ADVANCES IN TRIBOLOGY

Edited by **Pranav H. Darji**

Advances in Tribology
http://dx.doi.org/10.5772/61873
Edited by Pranav H. Darji

Contributors

David Johnson, Quan-Zhong Zhao, Zhuo Wang, Andreas Almqvist, Peter Wall, Shengmao Zhang, Yujuan Zhang, Pingyu Zhang, Guangbin Yang, Qiyin Lin, Baotong Li, Hong Zhao, Rana Afif Anaee, Majid Hameed Abdulmajeed, Binnur Sağbaş, José Cabral, Facundo Almeraya-Calderon, Citlalli Gaona, Carlos AgustÃn Poblano-Salas, Abraham Velasco Telles, David E. Sander, Hannes Allmaier, Hans-Herwig Priebsch, Mitra Akhtari Zavareh, Xiaoyan Liao, Changhong Wu

CBS, Edition **2017**

Published by InTech
Janeza Trdine 9, 51000 Rijeka, Croatia

Publishing Process Manager Edi Lipovic

Technical Editor SPi Global

Cover Designer InTech Design Team

Additional hard copies can be obtained from orders@intechopen.com

Advances in Tribology, Edited by Pranav H. Darji
p. cm.
Print ISBN 978-953-51-2742-0
Online ISBN 978-953-51-2743-7

INTECH
open science | open minds

Contents

Preface

Tribology is the science and technology of interacting surfaces in relative motion and of related subjects and practices. Understanding the nature of these interactions and solving the technological problems associated with the interfacial phenomena constitute the essence of tribology. The integrated study of friction, wear, and lubrication as constituent parts of tribological subject is now considered as a discrete discipline, having independent but direct effect on ultimate design, manufacture, and use of "customer-friendly" marketable products. In the major field of design and manufacturing of mechanical, production, automobile, and industrial engineering, typical and advanced methodologies and processes are implemented for the best performance of the product or machinery. Thus, the concept of tribology has come into practice for even better performance. The lack of consideration of tribological fundamentals in both fields is responsible for vast economic losses, including shortened life, excessive equipment downtime, and large expenditures of energy. Nowadays, it is very important that the tribological knowledge be implemented at each stage of design and manufacturing to minimize the frictional and wear losses, and ultimately these will serve as best preference for the economical growth of the nation. Currently, tribologists are playing a vital role in the same direction. By proper utilization of tribological knowledge, it is possible to improve the performance of existing product or machinery without affecting its conventional design and manufacturing method. This subject is equally important for analyzing the hazard rate, failure probability, and reliability from the viewpoint of systematic evaluation of the performance of product, process, and service for economic development of the world.

The purpose of research in tribology and application of tribology are understandably the minimization and elimination of losses resulting from friction and wear at all levels of technology where the rubbing of surfaces is involved. Thus, tribology has an important role to play in the more efficient utilization of energy and materials resources—through increased efficiency, enhanced reliability, and reduced maintenance costs, leading to increased productivity of engineering plant and machinery. In addition to its industrial applications, tribology is used for most of the household appliances/components, including medical instruments and practices. It is useful in almost all walks of life—from walking on the floor to flying in the air and so on. There are no mechanism, no machine, and no equipment that are not effected by tribological factors. What has become increasingly obvious in recent years is the inherently interdisciplinary nature of the tribologist's task; as well as involving practicing and academic engineers, advances in the subject have drawn upon the ingenuity and expertise of physicists, chemists, metallurgists, and material scientists. I hope that this volume may be of interest to research scholars, academicians, industrialists, professional engineers, and specialists in these related areas. This book would also be of immense help to various practicing engineers, technologists, managers, and supervisors engaged in the maintenance, operation, and upkeep of different machines, equipments, systems, and plants of various industries. Tribology is an area of active research, and the published chapters play a major role in continuation with the growth in this field.

This book contains original and innovative research studies on recent applications of tribology, contributed by the group of selected researchers describing the best of their work. Through its 11 chapters, the reader will have access to work in 3 major areas of tribology. These are surface engineering and coating, friction and wear mechanism, and lubrication technology. The first part of the book from Chapters 1 to 4 deals with the surface treatment and coating through which component life can be improved by reducing wear rate. The second part of the book from Chapters 5 to 7 deals with tribo-testing and tribo-system monitoring for friction and wear mechanism presented with real-life case studies. The third part from Chapters 8 to 11 discusses the advances in lubrication, which also includes the role of nanolubricants and lubrication additives.

I feel amazing pleasure to edit this book. I would like to express my sincere gratitude to all authors for their outstanding chapters. I also wish to acknowledge the InTech editorial staff, in particular, Mr. Edi Lipovic, publishing process manager, for indispensable technical assistance in book preparation and publishing. And of course, thanks to my parents, wife "Veera," and loving son "Fagun" for the support they always give to me. Thanks for being close to me during all my life.

Prof. (Dr.) Pranav H. Darji
Professor and Head, Department of Mechanical Engineering
C. U. Shah College of Engineering and Technology
Director, P. G. Studies and Research
Research, Development and Innovation Centre
C. U. Shah University
Wadhwan City, Gujarat, India

Section 1

Surface Engineering and Coating

Chapter 1

Manipulation of Tribological Properties of Metals by Ultrashort Pulsed Laser Micro-/Nanostructuring

Quan-Zhong Zhao and Zhuo Wang

Additional information is available at the end of the chapter

http://dx.doi.org/10.5772/64764

Abstract

Surface texturing as a means for controlling tribological properties of mechanical components is well known for many years. Various technologies have been developed for surface texturing. Among them, ultrashort pulsed laser surface texturing is one of the most promising ways to achieve micromachining in the field of tribological applications. Ultrashort pulsed laser technology can produce various micro-/nanostructures on the material surfaces to modulate their tribological properties. The aim of this chapter is to introduce the recent progress on ultrashort pulsed laser-induced frictional property change of metals and to demonstrate the potential applications of ultrashort pulsed laser-induced frictional property change of metal in various fields.

Keywords: ultrashort pulsed laser micro-/nanostructuring, surface texturing, tribological property, metals

1. Introduction

In the past decades, there has been a growing interest in the designing of surface structures in micro-/nanoscale to modulate tribological properties of materials [1–3]. Various surface modification technologies have been proposed to tune the tribological properties of various materials [1–3]. Surface texturing, an artificial topography on the surface of material, has been introduced to modulate the tribological properties of materials [1]. These textured surfaces can act as lubricant reservoirs for supplying lubricant, as traps for wear debris, and as microhydrodynamic bearings for improving load-carrying capacity [4–8]. Up till now, these textured surfaces have been utilized to improve the tribological performances of interfaces in various fields, such as piston rings, mechanical seals, thrust bearing, magnetic storage devices, cutting tools, and microelectro-mechanical system (MEMS) [9–14].

Many surface texturing technologies, including mechanical processing, lithography, ion beam texturing, and laser surface texturing have been developed for producing the micro-/nano-meter-sized structures on different material surfaces [15–18]. Compared to other techniques, laser structuring provides significant advantages including fast machining time, environment friendly, precise control of the geometric features of the patterns, and so on. Especially, the ultrashort pulsed laser with pulse width ranging from picosecond to femtosecond is one of the most promising ways to achieve micromachining in the field of tribological applications [19–37], which is owing to its ultrashort pulse width and ultrahigh peak power that can process almost all materials. Moreover, ultrashort pulsed lasers are expected to minimize the melt ejection and heat-affected effects for surface texturing in tribological applications.

Some researchers have proved that ultrashort pulsed laser structuring can modulate the friction behaviors of metal materials [23–37]. Surface textures not only reduce the friction coefficient, but also can raise the friction coefficient, depending on the geometry, scale, and subsequent modification. The aim of this chapter is to introduce the recent progress on ultrashort pulsed laser-induced frictional property change of metals and to demonstrate the potential applications of ultrashort pulsed laser-induced frictional property change of metal in various fields.

2. Ultrashort pulsed laser-induced micro-/nanostructures on the surface of metals: phenomena and mechanisms

Laser surface modification has been intensively studied since the invention of laser. The first step in any structural modification of a material by laser processing is the deposition of laser energy. The total laser energy and the spatial and temporal energy distribution determine what kind of final structure on the modified surface will be obtained [38].

During the interaction of laser pulses with metal targets, the laser energy is absorbed by free electrons through such mechanisms as inverse bremsstrahlung [39]. Then the evolution is the excitation of electrons from their equilibrium states to excited states by absorbing photons. The electronic excitation is followed by multiple secondary processes, ultimately ending in the final structural modification of the material. The time scales of various secondary processes can be identified as shown in **Figure 1** [38].

The primary electronic excitation is related to a quite short-lived coherent polarization of the metal material. The polarization is destroyed during the dephasing process within a time scale of about 10^{-14} s [40]. The pure dephasing changes the phase of the excited states but have no effect on the electronic energy distribution. The started distribution of excited electronic states corresponds to the set of states coupled by the optical transitions. The occupation of these primary states is promptly changed by the process of carrier-carrier interaction, and a quasi-equilibrium state is established among the electrons within a time scale of approximately

10^{-13}s. At this moment, the electron temperature (T_e) is greater than the lattice temperature (T_i). Subsequently, the quasi-equilibrium electrons cool down on a time scale of 10^{-13}–10^{-12} s by emission of phonons. The final stage of the thermalization process is the redistribution of the phonons over the entire Brillouin zone according to a Bose-Einstein distribution. It is generally accepted that a few picoseconds after the deposition of the laser energy the energy distribution is close enough to the thermal equilibrium. The thermal diffusion can take place on a time scale of the order of 10^{-11} s after the thermalization. The solid-liquid transition will take place at melting temperature when a sufficient amount of energy is deposited in the metal.

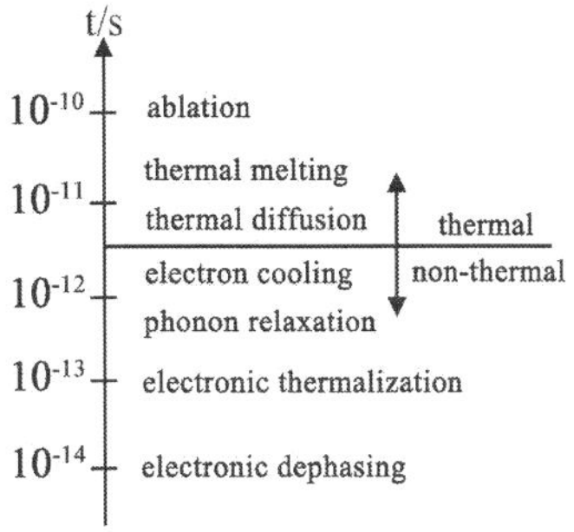

Figure 1. Time scale of the various secondary processes [38].

As mentioned above, it can be concluded that a distinct dividing line at about 10^{-12} s separate the regime of nonthermal processes and thermal processes. For the pulsed laser-metal interaction process, if the pulse duration (pulse width) is much longer than 10^{-12}s, the whole process will involve thermalization within the electron subsystem, energy transfer to the lattice, and energy losses due to the electron heat transport into target. Thus, the thermal mechanisms of laser ablation such as melting and boiling are likely to play an important role. While if the pulse duration is much shorter than 10^{-12}s, the thermal diffusion can be neglected, which avoids negative effects of energy transfer.

Chichkov et al. [41] compared holes drilled in 100 μm thick steel foils (in vacuum) with a pulse duration of 3.3 ns (1 ns = 10^{-9}s) laser and a pulse duration of 200 fs (1 fs = 10^{-15}s) laser. **Figure 2(a)** and **(c)** shows the schematic of nanosecond-pulse ablation and the surface morphology of holes drilled with a pulse duration of 3.3 ns and a fluence of F = 4.2 J/cm^2. The trace of the molten material can be seen in **Figure 2(c)**. Due to the longer pulse duration of nanosecond pulses, there is enough time for the thermal wave to propagate into the metal target and to create a relatively large molten layer. Thus, the solid-liquid and solid-vapor transformation take place in the process of laser ablation. The molten material is pushed out by the recoil pressure produced by the vaporization process, which lead to a "corona" around the hole as shown in **Figure 2(c)**. **Figure 2(b)** and **(d)** presents the schematic of femtosecond-pulse laser ablation and the hole drilled with a duration of 200 fs and a fluence of F = 0.5 J/cm^2. It can be seen that there is no trace of the molten material and only a vapor dust ring around the hole. As mentioned above, when the laser pulse duration is much shorter than time scale of thermal diffusion, nonthermal ablation mechanisms will occur. Due to the very short time scales involved in the ablation with femtosecond laser pulses, direct ionization and the formation of dense electron-hole plasmas can result in athermal phase transition, direct bond breaking, and explosive disintegration of the lattice through electronic repulsion. Therefore, the thermal conduction into metal can be neglected.

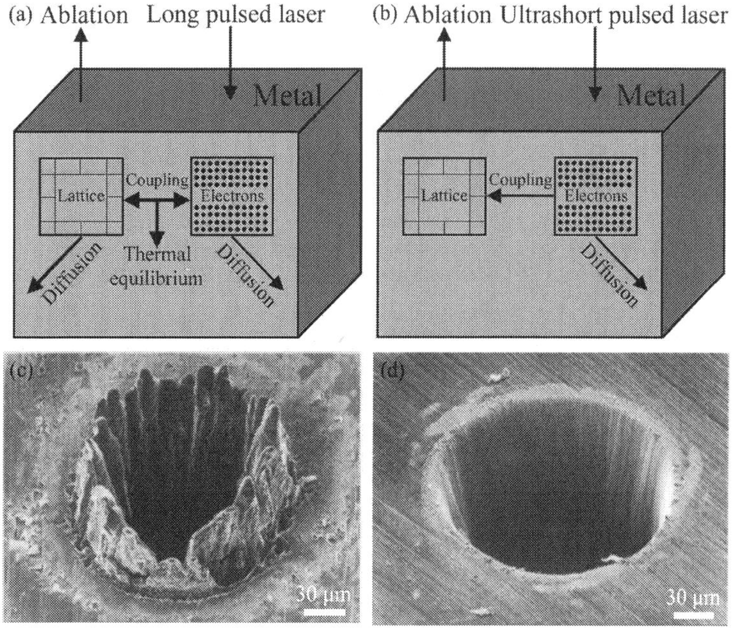

Figure 2. Schematic of nanosecond/femtosecond-pulse laser ablation and surface morphology of holes drilled in a 100 μm thick steel foil. (a) Schematic of nanosecond-pulse laser ablation, (b) schematic of femtosecond-pulse laser ablation, (c) surface morphology of holes drilled through a steel foil with a pulse duration of 3.3 ns and a fluence of $F = 4.2$ J/cm^2, (d) surface morphology of holes drilled through a steel foil with a pulse duration of 200 fs and a fluence of $F = 0.5$ J/cm^2 [41].

As early as 1965, a kind of near-wavelength periodic structures has been discovered in optical damages on semiconductor surfaces induced by a ruby laser [42]. Since then, these laser-induced periodic surface structures (LIPSS) or the so-called ripples have been studied extensively on various material surfaces. Recently, LIPSS as a kind of nanostructures have gained remarkable interest because they enable tuning of a wide range of properties, including wettability, colorization, and tribological properties [43–45]. In this section, we will introduce phenomena and mechanisms concerning ultrashort pulsed laser-induced periodic surface structures on the surface of metals.

Ultrashort pulsed laser-induced ripples on the surface of metals are formed as a result of light-matter interaction and their period (Λ), orientation, and morphology are strongly dependent on both the material properties and the laser irradiation conditions, including polarization, angles of incidence, laser energy fluence, incident light wavelength (λ), and pulse number/scanning speed [46–49]. Period is the most important ripples' parameter. Nowadays, ultrashort pulsed laser can easily induce subwavelength ripples on different metals. In general, the ripples can be separated into near-subwavelength ripples (NSRs, $0.4 < \Lambda/\lambda < 1$) and deep-subwavelength ripples (DSRs, $\Lambda/\lambda < 0.4$) in normal incidence [48, 50]. Orientation of ripples with respect to irradiation polarization is another parameter allowing to distinguish different ripple morphologies. For a linearly polarized laser beam, the origination of ripples is most often perpendicular to the incident laser polarization [51]. **Figure 3** shows uniform periodic nanostructures produced on the surface of stainless steel using an 800-nm femtosecond laser. It can be seen that the ripples are oriented perpendicular to the incident laser polarization (white double arrows direction). The period in the ripple structures is measured to be about 560 nm along the polarization direction, which is less than the wavelength of the employed femtosecond laser.

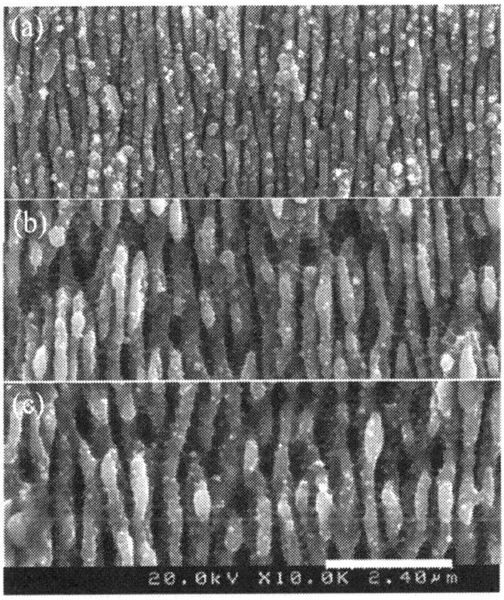

Figure 3. SEM images of the 304 stainless steel surfaces ablated by an 800-nm femtosecond laser. (a) DSRs, (b) NSRs, and (c) NSRs [48].

However, early studies showed that the period of ripples (Λ) is close to the incident light wavelength (λ) with a strong dependence on the angle of incidence (θ) when the ripples formed on the surface of metals using lasers with long pulse width (≥ns) [52].

$$\Lambda = \frac{\lambda}{1 \pm \sin\theta} \tag{1}$$

where the sign + or − depends on a scattered wave direction toward (+) or from (−) the incident beam. These classical ripples are widely accepted as a result of the interaction between the incident light and the surface scattering wave [52–54]. However, the period of ripple is generally found smaller compared to the laser wavelength when the ripple induced by using ultrashort pulsed lasers such as femtosecond lasers. Thus, the subwavelength ripples should not be ascribed to the classical ripples described by the scattering mode. Recently, some researchers propose that the interaction between the incident laser light and the excited surface plasma (SP) wave is responsible for the formation of ripples induced by femtosecond pulsed lasers on a metal surface [50]. They believed that the propagating SPs and the incident laser will interfere to form a fringe with a vector as shown in **Figure 4**:

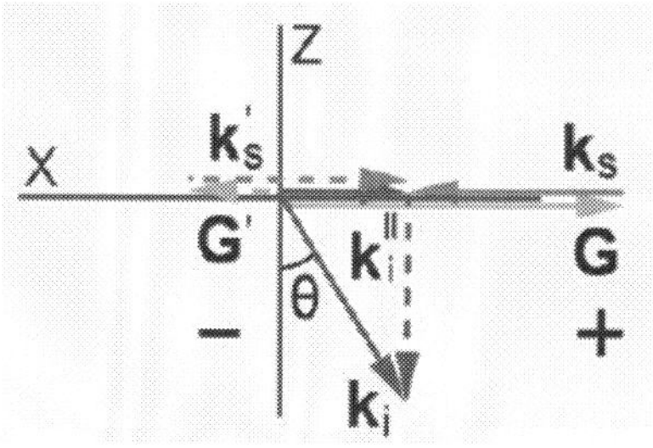

Figure 4. Schematic processes of SP-laser interactions [50].

$$\mathbf{G} = \mathbf{k}_i - \mathbf{k}_s \tag{2}$$

where $\mathbf{k}_i$ is the wavevector of incident laser, and $\mathbf{k}_s$ is the wavevector of SPs. Considering the components in the interface, they obtain

$$\Lambda = \frac{\lambda}{\frac{\lambda}{\lambda s} \pm \sin\theta} \tag{3}$$

where $\Lambda = 2\pi\ |\mathbf{G}|^{-1}$, $\lambda = 2\pi\ |\mathbf{k}_i|^{-1}$, $\lambda_s = 2\pi\ |\mathbf{k}_s|^{-1}$, and θ is the incident angle of laser. Assume $\varepsilon''<|\varepsilon'|$ (ε'' is the imaginary part of ε), the λ_s on a metal/dielectric interface is given by Raether [55]

$$\lambda_s = \lambda\left(\frac{\varepsilon' + \varepsilon_d}{\varepsilon'\varepsilon_d}\right)^{1/2} \tag{4}$$

where ε_d is the dielectric constant of the dielectric material (for air, $\varepsilon_d \approx 1$). Thus, the simple relationship $\Lambda = \lambda_s$ can be obtained in normal incidence. It means that in the situation with a destructible fluence, the interference fringes will induce permanent ripples on material surface with Λ equal to λ_s, which is always smaller than λ.

3. Modulation of tribological property of metals by ultrashort pulsed laser-induced microstructures

Lasers with long pulse width are used to produce surface textures to improve tribological properties of metal in the early stage [9, 11]. However, there is an obvious thermal effect during the long pulse width laser ablation process, which results in the reduction of machining accuracy. As above-mentioned, the ultrashort pulsed lasers, owing to their ultrashort pulse width and ultrahigh peak power, can safely process almost all materials with minimal heat effects for surface texturing. Therefore, the ultrashort pulsed laser surface texturing is expected to be a useful tool for improving tribological properties of materials (**Figure 5**).

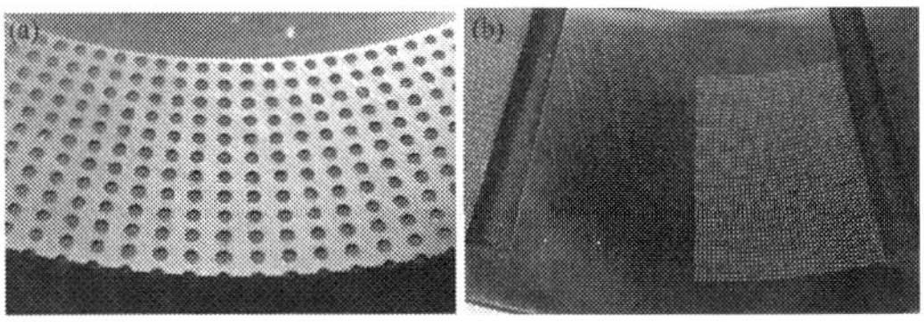

Figure 5. Laser textured surface. (a) Laser-textured stationary specimen of the seal. (b) Laser-textured thrust bearing [9, 11].

As early as 2004, Dumitru et al. [25] studied the effect of femtosecond laser ablation on tribological properties of coated WC-Co surfaces under dry condition. Surface textures with various shapes were found to occur after a given number of incident femtosecond laser pulses and the tribological tests demonstrated improved wear behaviors for the patterned coated WC-Co surfaces. **Figure 6** showed that the particle trap role of the femtosecond laser-induced pores is responsible for the improvement of tribological properties of coated WC-Co surfaces to a great extent.

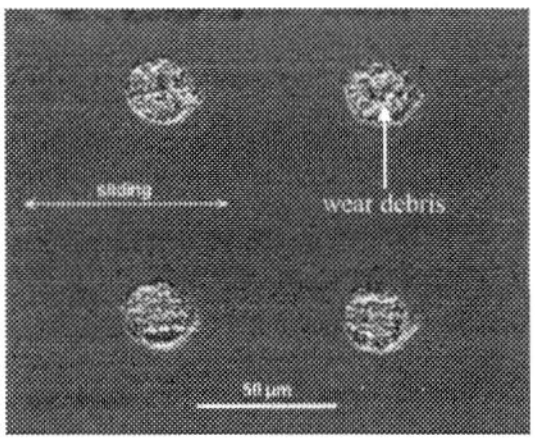

Figure 6. SEM image from wear track of the patterned coated WC-Co surface [25].

As mentioned above, ultrashort pulsed laser technology is one of the most promising ways to achieve micromachining in the field of tribological applications. In comparison of other laser-textured surface, there is very minimal resolidified and spatter particles on the ultrashort pulsed laser-textured surface due to its ablation mechanism. Bathe et al. [26] investigated the effect of laser-textured surface produced by different laser sources on the tribological behavior of gray cast iron under dry condition. The microtextured surfaces with dimple feature were produced on gray cast iron using millisecond (0.5 ms), nanosecond (40 ns), and femtosecond (120 fs) laser source. As shown in **Figure 7**, the femtosecond laser-textured surface exhibited the lowest steady-state coefficient and wear rate among the tested samples. The reason is ascribed to the absence or very minimal resolidified particles present on femtosecond laser-textured surface. During sliding wear test, the resolidified and spatter particles on the metal surface is collapsed readily, which is generally detrimental to the tribological properties. Moreover, the dimple can capture wear debris and reduce the plowing of metal surface. Consequently, the femtosecond laser-textured surfaces show a considerable lower friction coefficient and wear rates compared with the untextured and other long pulse width laser-textured surfaces. The SEM images of the wear track of the textured and untextured samples are shown in **Figure 8**. It can be seen that a spread of wear debris cover the worn surface of the untextured samples (**Figure 8a**). However, the wear track does not show much wear debris on surface (**Figure 8b–d**). This indicates that wear debris moves from the contact region and fills the dimples. In this case, dimples act as reservoirs of debris that leave free interface between the friction pairs, thus reducing friction and wear.

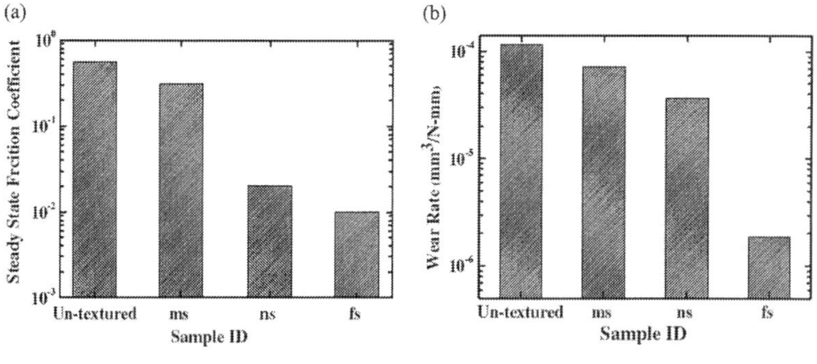

Figure 7. Ball-on-disk test performed at room temperature: (a) steady-state friction coefficient for untextured and textured samples, (b) wear rate for untextured and textured samples [26].

Femtosecond laser surface texturing not only reduces the friction coefficient, but also can raise the friction coefficient. The authors of this chapter have [27] found that tribological property of textured surfaces by femtosecond laser can be modulated. By producing regular arranged microgrooved textures with different spacing on the AISI 304L steel surfaces by an 800-nm femtosecond laser, they proved that the spacing of microgrooves had a significant impact on friction coefficient of textured surfaces under dry condition. As the spacing of the textured surfaces increases, the average friction coefficients first decrease and then increase (**Figure 9**).

And the structured surface with a microgroove spacing of 100 μm has the minimal average friction coefficient. When the period of microgroove is within 15–35 μm, the average friction coefficients of structured surfaces are more than that of the unstructured surface. However, when the period of microgroove is in the range of 50–300 μm, the average friction coefficients of structured surfaces are smaller than that of unstructured surface. In conclusion, the increase or decrease of average friction coefficient of textures surfaces depends on the microgroove spacing compared with untextured surface.

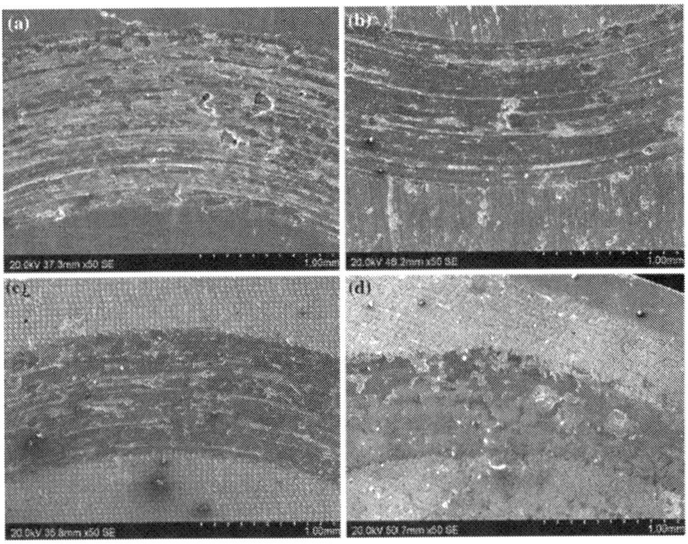

Figure 8. SEM images of wear track of (a) untextured, (b) millisecond laser textured, (c) nanosecond laser textured, and (d) femtosecond laser-textured samples [26].

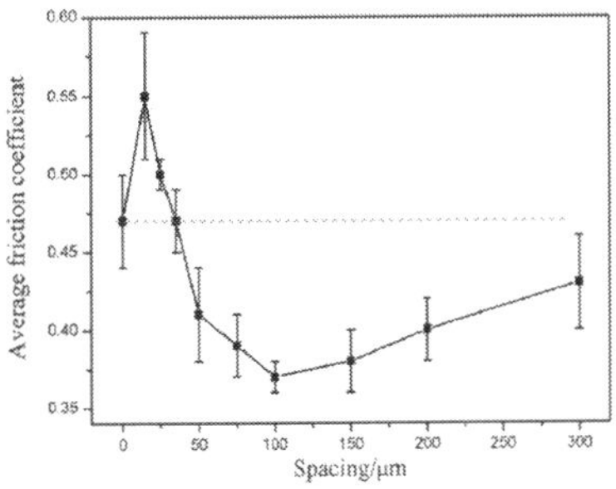

Figure 9. Average friction coefficients of untextured surface and textured surfaces with different microgroove spacing [27].

The result can be ascribed to the changing contributions of the three components of friction, i.e., due to the adhesion, to the plowing, and to the deformation [56]. The friction coefficient can be formulated by Eq. (5) under dry condition:

$$\mu = \mu_a + \mu_p + \mu_d \tag{5}$$

where μ_a is the adhesion friction coefficient component which is related to the real contact area [57], μ_p is the plowing friction coefficient component which is related to wear debris, and μ_d is the deformation friction coefficient component which is related to the surface roughness and contact stress, respectively [58]. As the increase of spacing of textured surfaces, the density of microgrooved textures decreases and the real contact area between two relative sliding surfaces increases correspondingly. The adhesion friction coefficient component depends on the real area of contact [57]. So, when the real area of contact increases, adhesion friction coefficient component increases. The plowing friction coefficient component also increases with the period of the microgroove as to the trapped effect becomes weaker. While the deformation component of friction coefficient decrease with the increase of spacing of textured surfaces due to the decrease of the contact stress. The small real contact corresponding to large contact stress may result in a transition from elastic deformation to the plastic deformation, usually causing a great increase of the friction force. Meanwhile, the deformation and collision of asperities lead to the ratchet coefficient which depends on the surface roughness. With increasing the period of microgrooves, the roughness of textured surfaces reduces, accordingly, the deformation friction coefficient component decreases. Therefore, there is a minimal friction coefficient caused by the changing contributions of three friction components. In summary, friction coefficients of AISI 304L steel can be controlled by changing the period of microgrooves of the textured surfaces.

Ultrashort pulsed laser surface texturing can reduce friction and wear under lubrication condition as well. In metal cutting with continuous chips, severe friction exists as the chip flows over the rake face of the cutting tool at high normal load and speed. In order to reduce friction and wear between the chip and the tool rake face, Lei et al. [28] utilized femtosecond laser surface texturing to produce microholes to form micropools filled with liquid or solid lubricants on the rake face of WC cutting inserts. They conducted a comparative investigation between micropool lubricated (surface-textured) cutting tools and flood-cooled (untextured) cutting tools. The cutting tests showed that the friction coefficient of the textured surface significantly reduced compared with that of untextured surface. The following two mechanisms can explain this phenomenon. First, the direct chip-tool contact area was reduced with embedded micropools that contain liquid or solid lubricants. Second, the lubricant may be squeezed out or spread over by the rubbing action of the chip flowing over the micropools to form a thin film at the chip-tool interface thus direct chip-tool contact can be further reduced. Consequently, the smaller direct contact area between the chip and tool rake face leaded to less friction coefficient. Ling et al. [29] also reported a research on utilization of surface textures to reduce adhesion and enhance drill bit life. The results from this study reveal that surface textures induced by picosecond laser on the margins of drill bits is a promising method for these purposes. The surface textures can serve as microreservoirs to retain some lubricant and release it under pressure or physical scribing. Therefore, properly designed microscale surface textures could provide positive impact on lubrication enhancement.

Surface texture will inevitably cause an increase in surface roughness which may destroy the oil films due to large local contact stress. Thus, a critical issue in application of surface texturing concerns the optimum surface texture under lubrication condition. The authors of this chapter [30] have investigated the effect of femtosecond laser-induced microgrooves on starved lubrication tribological properties of stainless steel. The results show that the average friction coefficients initially decrease and then increase as the spacing of the textured surfaces increases. Moreover, the average friction coefficient of textured surface is reduced by 30.9% with their optimum microgroove spacing (75 μm) compared with that of untextured surface. When the period of microgroove is very small, the reduction of actual contact area between friction pairs enhances the average contact stress at the sliding surface, which makes the oil film easily collapsible in the untreated region. When the period of microgroove is very large, the grooves as oil reservoirs (secondary lubrication effect) cannot be sufficiently provided to the sliding contact surface, and the wear debris cannot be effectively trapped by the grooves (wear debris-trapped effect). Therefore, the changing contributions of the three effects of friction lead to the existence of the optimum friction property for the textured samples with the increase of microgroove spacing. In a word, femtosecond laser surface texturing has a marked potential for reducing friction and wear under lubrication condition if the surface textures were distributed in an appropriate manner.

4. Reduction of friction of metals by ultrashort pulsed laser-induced periodic surface nanostructures

Laser-induced periodic surface structures (LIPSS) or the so-called ripple structures exhibit several amazing properties such as the capability of tuning wettability, reflectivity, and tribological properties. The modification of tribological properties has specifically attracted attention because of its potential application in industrial fields such as magnetic recording devices, cutting tools, and microelectro-mechanical system (MEMS) [12–14].

As the density of magnetic recording on computer hard disk drive increases rapidly and approaches the level of 1 Tb/in^2, the flying height of the magnetic head (the slider), or hard disk interface, has to be reduced to about 5 nm. Such ultralow hard disk interface can only be realized on super smooth surfaces of slider and disk; however, this will result in serious stiction at the hard disk interfaces. Therefore, reduction of the interfacial adhesion and stiction has attracted great attentions in recent years. Some researchers found that the surface texturing is an effective method to reduce stiction and adhesion at the hard disk interfaces. Hanchi et al. [59] reported that the textured sliders can prevent the catastrophic failure caused by a sudden rise of friction from super smooth surface. Wang et al. [60] also showed that the textured sliders exhibited less lubricant depletion and smaller vibrations compared with untextured sliders. Tagawa et al. [31] proposed a novel concept of contact sliders with nanotextures produced by femtosecond laser processing. Contact sliders experiments were carried out using the contact sliders with and without nanotextures. It was found that the nanotextured sliders facilitate the development of contact hard disk interface with lower friction, low wear of the contact slider surfaces, and low contact sliders bouncing vibration.

One of the successful applications of laser surface texturing is in cutting tools. Surface textures with micrometer scales on the tool surface have been reported to improve tribological characteristics [32, 33]. The nanoscale textures are also the ideal candidate for improving the properties of cutting tools. Kawasegi et al. [32] developed WC-Co cutting tools that had nanoscale textures on their surfaces using interference inscription of a femtosecond laser. The effect of nanoscale textures on the machinability of an aluminum alloy was investigated with a turning experiment applying the minimum quantity lubrication method. The tested results indicated that the ripple structure decreased the cutting force due to the corresponding reduction in the friction on the rake face (**Figure 10**).

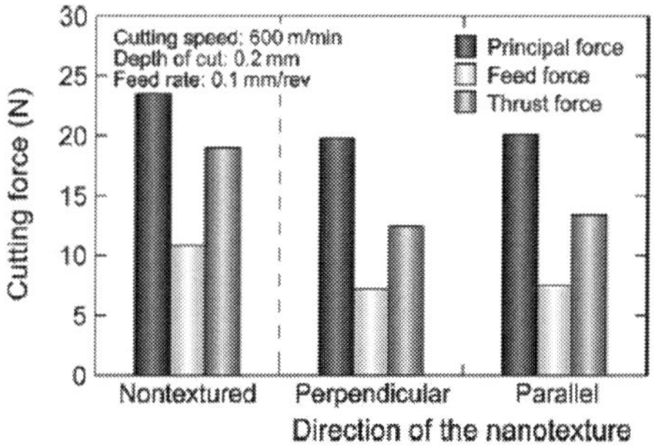

Figure 10. Comparison of the cutting forces required for the nanotextured tools with various texture directions [32].

Using cutting fluids is one of the effective methods to alleviate the severe friction and wear conditions in metal cutting operations. However, the lubrication effectiveness of cutting fluids reduces because of its difficult infiltration into the tool-chip interface during high-speed machining. Moreover, most of the fluids with environmentally harmful compositions are hard to dispose and expensive to recycle. Based on the above reasons, research on metal machining under dry condition has caused more attention for its positive role in reducing environment pollution and production costs. When machining under dry condition, the friction and adhesion between the chip and tool becomes higher, resulting in higher temperature and wear rates. It was thought that a solution to these problems could be achieved by developing new cutting tools with the purpose of reducing the heat generated by lowering the friction. In view of the above, nanoscale surface texturing was made on the rake face of the cutting tools with femtosecond laser pulses, which are expected to decrease friction and wear due to the reduced contact length at the tool-chip interface of the nanotextured tools. Deng et al. [34] fabricated nanoscale surface textures on the rake face close to the main cutting edge of the WC/TiC/Co carbide tools by femtosecond laser surface texturing. Dry cutting texts were carried out with the rake face-textured tools (TT) and the conventional carbide tools (CT). Results indicate that the cutting forces, the cutting temperature and the friction coefficient at the tool-chip interface of the TT were significantly reduced in comparison with that of the conventional CT:

$$F_f = a_w l_f \tau_c \tag{6}$$

where a_w is the cut width, l_f is the tool-chip contact length, and τ_c is the shear strength at the tool-chip interface, respectively. As shown in **Figure 11**, the nanoscale surface textures decrease the tool-chip contact length l_f, which result in the reduction of friction between the tool-chip interfaces.

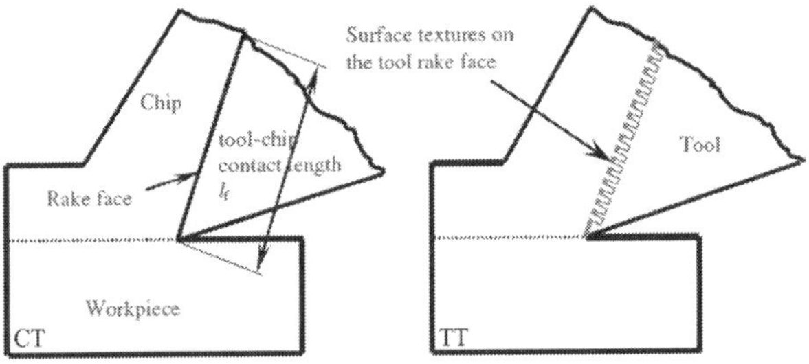

Figure 11. Schematic diagram of the tool-chip contact length for CT and TT. In the case of dry cutting, the friction force (Ff) at tool rake face can be calculated as [34].

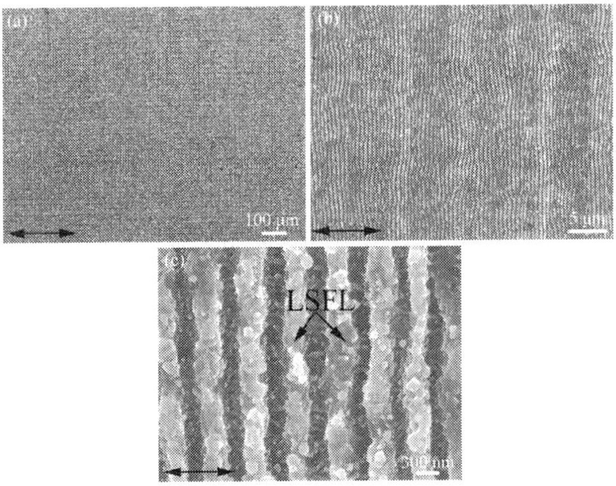

Figure 12. SEM micrographs of the LIPSS (laser-induced periodic surface structures) on the AISI 304L steel surfaces. Note the different magnifications used in (a)–(c). The black double-headed arrow indicates the polarization direction of the femtosecond laser pulses [35].

The authors of this chapter also [35] found the effect of femtosecond-laser-induced periodic surface structures (LIPSS) on the tribological properties of stainless steel in the conditions of starved lubrication and in dry contacts, respectively. By utilizing an 800-nm femtosecond laser to produce uniform ripple structures on the stainless steel surface are shown in **Figure 12**. The tribological properties of original surface and nanotextured surfaces with LIPSS were investigated under both dry and starved oil lubricated conditions. The friction coefficient of nanotextured surfaces with LIPSS has shown a lower value than that of the original surface under both dry and starved oil lubricated conditions. This finding may suggest applications in field such as microelectro-mechanical systems.

5. Manipulation of tribological properties of metals by ultrashort pulsed laser texturing and quenching

In the past, several classes of tribological experiments were developed in order to investigate the benefits of laser surface texturing in terms of transition between different lubrication regimes, reduction of friction coefficients, and reduction of wear rates. As a matter of fact, there exist possible local quenched effects accompanying with laser surface texturing process as a consequence of the interaction between laser beam and metal materials. As mentioned in Section 2, the time scale for a considerable energy transfer from electronics to the lattice is about 1 ps (10^{-12} s) during the interaction of pulse laser radiation with metal targets [41]. Thus, the thermal diffusivity cannot be neglected during the nanosecond or picosecond laser surface texturing process. The high energy deposited only in a very thin layer within picosecond time scale will result in the formation of a laser quenched layer on the textured surfaces. The laser quenching effects can lead to a reduction in grain size, phase transition, and change surface chemistry, which are possible to enhance tribological properties [59, 60].

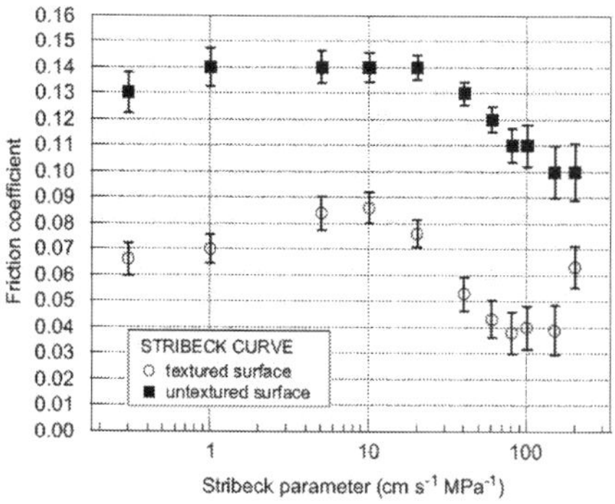

Figure 13. Friction coefficients against Stribeck parameters calculated as the ratio between mating speeds (range from 0.06 to 40 cm/s) and nominal contact pressures (maintained constant: roughly 0.2 MPa) [59].

Gualtieri et al. [59] fabricated microdimples on the 30NiCrMo12 nitride steel by nanosecond laser surface texturing. They also found that laser ablation in creating microdimples is accompanied to local quenching which caused grains size reduction and local hardening in micrometric subsurface areas near the microdimple edges. Tribological tests showed a reduction of friction coefficient due to the well-known hydrodynamic lift effect ensured by microdimples in full lubrication configuration (**Figure 13**). The local hardening induced by laser quenching may significantly improve the wear resistance of nitride steel during friction process.

Laser interference metallurgy is one possible approach to create well-defined surface topographies on the microscale accompanied with metallurgical effects including melting, resolidification, and the formation of intermetallic phases. Gachot et al. [60] applied the aforementioned technique to pattern both interacting surfaces (steel substrate and tribometer ball) to control the involved contact geometries under dry condition. They found that the dry friction between two laser-structured solids depends not only on the geometric characteristics of textured patterns but also on the surface chemistry and mechanical properties of textured surfaces.

Figure 14 shows the change of average friction coefficients of textured surfaces with different spacing in the process of picosecond laser surface texturing under dry condition. It can be seen that the average friction coefficients initially increase then decrease as the spacing of the microgroove increases. Moreover, the picosecond laser surface texturing can reduce the average friction coefficient if the microgrooves are distributed in an appropriate manner. The well-controlled tribological properties are attributed to the combined effects of laser surface texturing and laser quenching [61].

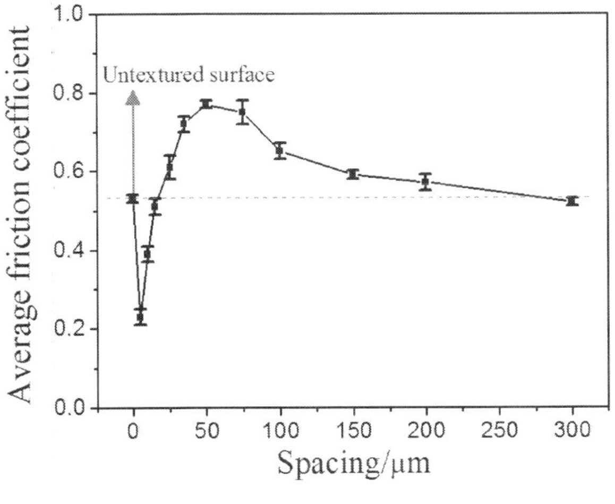

Figure 14. Average friction coefficients of untextured and textured surfaces with different spacing under dry condition [61].

6. Reduction of friction of metals by combining laser-induced periodic surface nanostructures and coating techniques

As mentioned above, laser-induced periodic surface nanostructures have marked potential to improve tribological properties of metal materials. However, only laser surface-texturing technology cannot meet strict requirements under harsh condition. As well known, coating materials surfaces with hard or soft solid lubricant has been considered to be a feasible method to enhance the tribological properties of materials. Thus, a combination of laser surface texturing and solid lubricating coatings are expected to significantly enhance the tribological properties of material substrates.

Lian et al. [36] developed an effective cutting tool named tungsten disulfide (WS_2) soft-coated nanotextured self-lubricating tool which is fabricated by two steps. First, nanoscale surface textures were produced on the tool-chip interface of rake face of uncoated YS8 (WC + TiC + Co) cemented carbide cutting inserts by femtosecond laser surface texturing. Then the nanotextured tools were deposited with WS_2 soft coatings by medium-frequency magnetron sputtering, multiarc ion plating, and ion beam-assisted deposition technique. From **Figure 15**, it can be seen that the friction coefficient at the tool-chip interface of the nanotextured tools (CFT) was reduced compared with that of the conventional one (YS8). Moreover, the WS_2 soft-coated nanotextured self-lubricating tool (CFTWS) exhibits the lowest friction coefficient among all the tools under the same tested condition. Zhang et al. [37] also studied the synergic effect of the laser-induced periodic surface nanostructures and the $Ti_{55}Al_{45}N$ hard coating on dry cutting property of WC/Co cutting tool. They developed two kinds of WC/Co-based $Ti_{55}Al_{45}N$ coated tools by changing processing sequence of femtosecond laser surface texturing and physical vapor deposition method as shown in **Figure 16**. Both the first $Ti_{55}Al_{45}N$-coated and then nanoscale-textured tools (CNT) and the first nanoscale-textured and then $Ti_{55}Al_{45}N$-coated tools (NCT) have lower friction coefficient compared with the conventional coated tool (CCT).

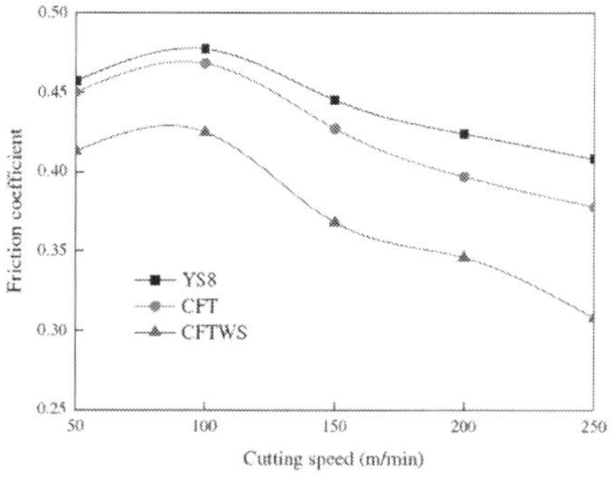

Figure 15. Friction coefficient between the tool-chip interface of the YS8, CFT, and CFTWS tools at different cutting speed (ap = 0.3 mm, f = 0.1 mm/r) [36].

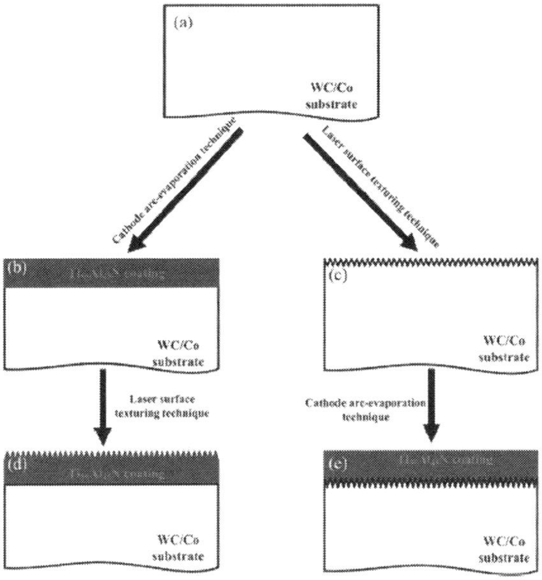

Figure 16. Schematic diagram showing the surface treatment procedure on WC/Co cemented carbide substrate: (a) polished, (b) coated, (c) textured, (d) first coated and then textured, and (e) first textured and then coated specimens [37].

7. Summary and outlook

There is an increasing need to control friction and wear in order to extend the lifetime of mechanical systems, to improve their reliability, and to conserve resources and energy. Previous investigations and applications have been demonstrated that tribological properties of metal materials can be improved by surface-texturing technology. However, ultrashort pulsed laser technology is considered as one of the most promising ways to achieve micromachining in the field of tribological applications, which is owing to its ultrashort pulse width and ultrahigh peak power that can process almost all materials. Moreover, ultrashort pulsed lasers are expected to minimize the heat-affected effects for surface texturing in tribological applications. To meet different requirements, various micro-/nanostructures can be produced by ultrashort pulsed laser surface texturing on the material surfaces to tune their tribological properties. Combining LIPSS with some solid lubricant coatings is also a good idea to obtain a desired reduction in friction and wear. What needs illustrating is that the surface textures are a kind of hierarchical structures consisting of micro-/nanostructures in the real application. Sugihara et al. [62] developed a cutting tool with a banded nano-/microtextured surface and it was revealed that the surface significantly improved the antiadhesiveness and lubricity. In a word, ultrashort pulsed laser surface texturing has a marked potential to modulate tribological properties of metal materials.

Author details

Quan-Zhong Zhao[1*] and Zhuo Wang[2]

*Address all correspondence to: zqz@siom.ac.cn

1 State Key Laboratory of High Field Laser Physics, Shanghai Institute of Optics and Fine Mechanics, Chinese Academy of Sciences, Shanghai, China

2 School of Materials Science and Engineering, Shanghai Jiao Tong University, Shanghai, China

References

[1] Etsion I. State of the art in laser surface texturing. J. Tribol-T. ASME. 2005; 127(1): 248–253. DOI: 10.1115/1.1828070

[2] Han Z, Zhang YS, Lu K. Friction and wear behavior of nanostructured metals. J. Mater. Sci. Technol. 2008; 24(4): 483–494.

[3] Fox-Rabinovich GS, Yamamoto K, Beake BD, Gershman IS, Kovalev AI, Veldhuis SC, Aguirre MH, Dosbaeva G, Endrino JL. Hierarchical adaptive nanostructured PVD coatings for extreme tribological applications: the quest for nonequilibrium states and emergent behavior. Sci. Technol. Adv. Mat. 2012; 13(4): 1–26. DOI: 10.1088/1468-6996/13/4/043001

[4] Vihena LM, Sedlacek M, Podgornik B. Surface texturing by pulsed Nd: YAG laser. Tribol. Int. 2009; 42(10): 1496–1504. DOI: 10.1016/j.triboint.2009.06.003

[5] Xing YQ, Deng JX, Wu Z, Cheng HW. Effect of regular surface textures generated by laser on tribological behavior of Si3N4/TiC ceramic. Appl. Surf. Sci. 2013; 265: 823–832. DOI: 10.1016/j.apsusc.2012.11.127

[6] Li KM, Yao ZQ, Hu YX, Gu WB. Friction and wear performance of laser peen textured surface under starved lubrication. Tribol. Int. 2014; 77: 97–105. DOI: 10.1016/j.triboint.2014.04.017

[7] Kovalchenko A, Ajayi O, Erdemir A, Fenske G, Etsion I. The effect of laser texturing of steel surfaces and speed-load parameters on the transition of lubrication regime from boundary to hydrodynamic. Tribol. Trans. 2004; 47(2): 299–307. DOI: 10.1080/05698190490440902

[8] Costa HL, Hutchings IM. Hydrodynamic lubrication of textured steel surfaces under reciprocating sliding conditions. Tribol. Int. 2007; 40(8): 1227–1238. DOI: 10.1016/j.triboint.2007.01.014

[9] Qiu Y, Khonsari MM. Experimental investigation of tribological performance of laser textured stainless steel rings. Tribol. Int. 2011; 44(5): 635–644. DOI: 10.1016/j.triboint.2011.01.003

[10] Ryk G, Etsion I. Testing piston rings with partial laser surface texturing for friction reduction. Wear. 2006; 261(7–8): 792–796. DOI: 10.1016/j.wear.2006.01.031

[11] Marian VG, Gabriel D, Knoll G, Filippone S. Theoretical and experimental analysis of a laser textured thrust bearing. Tribol. Lett. 2011; 44(3): 335–343. DOI: 10.1007/s11249-011-9857-8

[12] Wu Z, Deng JX, Chen Y, Xing YQ, Zhao J. Performance of the self-lubricating textured tools in dry cutting of Ti-6Al-4V. Int. J. Adv. Manuf. Tech. 2012, 62(9–12): 943–951. DOI: 10.1007/s00170-011-3853-x

[13] Raeymaekers B, Etsion I, Talke FE. Enhancing tribological performance of the magnetic tape/guide interface by laser surface texturing. Tribol. Lett. 2007; 27(1): 89–95. DOI: 10.1007/s11249-007-9211-3

[14] Komvopoulos K. Adhesion and friction forces in microelectromechanical system: mechanisms, measurement, surface medication techniques, and adhesion theory. J. Adhes. Sci. Technol. 2003; 17(4): 477–517. DOI: 10.1163/15685610360554384

[15] Pettersson U, Jacobson S. Textured surfaces for improved lubrication at high pressure and low sliding speed of roller/piston in hydraulic motors. Tribol. Int. 2007; 40(2): 355–359. DOI: 10.1016/j.triboint.2005.11.024

[16] Pettersson U, Jacobson S. Influence of surface texture on boundary lubricated sliding contacts. Tribol. Int. 2003; 36(11): 857–864. DOI: 10.1016/S0301-679X(03)00104-X

[17] Wang X, Kato K. Improving the anti-seizure ability of SiC seal in water with RIE texturing. Tribol. Lett. 2002; 14(4): 275–280. DOI: 10.1023/A:1022650813314

[18] He DQ, Zheng SX, Pu JB, Zhang GG, Hu LT. Improving tribological properties of titanium alloys by combining laser surface texturing and diamond-like carbon film. Tribol. Int. 2015; 82: 20–27. DOI: 10.1016/j.triboint.2014.09.017

[19] Chen CY, Wu BH, Chung CJ, Li WL, Chien CW, Wu PH, Cheng CW. Low-friction characteristics of nanostructured surfaces on silicon carbide for water-lubricated seals. Tribol. Lett. 2013; 51(1): 127–133. DOI: 10.1007/s11249-013-0153-7

[20] Chen CY, Chung CJ, Wu BH, Li WL, Chien CW, Wu PH, Cheng CW. Microstructure and lubricating property of ultra-fast pulse textured silicon carbide seals. Appl. Phys. A-Mater. 2012; 107(2): 345–350. DOI: 10.1007/s00339-012-6822-9

[21] Kietzig AM, Hatzikiriakos SG, Englezos P. Ice friction: the effects of surface roughness, structure, and hydrophobicity. J. Appl. Phys. 2009; 106(2): 0243031–0243037. DOI: 10.1063/1.3173346

[22] Sakai T, Nedyalkov N, Obara M. Friction characteristics of submicrometre-structured surfaces fabricated by particle-assisted near-field enhancement with femtosecond laser. J. Phys. D-Appl. Phys. 2007; 40(23): 7485–7491. DOI: 10.1088/0022-3727/40/23/035

[23] Garrelie F, Loir AS, Donnet LC, Rogemond F, Le Harzic R, Belin M, Audouard E, Laporte P. Femtosecond pulsed laser deposition of diamond-like carbon thin films for tribological application. Surf. Coat. Tech. 2003; 163: 306–312. DOI: 10.1016/S0257-8972(02)00481-4

[24] Arslan A, Masjuki HH, Varman M, Kalam MA, Quazi MM, Al Mahmud KAH, Gulzar M, Habibullah M. Effect of texture diameter and depth on the tribological performance of DLC coating under lubricated sliding condition. Appl. Surf. Sci. 2015; 356: 1135–1149. DOI: 10.1016/j.apsusc.2015.08.194

[25] Dumitru G, Romano V, Weber HP, Gerbig Y, Haefke H, Bruneau S, Hermann J, Sentis M. Femtosecond laser ablation of cemented carbides: properties and tribological applications. Appl. Phys. A-Mater. 2004; 79(3): 629–632. DOI: 10.1007/s00339-004-2675-1

[26] Bathe R, Krishna VS, Nikumb SK, Padmanabham G. Laser surface texturing of gray cast iron for improving tribological behavior. Appl. Phys. A-Mater. 2014; 117(1): 117–123. DOI: 10.1007/s00339-014-8281-y

[27] Wang Z, Zhao QZ, Wang CW, Zhang Y. Modulation of dry tribological property of stainless steel by femtosecond laser surface texturing. Appl. Phys. A-Mater. 2015; 119(3): 1155–1163. DOI: 10.1007/s00339-015-9085-4

[28] Lei ST, Devarajan S, Chang ZH. A study of micropool lubricated cutting tool in machining of mild steel. J. Mater. Process. Tech. 2009; 209(3): 1612–1620. DOI: 10.1016/j.jmatprotec.2008.04.024

[29] Ling TD, Liu PZ, Xiong SW, Grzina D, Cao J, Wang QJ, Xia ZC, Talwar R. Surface texturing of drill bits for adhesion reduction and tool life enhancement. Tribol. Lett. 2013; 52(1): 113–122. DOI: 10.1007/s11249-013-0198-7

[30] Wang Z, Li YB, Bai F, Wang CW, Zhao QZ. Angle-dependent lubricated tribological properties of stainless steel by femtosecond laser surface texturing. Opt. Laser Technol. 2016; 81: 60–66. DOI:10.1016/j.optlastec.2016.01.034

[31] Tagawa N, Takada M, Mori A, Sawada H, Kawahara K. Development of contact sliders with nanotextures by femtosecond laser processing. Tribol. Lett. 2006; 24(2): 143–149. DOI: 10.1007/s11249-006-9142-4

[32] Kawasegi N, Sugimori H, Morimoto H, Morita N, Hori I. Development of cutting tools with microscale and nanoscale textures to improve frictional behavior. Precis. Eng. 2009; 33(3): 248–254. DOI: 10.1016/j.precisioneng.2008.07.005

[33] Zhang KD, Deng JX, Sun JL, Jiang C, Liu YY, Chen SA. Effect of micro/nano-scale textures on anti-adhesive wear properties of WC/Co-based TiAlN coated tools in AISI

316 austenitic stainless steel cutting. Appl. Surf. Sci. 2015; 335: 602–614. DOI: 10.1016/j.apsusc.2015.07.132

[34] Deng JX, Lian YS, Wu Z, Xing YQ. Performance of femtosecond laser-textured cutting tools deposited with WS2 solid lubricant coatings. Surf. Coat. Tech. 2013; 222: 135–143. DOI: 10.1016/j.surfcoat.2013.02.015

[35] Wang Z, Zhao QZ, Wang CW. Reduction of friction of metals using laser-induced periodic surface nanostructures. Micromachines 2015; 6(11): 1606–1616. DOI: 10.3390/mi6111444

[36] Lian YS, Deng JX, Yan GY, Cheng HW, Zhao J. Preparation of tungsten disulfide (WS2) soft-coated nano-textured self-lubricating tool and its cutting performance. Int. J. Adv. Manuf. Tech. 2013; 68(9–12): 2033–2042. DOI: 10.1007/s00170-013-4827-y

[37] Zhang KD, Deng JX, Meng R, Gao P, Yue HZ. Effect of nano-scale textures on cutting performance of WC/Co-based Ti55Al45N coated tools in dry cutting. Int. J. Refract. Met. H. 2015; 51: 35–49. DOI: 10.1016/j.ijrmhm.2015.02.011

[38] Von der linde D, Sokolowski-Tinten, K, Bialkowski J. Laser-solid interaction in the femtosecond time regime. Appl. Surf. Sci. 1997; 109: 1–10. DOI: 10.1016/S0169-4332(96)00611-3

[39] Momma C, Nolte S, Chichkov BN, Alvensleben FV, Tunnermann A. Precise laser ablation with ultrashort pulses. Appl. Surf. Sci. 1997; 109: 15–19. DOI: 10.1016/S0169-4332(96)00613-7

[40] Becker PC, Fragnito HL, Brito Cruz CH, Fork RL, Cunningham JE, Henry JE, Shank CV. Femtosecond photon echoes from band-to-band transitions in GaAs. Phys. Rev. Lett. 1988; 61(14): 1647–1649. DOI: 10.1103/PhysRevLett.61.1647

[41] Chichkov BN, Momma C, Nolte S, Alvensleben FV, Tunnermann A. Femtosecond, picosecond and nanosecond laser ablation of solids. Appl. Phys. A-Mater. 1996; 63(2): 109–115.

[42] Birnbaum M. Semiconductor surface damage produced by ruby lasers. J. Appl. Phys. 1965; 36: 3688–3689.

[43] Wu B, Zhou M, Li J, Ye X, Li G, Cai L. Superhydrophobic surfaces fabricated by microstructuring of stainless steel using a femtosecond laser. Appl. Surf. Sci. 2009; 256(1): 61–66. DOI: 10.1016/j.apsusc.2009.07.061

[44] Ahsan MS, Ahmed F, Kim YG, Lee MS, Jun MBG. Colorizing stainless steel surface by femtosecond laser induced micro/nano-structures. Appl. Surf. Sci. 2011; 257(17): 7771–7777. DOI: 10.1016/j.apsusc.2011.04.027

[45] Yasumaru N, Miyazaki K, Kiuchi J. Control of tribological properties of diamond-like carbon films with femtosecond-laser-induced nanostructuring. Appl. Surf. Sci. 2008; 254(8): 2364–2368. DOI: 10.1016/j.apsusc.2007.09.037

[46] Zhao QZ, Malzer S, Wang LJ. Formation of subwavelength periodic structures on tungsten induced by ultrashort laser pulses. Opt. Lett. 2007; 32(3): 1932–1934. DOI: 10.1364/OL.32.001932

[47] Mannion PT, Magee J, Coyne E, O'Connor GM, Glynn TJ. The effect of damage accumulation behavior on ablation thresholds and damage morphology in ultrafast laser micro-machining of common metals in air. Appl. Surf. Sci. 2004; 233(1–4): 275–287. DOI: 10.1016/j.apsusc.2004.03.229

[48] Yasumaru N, Sentoku E, Miyazaki K, Kiuchi J. Femtosecond-laser-induced nanostructure formed on nitride stainless steel. Appl. Surf. Sci. 2013; 264(12): 611–615. DOI: 10.1016/j.apsusc.2012.10.076

[49] Bizi-bandoki P, Valette S, Audouard E, Benayoun S. Effect of stationary femtosecond laser irradiation on substructures' formation on a mold stainless steel surface. Appl. Surf. Sci. 2013; 270: 197–204. DOI: 10.1016/j.apsusc.2012.12.168

[50] Huang M, Zhao FL, Cheng Y, Xu NS, Xu ZZ. Origin of laser-induced near-subwavelength ripples: interference between surface plasmons and incident laser. ACS Nano 2009; 3(12): 4062–4070. DOI: 10.1021/nn900654v

[51] Buividas R, Mikutis M, Juodkazis S. Surface and bulk structuring of materials by ripples with long and short laser pulses: recent advances. Prog. Quant. Electron. 2014; 38(3): 119–156. DOI: 10.1016/j.pquantelec.2014.03.002

[52] Zhou GS, Fauchet PM, Siegman AE. Growth of spontaneous periodic surface structures on solids during laser illumination. Phys. Rev. B 1982; 26(10): 5366–5381. DOI: 10.1103/PhysRevB.26.5366

[53] Soileau MJ. Ripple structures associated with ordered surface defects in dielectrics. IEEE J. Quantum Elect. 1984; 20(5): 464–467. DOI: 10.1109/JQE.1984.1072422

[54] Emmony DC, Howson RP, Willis LJ. Laser mirror damage in germanium at 10.6 μm. Appl. Phys. Lett. 1973; 23(11): 598–600. DOI: 10.1063/1.1654761

[55] Raether H. Surface plasmons on smooth and rough surfaces and on gratings. Springer; 1988. DOI: 10.1007/BFb0048317

[56] Suh NP, Sin HC. The genesis of friction. Wear 1981; 69(1): 91–114. DOI: 10.1016/0043-1648(81)90315-X

[57] Kim DE, Suh NP. On microscopic mechanism of friction and wear. Wear 1991; 149(1–2): 199–208. DOI:10.1016/0043-1648(91)90373-3

[58] Ding Q, Wang LP, Hu LT, Hu TC, Wang Y. The pairing-dependent effects of laser surface texturing on micro tribological behavior of amorphous carbon film. Wear 2012; 274: 43–49. DOI: 10.1016/j.wear.2011.08.008

[59] Gualtieri E, Borghi A, Calabri L, Pugno N, Valeri S. Increasing nanohardness and reducing friction of nitride steel by laser surface texturing. Tribol. Int. 2009; 42(5): 699–705. DOI: 10.1016/j.triboint.2008.09.008

[60] Gachot C, Rosenkranz A, Reinert L, Ramos-Moore E, Souza N, Muser MH, Mucklich F. Dry friction between laser-patterned surfaces: role of alignment, structural wavelength and surface chemistry. Tribol. Lett. 2013; 49(1): 193–202. DOI: 10.1007/s11249-012-0057-y

[61] Wang Z, Wang CW, Wang M, Zhao QZ. Manipulation of tribological properties of stainless steel by picosecond laser texturing and quenching. To be published on Tribol. Int.

[62] Sugihara T, Enomoto T. Development of a cutting tool with a nano/micro-textured surface-improvement of anti-adhesive effects by considering the texture patterns. Precis. Eng. 2009; 33(4): 425–429. DOI: 10.1016/j.precisioneng.2008.11.004

Chapter 2

Tribology of Slip Surfaces in Journal Bearings

Qiyin Lin, Baotong Li and Hong Zhao

Additional information is available at the end of the chapter

http://dx.doi.org/10.5772/63532

Abstract

Boundary slip can be triggered by certain engineered surfaces. Journal bearings with slip surfaces have markedly different performances compared with traditional journal bearings. In order to reasonably utilise boundary slip to improve the tribological performances of journal bearings, including hydrodynamic and hybrid journal bearings, it is necessary to draw out the impact laws of boundary slip on the performances of journal bearings. The design criterion of slip surfaces is presented in this chapter; only a well-designed slip surface could improve the tribological performances of journal bearings. This chapter could provide a valuable guide for the design of slip surfaces in journal bearings.

Keywords: boundary slip, journal bearing, tribological performance, impact law, design criterion

1. Introduction

Tribological performances of lubricated contacts are strongly governed by the boundary conditions of fluid flows that provide lubrication. Traditional textbooks always assume that the immediate layer of liquid next to a solid surface moves with the same tangential velocity as the solid surface itself, which is the well-known 'no-slip' assumption; but for over a century, there have been persistent doubts about its validity.

In recent years, a number of experiments have shown that for certain engineered surfaces, the no-slip boundary condition is not a valid one, and boundary slip might occur at the fluid-solid interface [1–3]. Such slip surfaces can be obtained by modifying the geometrical micro- or nanostructure of the surfaces and controlling the surface energy. The micro- or nanostructured patterns on solid surfaces can be created using micro-nano fabrication techniques such as plasma etching, lithographic techniques, electro chemical etching, laser texturing and so on.

And the surface energy of a solid surface can be controlled by such techniques as film or molecule deposition, solution coating, or self-assembly of hydrophobic layers.

A large number of literatures reported that the boundary slip phenomenon is very obvious and remarkable when liquids flow over superhydrophobic surfaces (not-wettable surfaces). It is remarkable that when boundary slip is introduced into lubricated contacts, their tribological performances would present significant differences comparing with that without slip [4–7]. Slip surfaces will significantly influence the tribological performances of lubricated contacts [5–12]. Thus, it is necessary and valuable to draw out these changes and differences with and without boundary slip. This chapter will look into the tribology of slip surfaces in journal bearings and reveal their impact laws on the tribological performances of hydrodynamic and hybrid journal bearings.

2. Slip numerical models

Nowadays, there are mainly three numerical models to describe the boundary slip phenomenon, i.e. slip length model, limiting shear stress slip model and slip intensity model.

The slip length model, also named Navier slip model, states that the slip velocity, U_s, is proportional to the surface shear rate, $\partial u/\partial y$, and slip length, b, seeing Eq. (1).

$$U_s = b \cdot \frac{\partial u}{\partial y} \tag{1}$$

The slip length, b, is defined as the fictive distance below the solid surface where the velocity extrapolates linearly to zero, as shown in **Figure 1**.

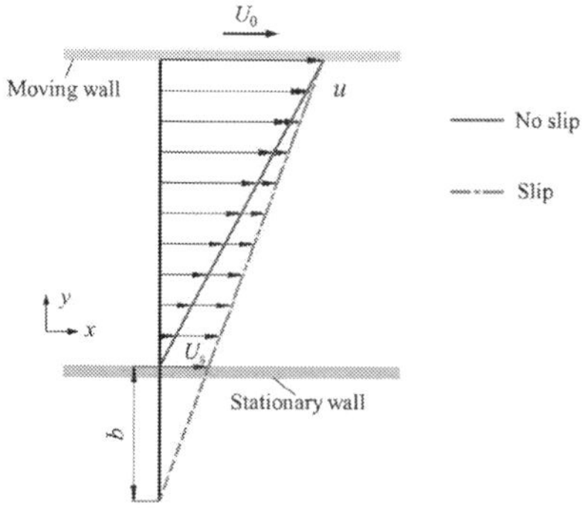

Figure 1. Slip length model.

The limiting shear stress slip model assumes that there is a critical shear stress, τ_c, at the fluid-solid interface and the wall slip occurs only when the wall shear stress, τ_0, reaches the critical value, τ_c. If slippage occurs, the surface shear stress, τ_s, is equal to the critical value, i.e.

$$\tau_s = \tau_c \quad (\tau_0 \geq \tau_c)$$

and others,

$$\tau_s = \tau_0 \quad (\tau_0 < \tau_c)$$

where

$$\tau_0 = \mu \cdot \frac{\partial u}{\partial y}$$

The reported slip length, b, ranges from the order of the mean free path of fluid molecular to micrometer, even centimeter; the reported critical shear stress, τ_c, still exists in an over-broad range; consequently, it is difficult to qualify the exacting value of slip length and critical shear stress.

Not only the slip length model but also the limiting shear stress slip model indicates that the slip velocity, U_s, is always related to fluid velocity, U_i, at the nearest region close to this fluid-solid interface. For the convenience, to determine the slip intensity, a slip-intensity factor, γ, is introduced and restricted from 0 to 1, which seems to be much easier to qualify, and the slip intensity increases with the value of slip-intensity factor, γ; thus, a slip intensity model is derived from the limiting shear stress slip model then and is expressed as

$$U_s = \gamma \cdot (U_i - n \cdot < U_i \cdot n >) \tag{2}$$

where n is the surface normal vector; the dot produce, $U_i \cdot n$, is the projection value of fluid velocity, U_i, projecting to the surface normal direction; the minus sign, –, in the formula represents that the angle between the surface normal vector, n, and the velocity vector, U_i, is an obtuse angle; and the dot produce, $n \cdot <U_i \cdot n>$, represents the velocity component of, U_i, in the surface normal direction.

Assuming that the three components of slip velocity, U_s, are u_s, v_s, w_s, and those three components of fluid velocity, U_i, are u_i, v_i, w_i, respectively, under the coordinate system x_n-y_n-z_n, where the direction of surface normal vector n is -z_n, as shown in **Figure 2**, Eq. (2) can be rewritten as

$$\begin{cases} u_s = \gamma \cdot u_i \\ v_s = \gamma \cdot v_i \\ w_s = 0 \end{cases}$$

Consequently, the slip intensity model indicates that the boundary slip only occurs in the tangential direction. When $\gamma = 1$, this slip model becomes the limiting shear stress slip model with $\tau_c = 0$, i.e. a shear free condition, also called perfect slip condition. When $\gamma = 0$, this slip model regresses to a no-slip boundary condition with zero velocity (stationary wall). For the convenience to identify the slip intensity and analyse its influences on the performances of journal bearings, the slip intensity model is utilised to represent the boundary slip in this chapter.

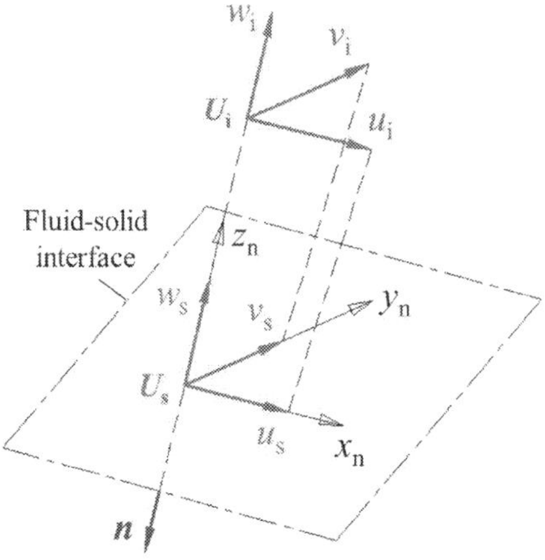

Figure 2. Schematic diagram of velocity.

3. Governing equations and cavitation model

In order to reduce the generation of frictional heat in high-speed journal bearing systems, lubricant with low-viscosity, such as water, is usually utilised, and water is chosen as the lubricant in this chapter. Assuming that water is a continuous isoviscous incompressible fluid medium, so continuity equation and full Navier–Stokes equations are used to predict the performances of the fluid domain. For steady-state study, these governing equations are expressed as

$$\frac{\partial u}{\partial x} + \frac{\partial v}{\partial y} + \frac{\partial w}{\partial z} = 0 \tag{3}$$

$$\rho\left(u\frac{\partial u}{\partial x} + v\frac{\partial u}{\partial y} + w\frac{\partial u}{\partial z}\right) = -\frac{\partial p}{\partial x} + \mu\left(\frac{\partial^2 u}{\partial x^2} + \frac{\partial^2 u}{\partial y^2} + \frac{\partial^2 u}{\partial z^2}\right) \tag{4}$$

http://dx.doi.org/10.5772/63532

$$\rho\left(u\frac{\partial v}{\partial x}+v\frac{\partial v}{\partial y}+w\frac{\partial v}{\partial z}\right)=-\frac{\partial p}{\partial y}+\mu\left(\frac{\partial^2 v}{\partial x^2}+\frac{\partial^2 v}{\partial y^2}+\frac{\partial^2 v}{\partial z^2}\right) \tag{5}$$

$$\rho\left(u\frac{\partial w}{\partial x}+v\frac{\partial w}{\partial y}+w\frac{\partial w}{\partial z}\right)=-\frac{\partial p}{\partial z}+\mu\left(\frac{\partial^2 w}{\partial x^2}+\frac{\partial^2 w}{\partial y^2}+\frac{\partial^2 w}{\partial z^2}\right) \tag{6}$$

where $u/v/w$ are velocity components, $x/y/z$ are Cartesian coordinates, ρ is fluid density, p is fluid pressure, μ is viscosity.

Additionally, gas may escape from the lubricant or lubricant may vapour, when the pressure drops to negative value, which means that the fluid film of journal bearings ruptures and a cavitation phenomenon occurs. As it was well known, some numerical models had been developed to describe the cavitation phenomenon, for example, the half-Sommerfeld boundary condition, Reynolds boundary condition and the Jakobsson-Floberg-Olsson (JFO) formulation. In order to consider the generating mechanism of vapour and the transitions between liquid and vapour, the cavitation phenomenon in fluid domain is governed by a phase change model based on pressure change, i.e. phase transitions between liquid and vapour resulting from pressure change.

The transition from liquid phase to vapour phase is triggered when the pressure is less than the saturation vapour pressure, P_{Sat}. All the variables in the governing equations adopt the corresponding value of mixture phase (i.e. liquid phase + vapour phase). The physical properties of the mixture phase are scaled by the liquid volume fraction, f. The liquid volume fraction, f, is defined as the ratio of liquid volume to the total volume (liquid volume + vapour volume). So, the physical properties of mixture phase α are defined as

$$\alpha = f\cdot\alpha_{\text{l}} + (1-f)\cdot\alpha_{\text{v}} \tag{7}$$

The subscripts l and v represent the liquid phase and vapour phase, respectively. As the pressure is higher than the saturation vapour pressure, P_{Sat}, the liquid volume fraction f is equal to 1. As it is well known, the ambient pressure (its absolute value is 1 atm) is usually used to study the cavitation phenomenon in tribological analysis. The saturation vapour pressure, P_{Sat}, is equal to 1 atm in this chapter. When there is a liquid-vapour transition as the pressure changes, mass conservation is satisfied and the mass transfer rate m is to be modelled as

$$\begin{cases} m^{+} = \dfrac{2C_{\text{c}}}{U_{\text{m}}^{2}t_{\text{m}}}\cdot(1-f)\cdot\max(0, p-P_{\text{Sat}}) \\ m^{-} = \dfrac{2C_{\text{v}}\rho_{\text{l}}}{U_{\text{m}}^{2}t_{\text{m}}\rho_{\text{v}}}\cdot f\cdot\min(0, p-P_{\text{Sat}}) \end{cases} \tag{8}$$

where C_c, C_v, U_m and t_m are empirical constants based on the mean flow. The superscripts, + and –, represent the phase change from vapour to liquid (condensation) and the phase change from liquid to vapour (vapourisation), respectively.

4. Influences of boundary slip on hydrodynamic journal bearings

4.1. Hydrodynamic journal bearing model

The hydrodynamic journal bearing is presented in **Figure 3**. The journal radius, *r*, is 25 mm; the bearing width, *B*, is 25 mm; the radial clearance, *c*, is 50 μm; and the eccentricity is 0.55. The density of lubricant (water), ρ, is 998.2 kg·m^{-3}, and its viscosity, μ, is 1.003 mPa·s. The thickest and thinnest positions of the lubrication film are located at 90° and 270° in the circumferential direction, respectively. The rotational speed of journal ω is 10^4 rpm, about 1047.2 rad·s^{-1}. The pressures at the two end surfaces of lubricant domain in axial direction are equal to ambient pressure.

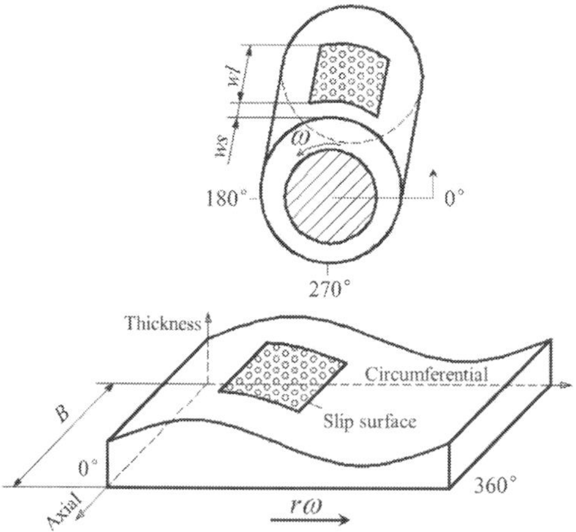

Figure 3. Hydrodynamic journal bearing with a slip surface.

The slip surface is designed on the internal surface of bearing bush, as shown in **Figure 3**. The distance from the bearing end surface to the starting position of slip surface in axial direction is marked as *ws*. The ratio of *ws* to the bearing width, *B*, is defined as the dimensionless slip-surface starting position, WS, i.e. WS = *ws*/*B*. And the dimensionless slip-surface width, WL, is defined as the ratio of the slip-surface width, *wl*, in axial direction to the bearing width, *B*, i.e. WL = *wl*/*B*. The slip intensity model is applied to slip region, and the traditional no-slip boundary condition is applied to other regions.

http://dx.doi.org/10.5772/63532

4.2. Slip intensity's influences

The influences of slip intensity on tribological performance of hydrodynamic journal bearing are investigated first. Two bearing models with different slip surfaces are analysed: one slip surface is located at 160–200° in circumferential direction, the other slip surface is located at 250–260°; their axial location is both defined by WL = 1.0 (which implies WS = 0).

Boundary slip not only could improve the load-carrying capacity of hydrodynamic journal bearing but also could reduce it, which is related to the position of slip surface, as shown in **Figure 4**, and we will analyse in detail later. The load-carrying capacity, w, is obtained from the pressure integral over the journal surface, and the dimensionless load-carrying capacity is defined as $W = w \cdot c^2 / (\mu \cdot r^3 \cdot \omega \cdot B)$.

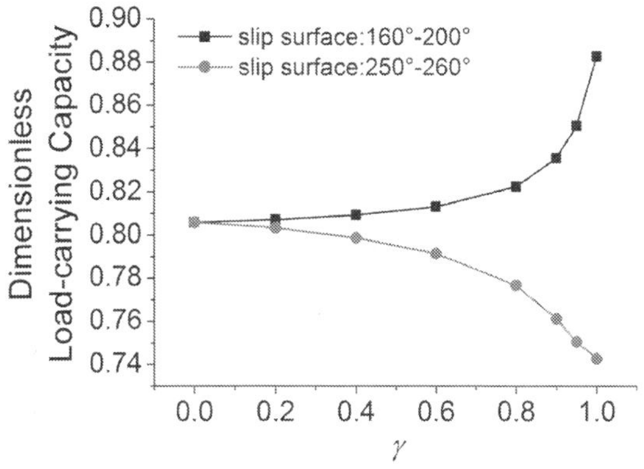

Figure 4. Influences of slip intensity.

If boundary slip could enhance the load-carrying capacity, its beneficial influence increases with the slip intensity, i.e. the value of slip-intensity factor, γ. Otherwise, if boundary slip has adverse effect on load-carrying capacity, its negative influence will also increase with the slip intensity. In a word, the influence of slip intensity is monotonic. For focusing on the influences of the position and size/area of slip surface, a same slip intensity, $\gamma = 1$, is utilised in the following analysis.

4.3. Influences of circumferential positions and sizes of slip surfaces

The influences of location and size of slip surface in the circumferential direction on pressure and load-carrying capacity are investigated. In this section, the impacts of location and size of slip surface in axial direction are not taken into account, so the slip region covers the whole internal surface of bearing bush in axial direction. The dimensionless pressure is defined as $P = p \cdot c^2 / (\mu \cdot r \cdot \omega \cdot B)$.

4.3.1. Slip surfaces in cavitation zone

The pressure distribution in the medium cross section in axial direction is illustrated in **Figure 5**, for these two situations, namely the size of slip surface is zero and the slip surface is located from 320° to 340°. For this situation, the size of slip surface is zero, i.e. there is no slip surface, and the maximum pressure is located at 234.5°. This pressure peak results from the fluid hydrodynamic action while the lubricant is flowing into a convergence region (bearing bush and journal shaft are not concentric). The hydrodynamic pressure rising zone is located from 94.5° to 234.5°, where $\partial p/\partial \theta > 0$, θ represents the circumferential angle. The pressure drop zone is located from 234.5° to 283.6°, where $\partial p/\partial \theta < 0$. The cavitation zone is located at 0–94.5° and 283.6–360°.

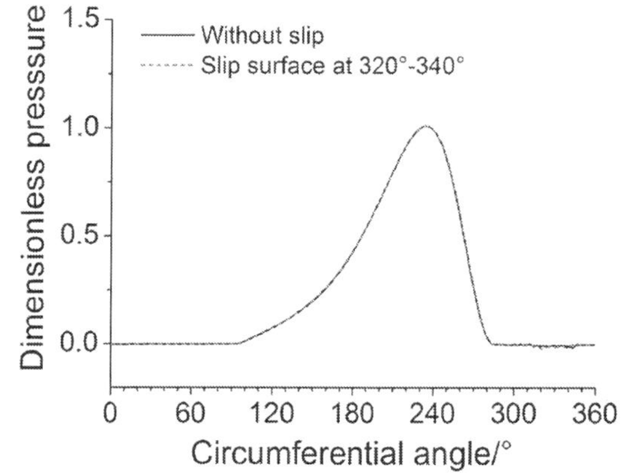

Figure 5. Pressure distribution for without slip surface and slip surface located at 320–340°.

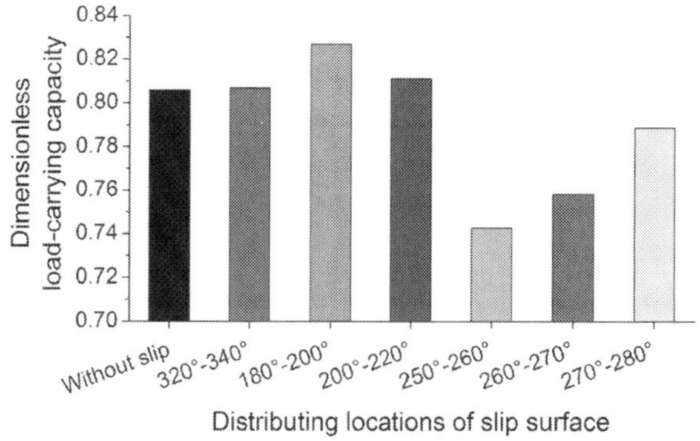

Figure 6. Load-carrying capacity for different slip surfaces.

It can be seen from **Figure 5**, for this situation, the slip surface is located from 320° to 340°, namely it's just located at the cavitation zone, its pressure distribution is the same as that when there is no boundary slip. The difference in load carrying capacity between these two situations is also very small and can be neglected, as shown in **Figure 6**. The reason lies on the pressure in cavitation zone. When fluid pressure drops below the saturation vapour pressure, phase change occurs and liquid is converted into vapour; then the pressure in these vapour regions (cavitation regions) is equal to the saturation vapour pressure. Consequently, if slip surface is located in the cavitation zone, it would have no influence on pressure.

4.3.2. Slip surfaces in pressure rising zone

Pressure distribution along circumferential direction for these three situations, namely there is no slip surface and the slip surface is from 180° to 200° and from 200° to 220°, is presented in **Figure 7**. These three situations all have a pressure peak, which is located at the same position near 234.5°. This pressure peak is induced by the fluid hydrodynamic action when the lubricant flows into a convergence region, because the bearing bush and rotary shaft are not concentric.

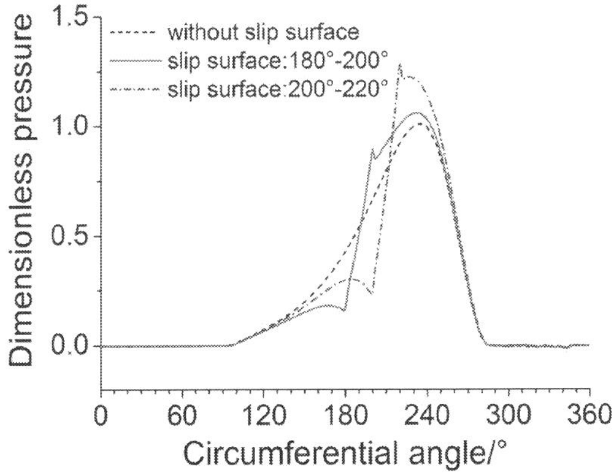

Figure 7. Pressure distribution for slip surfaces located at pressure rising zone.

For these two situations, namely slip surface is located at 180–200° and 200–220°, there is one other pressure peak, which is located at 200° and 220°, respectively. These two additional pressure peaks are just located at the end line of the downstream zone of slip surface, which is induced by slip surface. This is because when the fluid (lubricant) flows from a slip region into a no-slip region, its speed will decrease, i.e. the kinetic energy of lubricant will decrease, and the kinetic energy will transform into pressure energy; thus, the slip surface will produce a fluid hydrodynamic action in its downstream zone.

It can be seen from **Figure 7**, both these two slip situations also have a valley point in the curves of pressure distribution, which are located at 180° and 200°, i.e. the starting positions of slip surfaces; that is, to say that there is a negative fluid hydrodynamic action in the upstream zone of slip region. These valley points of pressure are induced by slip surface too. This is because when the fluid (lubricant) flows from a no-slip region to a slip region, the speed of fluid increases, i.e. the kinetic energy of lubricant is increased, and the increased part of kinetic energy is transformed from the pressure energy, so the pressure is decreased.

Taking the highest pressure of this case without slip surface, namely the pressure value at 234.5°, as a reference pressure, when the slip surface is located at 180–200°, the value of pressure peak at 200° is about 88.5% of reference pressure and the pressure value at 234.5° is 1.1 times of reference pressure. When the slip surface is located at 200–220°, the value of pressure peak at 220° is about 1.28 times of reference pressure and the pressure value at 234.5° is 1.21 times of reference pressure. Consequently, slip surfaces could enhance fluid hydrodynamic action. Slip surfaces at 180–200° and 200–220° are located at the hydrodynamic pressure rising zone, namely the region from 94.5° to 234.5°. It also can be seen form **Figure 6**, the load-carrying capacities of these two situations, i.e. slip surfaces located at 180–200° and 200–220°, are larger than that without slip surface. These indicate that the fluid hydrodynamic actions produced by boundary slip and convergence structure would promote each other when the slip surface is located in pressure rising zone, and the pressure of lubrication film will increase, and then it also results in an increase in load-carrying capacity.

The pressure distribution for these slip surfaces whose starting positions are fixed at 160° is illustrated in **Figure 8**. Due to the negative fluid hydrodynamic action induced by slip surfaces in their upstream zones, there is a valley value located at 160° for each pressure distribution curve. It can be seen clearly from **Figure 8**, for these three situations that slip surface is located at 160–200°, 160–220° and 160–240°, and the fluid films have ruptured lightly, meaning that a little cavitation phenomenon occurs.

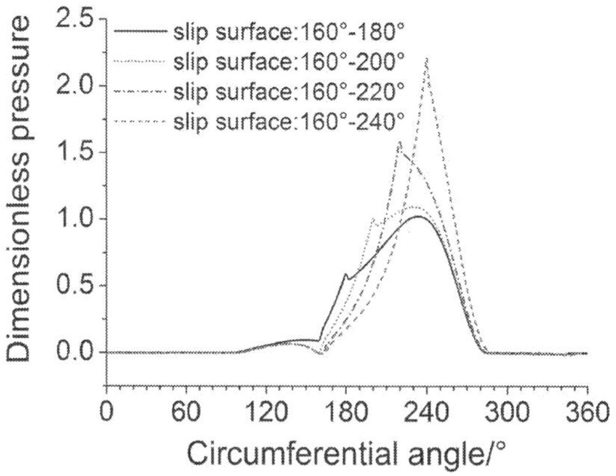

Figure 8. Pressure distribution for slip surfaces starting at 160°.

The pressure distribution curve for slip surface located at 200–220° has two pressure peaks, one located at 220° and the other located at 234.5°; while it has only one pressure peak located at 220°, for slip surface located at 160–220°. When slip surfaces' end positions are located near 234.5°, the two pressure peaks induced by slip surface and convergence structure, respectively, may merge into a single peak, for example, the slip surfaces located at 160–220° and 160–240°. The size/area of slip surface located at 160–220° is bigger than that located at 200–220°; thus, the fluid hydrodynamic action induced by slip surface located at 160–220° is correspondingly stronger than that for the slip surface located at 200–220°; then, the pressure peak due to slip surface located at 160–220° covers the pressure peak due to convergence structure, while the pressure peak due to slip surface located at 200–220° cannot cover the pressure peak due to convergence structure.

The load-carrying capacity for these slip surfaces whose starting positions are fixed at 160° is illustrated in **Figure 9**, and it shows that the load-carrying capacity increases with the size/area of slip surface. When the end position of slip surface moves from 180° to 240°, correspondingly the circumferential length of slip surface increases from 20° to 80°, and the improving ratio of load-carrying capacity, comparing with the load-carrying capacity without slip surface, increases from 3.2 to 27.9%. It can be concluded that the load-carrying capacity increases with the size/area of slip surface located at the hydrodynamic pressure rising zone.

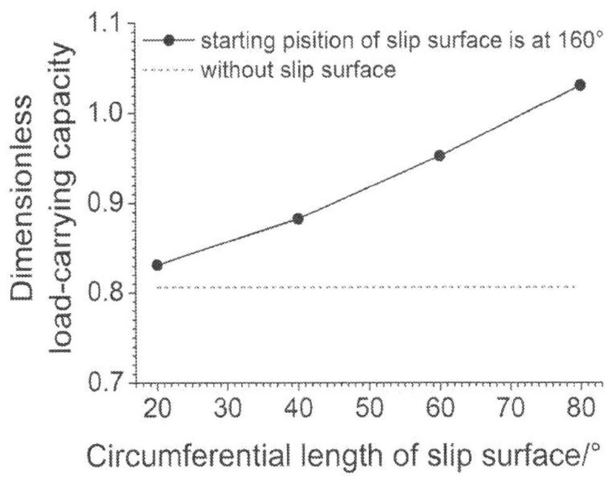

Figure 9. Load-carrying capacity for slip surfaces starting at 160 °.

4.3.3. Slip surfaces in pressure drop zone

The pressure distribution for slip surfaces located at pressure drop zone, namely 250–260°, 260–270° and 270–280°, is shown in **Figure 10**. Due to the negative fluid hydrodynamic action induced by slip surfaces at their upstream zones, these three pressure distribution curves have a valley value located at each starting position of slip surfaces, i.e. at 250°, 260° and 270°. There still is a pressure peak located at each end position of slip surfaces, i.e. at 260°, 270° and 280°. These pressure peaks are induced by slip surfaces. Because these positions of the negative fluid hydrodynamic action induced by slip surfaces located near the high-pressure zone of the fluid hydrodynamic action induced by convergence structure, the fluid hydrodynamic action induced by convergence structure will be inhibited and its intensity correspondingly becomes much lower. Sometimes, the fluid film at the upstream zone of slip surface will be broken up, for example, for these situations when slip surfaces are located at 260–270° and 270–280°; and it is meant that the cavitation phenomenon is enhanced. This is the reason why the load-carrying capacity for those slip surfaces located at pressure drop zone is smaller than that without slip surface. It can be clearly seen from **Figure 6**, the load-carrying capacity, when slip surface locates at 250–260°, 260–270° and 270–280°, is much smaller compared with that without slip surface.

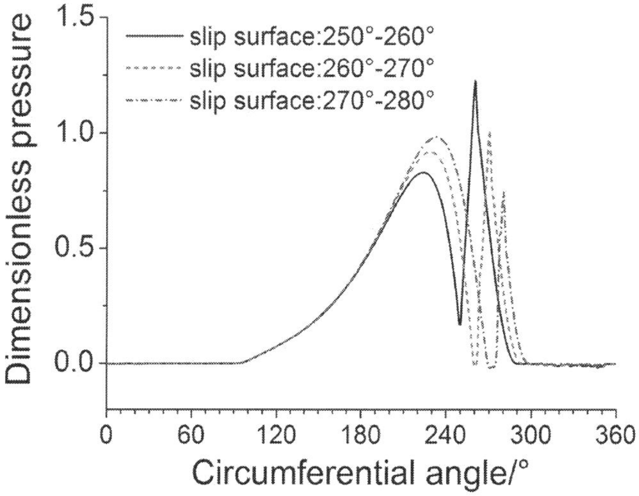

Figure 10. Pressure distribution for slip surfaces located at pressure drop zone.

The pressure distribution for these slip surfaces whose starting positions are fixed at 240° is illustrated in **Figure 11**; these four slip surfaces all are located at the pressure drop zone. Due to the negative fluid hydrodynamic action induced by slip surfaces in their upstream zones, there is a valley value located at 240° for each pressure distribution curve. It can be seen clearly, for these three situations that slip surface located at 240–260°, 240–270° and 240–280°, the fluid films at the upstream zones of slip surfaces have ruptured, meaning that cavitation phenomenon occurs. Because the additional cavitation zones are located at the high-pressure zone of the fluid hydrodynamic action induced by convergence structure, there is an adverse influence on load-carrying capacity. As shown in **Figure 12**, when the starting positions of slip surfaces fix at 240° and their end positions move from 250° to 280°, their load-carrying capacities are all smaller than that without slip surface; as the circumferential length of slip surface increases from 10° to 40°, the decreasing ratio of load-carrying capacity, comparing with the load-carrying capacity without slip surface, increases from 5.9 to 17.0%. It can be concluded that the load-carrying capacity decreases with the size/area of slip surface located at hydrodynamic pressure drop zone.

http://dx.doi.org/10.5772/63532

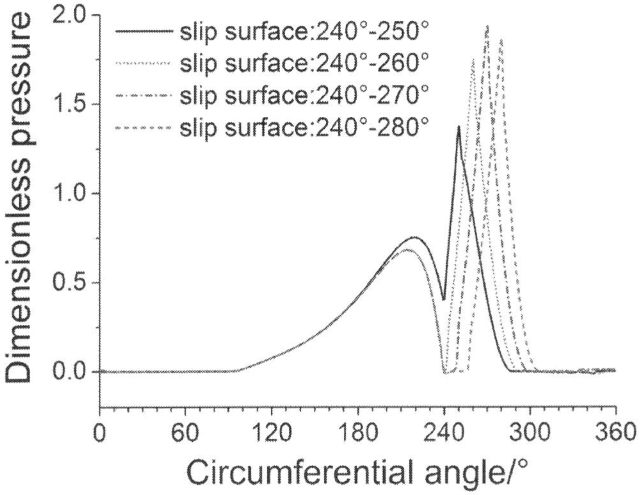

Figure 11. Pressure distribution for slip surfaces starting at 240°.

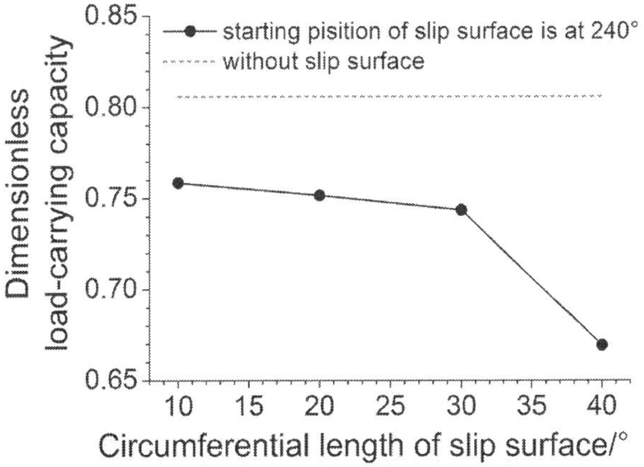

Figure 12. Load-carrying capacity for slip surfaces starting at 240°.

4.4. Influences of axial positions and sizes of slip surfaces

The influences of position and size of slip surface in the axial direction on load-carrying capacity are investigated in the following section. To focus on the impact of the parameters of slip surface in axial direction, the positions and sizes of slip surfaces in the circumferential direction are kept unchanged.

The three-dimensional distribution of local pressure for slip surfaces with different positions and areas in axial direction is illustrated in **Figure 13**; all these slip surfaces are located at 100–120° in the circumferential direction. The pressure peaks and valleys in the downstream and upstream zones of slip surfaces, respectively, are very obvious. It can be seen from **Figure 13 (a), (c)** and **(e)**, namely cases for slip surfaces with a same size (WL = 0.2), if the axial widths of slip surfaces are equal, the widths of pressure peak zones, as well as pressure valley zones, also kept equal, even though the axial positions of slip surfaces are different. As shown in **Figure 13 (b), (d)** and **(f)**, namely cases for slip surfaces with a same starting position (WS = 0.3), as the dimensionless slip-surface width increase from 0.3 to 0.5, the widths/sizes of pressure peak and valley zones still increase correspondingly. It can be concluded that the sizes of pressure peak and pressure valley zones induced by slip surfaces increase with the axial sizes of slip surfaces, but have no relationship with the axial positions of slip surfaces.

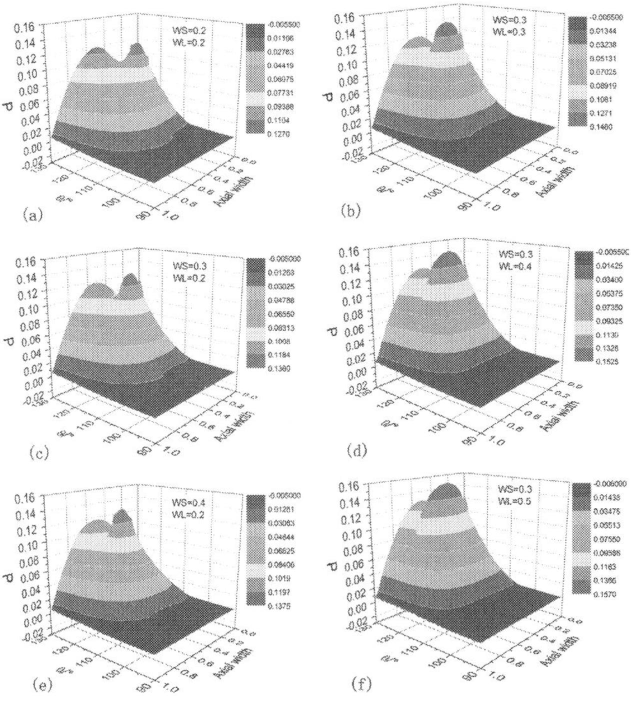

Figure 13. Three-dimensional distribution of local pressure for different slip surfaces.

The load-carrying capacity for slip surfaces with different positions and sizes in the axial direction is presented in **Figure 14**. These slip surfaces are located at 160–180° in the circumferential direction. The load-carrying capacity increases with the axial size of slip surface. For the case that the dimensionless starting position of slip surface, WS, is equal to 0.2, there is an increase of 3.8% in the load-carrying capacity when the dimensionless slip-surface width, WL, increases from 0.1 to 0.6.

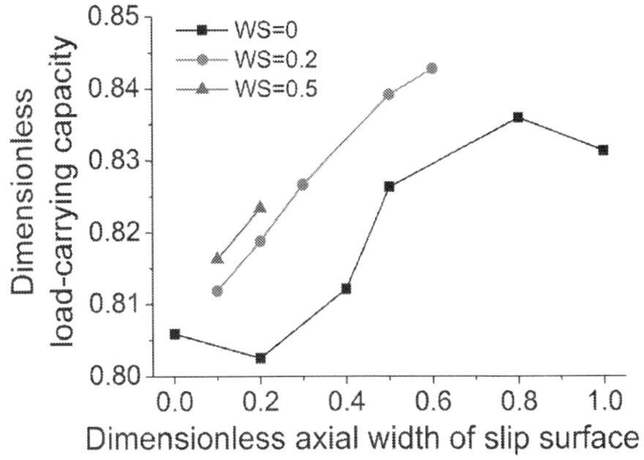

Figure 14. Load carrying-capacity for different slip surfaces.

The distribution curve of load-carrying capacity for WS = 0.5 lies above the curve for WS = 0.2, and the curve for WS = 0.2 lies above the curve for WS = 0, as shown in **Figure 14**. When the axial width of slip surface keeps unchanged (WS = 0.2), moving the starting position of slip surface from outlet (WS = 0) to bearing centre (WS = 0.5), there is an increase of 1.7% in load-carrying capacity. This indicates that the enhanced impacts of slip surfaces on load-carrying capacity are much stronger if the slip surface is located near the bearing centre, under the same axial width.

It can be seen from **Figure 14** the load-carrying capacity for the situation that WS = 0 and WL = 0.2 is smaller than that when WS = 0 and WL = 0. The load-carrying capacity for the situation WS = 0 and WL = 1.0 is also smaller than that when WS = 0 and WL = 0.8. When WS =0, the starting position of slip surface is just located at the bearing axial outlet. When WS = 0 and WL = 1.0, both the starting and the end positions of slip surface are just located at the bearing axial outlet. This situation, WS = 0 and WL = 0, means the size of slip surface is 0, namely there is no slip surface in bearing bush surface. Thus, it can be concluded that slip surface located at bearing axial outlet has an adverse effect on load-carrying capacity.

5. Influences of boundary slip on hybrid journal bearings

5.1. Hybrid journal bearing model

The hybrid journal bearing with two rectangular recesses is presented in **Figure 15**. The journal radius, r, is 25 mm; the bearing width, B, is 25 mm; the radial clearance, c, is 50 μm; and the eccentricity is 0.55. The density of lubricant (water), ρ, is 998.2 $kg{\cdot}m^{-3}$, and its viscosity, μ, is

1.003 mPa·s. The axial width of rectangular recess is 15 mm, its circumferential length is 30°, and its radial depth is 50 μm. These two rectangular recesses are located at 165–195° and 345–360°(0°)–15° along the circumference. The rotational speed of journal ω is 10^4 rpm, about 1047.2 rad·s^{-1}. The supply pressure is 15 kPa. The pressures at the two end surfaces of lubricant domain in axial direction are equal to ambient pressure. The thickest and thinnest positions of the lubrication film are located at 130° and 310° in the circumferential direction, respectively, seeing **Figure 15 (c)** which illustrates the thickness distribution and pressure distribution without slip surface.

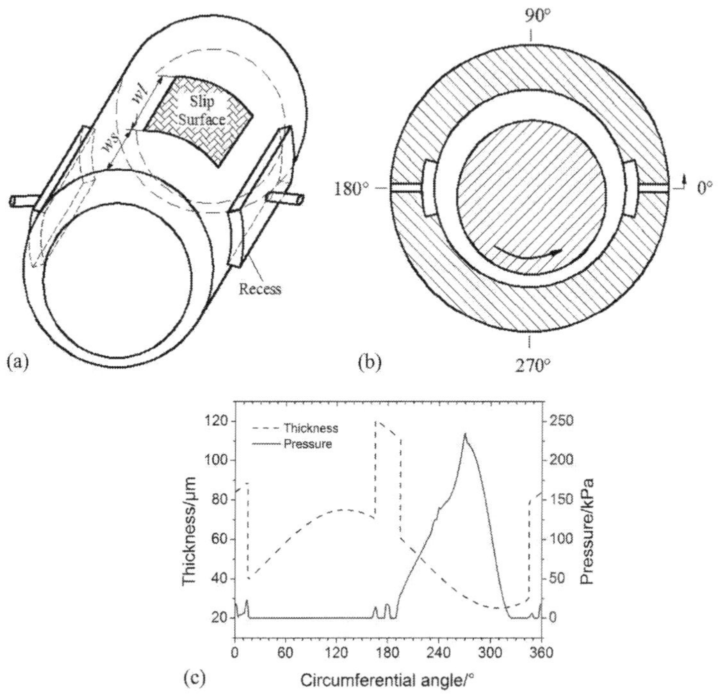

Figure 15. Schematic diagram of hybrid journal bearing: (a) complete view, (b) medium cross section in axial direction, (c) lubrication-film thickness and pressure distribution.

The slip surface is designed on the internal surface of bearing bush, as shown in **Figure 15 (a)**. The distance from the bearing end surface to the starting position of slip surface in axial direction is marked as ws. The ratio of ws to the bearing width, B, is defined as the dimensionless slip-surface starting position, WS, i.e. WS = ws/B. And the dimensionless slip-surface width, WL, is defined as the ratio of the slip-surface width, wl, in axial direction to the bearing width, B, i.e. WL=wl/B. The slip intensity model is applied to slip region, and the traditional no-slip boundary condition is applied to other regions.

5.2. Slip intensity's influences

The influences of slip intensity on tribological performance of hydrodynamic journal bearing are investigated first. Two bearing models with different slip surfaces are analysed: one slip surface is located at 210–240° in circumferential direction, the other slip surface is located at 270–300°; their axial location is both defined by WL = 1.0 (which implies WS = 0).

Boundary slip not only could improve the load-carrying capacity of hydrodynamic journal bearing but also could reduce it, which is related to the position of slip surface, as shown in **Figure 16**. This phenomenon is similar with that for hydrodynamic journal bearings with slip surfaces and we will analyse in detail later.

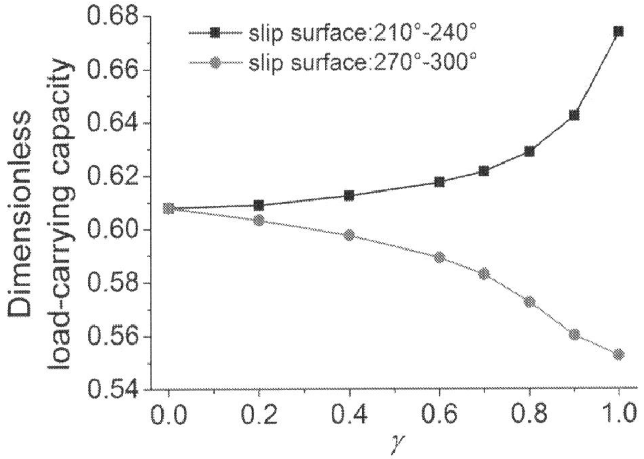

Figure 16. Influences of slip intensity.

If the boundary slip could enhance the load-carrying capacity, its beneficial influence increases with the slip intensity, i.e. the value of slip-intensity factor, γ. Otherwise, if the boundary slip has adverse effect on load-carrying capacity, its negative influence will increase with the slip intensity. In a word, the influence of slip intensity is monotonic. For focusing on the influences of the position and size/area of slip surface, a same slip intensity, $\gamma = 1$, is utilised in the following analysis.

5.3. Influences of circumferential positions and sizes of slip surfaces

The influences of location and size of slip surface in the circumferential direction on pressure and load-carrying capacity are investigated. In this section, the impacts of location and size of slip surface in axial direction are not taken into account, so the slip region covers the whole internal surface of bearing bush in axial direction.

5.3.1. Slip surfaces in cavitation zone

The pressure distribution in the medium cross section in axial direction is illustrated in **Figure 17**, for these two situations, namely the size of slip surface is zero and the slip surface is located from 60° to 90°. For this situation, the size of slip surface is zero, i.e. there is no slip surface, and the maximum pressure is about located at 270°. This pressure peak results from the fluid hydrodynamic action while the lubricant is flowing into a convergence region (bearing bush and journal shaft are not concentric). The hydrodynamic pressure rising zone is located from 195° to 270°, where $\partial p/\partial\theta > 0$, θ represents the circumferential angle. The pressure drop zone is located from 270° to 323°, where $\partial p/\partial\theta < 0$. The cavitation zone is located at 15–165° and 323–345°.

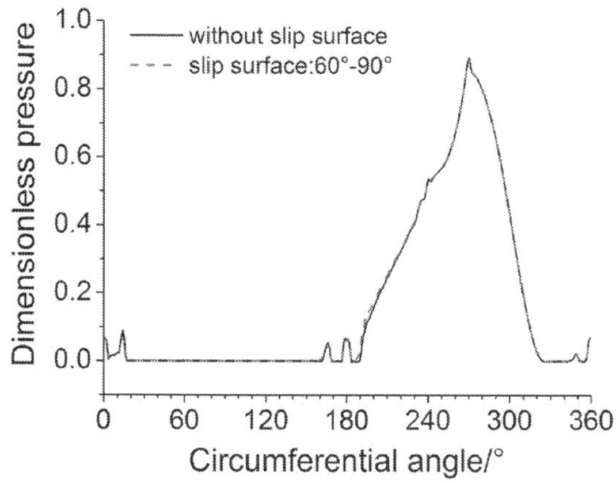

Figure 17. Pressure distribution for without slip surface and slip surface located at 60–90°.

It can be seen from **Figure 17**, for this situation, the slip surface is located from 60° to 90°, namely it's just located at the cavitation zone, its pressure distribution is the same as that when there is no boundary slip. The reason lies on the pressure in cavitation zone. When fluid pressure drops below the saturation vapour pressure, phase change occurs and liquid is converted into vapour, then the pressure in these vapour regions (cavitation regions) is equal to the saturation vapour pressure. Consequently, if slip surface is located in the cavitation zone, it would have no influence on pressure. Their difference in load-carrying capacity is also very small and can be neglected.

5.3.2. Slip surfaces in pressure rising zone

The pressure distribution in the medium cross section in axial direction is illustrated in **Figure 18**, for these four situations, namely the size of slip surface is zero and the slip surface located at 240–270°, 210–270° and 195–270°. These three slip surfaces, namely located at 240–270°, 210–270° and 195–270°, are located in the pressure rising zone.

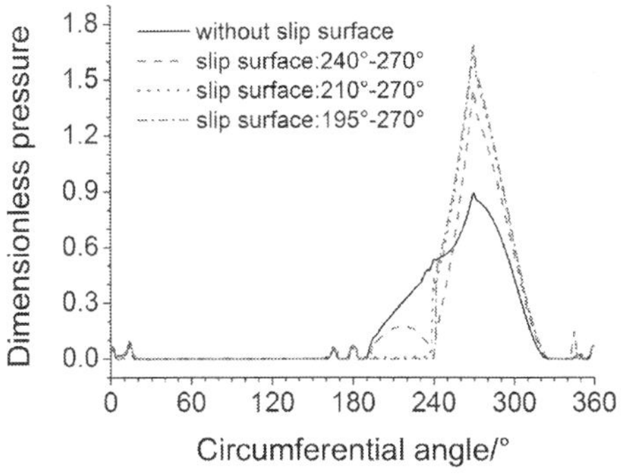

Figure 18. Pressure distribution for slip surfaces located in pressure rising zone.

For the three situations, namely when the slip surfaces are located at 240–270°, 210–270° and 195–270°, i.e. the end positions of slip surfaces are fixed at 270°, their maximum pressures are all higher than that without slip surface. These slip surfaces could enhance the fluid hydrodynamic action. The reason is that when the lubricant flows from a slip surface into a non-slip surface, its speed will decrease, i.e. the kinetic energy of the lubricant will decrease and then the kinetic energy will transform into pressure energy; as a result, the slip surface will produce a fluid hydrodynamic action in its downstream zone. The position of maximum pressure is just located at the end line of the downstream zone of the slip surface, as also clearly shown in **Figure 20**. Thus, the load-carrying capacity increases with the size of slip surface in the pressure rising zone, as shown in **Figure 19**.

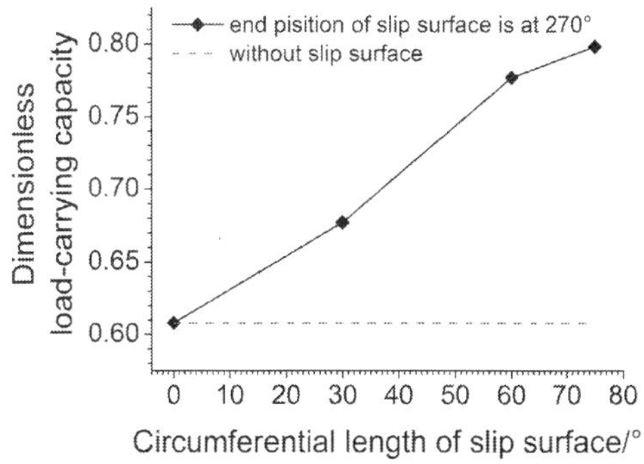

Figure 19. Load-carrying capacity for slip surfaces located in pressure rising zone.

Comparing with the situation without the slip surface, there is an increase of 11.4% in the load-carrying capacity when the slip surface is located from 240–270°. When the starting position of slip surface moves from 240° to 195°, the size of slip surface correspondingly increases from

30° to 75°, and the rate of load-carrying capacity comparing with the case without slip surface increases to 31.3%.

Figure 18 also shows that there is a negative fluid hydrodynamic action in the upstream zone of slip surface, indicating that there is a decrease in pressure. Because when the lubricant flows from a non-slip surface to a slip surface, the speed of fluid increases, i.e. the kinetic energy of the lubricant is increased, and the increased part of kinetic energy is transformed from the pressure energy, thereby decreasing the pressure. Thus, in the upstream zone of slip surface, the fluid hydrodynamic action induced by convergence structure will break down. In particular, when the slip surface is located in the pressure drop zone, the high-pressure zone induced by convergence structure, namely the main load-carrying zone, maybe broken up by the negative fluid hydrodynamic action due to slip surface, and the load-carrying capacity also may decrease, which will be discussed in the following section.

5.3.3. Slip surfaces in pressure drop zone

Pressure distribution and load-carrying capacity for slip surfaces located in pressure drop zone are presented in **Figures 20** and **21**, respectively. **Figure 21** indicates that the load-carrying capacity decreases with increasing size of the slip regions in the pressure drop zone. These four slip regions have the same starting location, located at 270°, and their end locations are located at 300°, 305°, 310° and 315°. Compared with the situation without slip surface, there is a decrease of 9.1% in load-carrying capacity when the slip surface is from 270° to 300°. When extending the end position of slip surface to 315°, the load-carrying capacity has a further decrease and its decrease rate, comparing with the case without slip surface, increases to 57.2%.

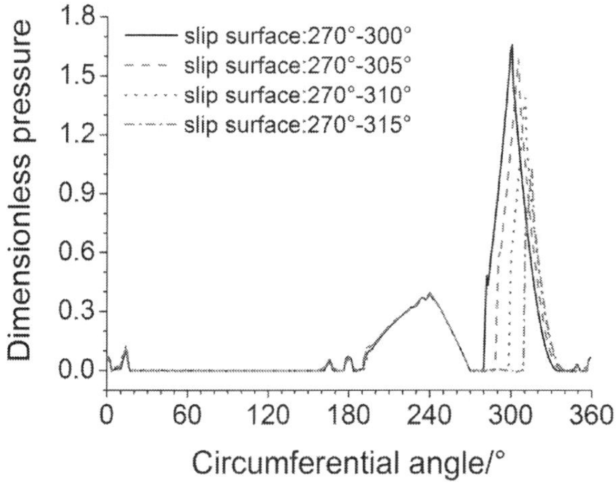

Figure 20. Pressure distribution for slip surfaces located in pressure drop zone.

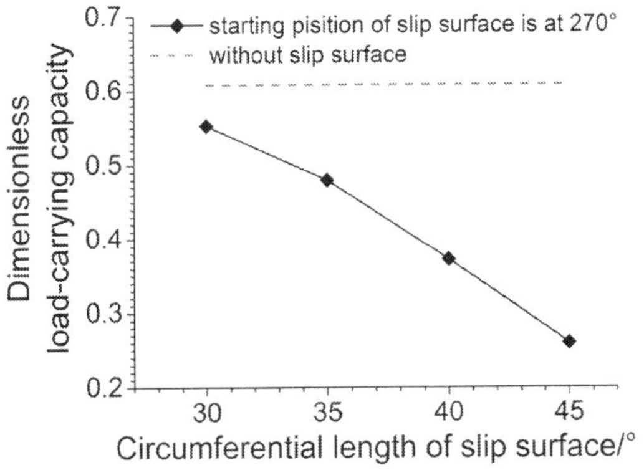

Figure 21. Load-carrying capacity for slip surfaces located in pressure drop zone.

5.4. Influences of axial positions and sizes of slip surfaces

The influences of position and size of slip surface in the axial direction on load-carrying capacity are investigated in the following section. To focus on the impact of the parameters of slip surface in axial direction, the positions and sizes of slip surfaces in the circumferential direction are kept unchanged, and these slip surfaces are located from 210° to 240°. The load-carrying capacity for slip surfaces with different positions and sizes in the axial direction is presented in **Figure 22**. The load-carrying capacity increases with the axial size of slip surface. For the case that the dimensionless starting position of slip surface, WS, is equal to 0.15, there is an increase of 8.4% in the load-carrying capacity when the dimensionless slip-surface width, WL, increases from 0.25 to 0.7.

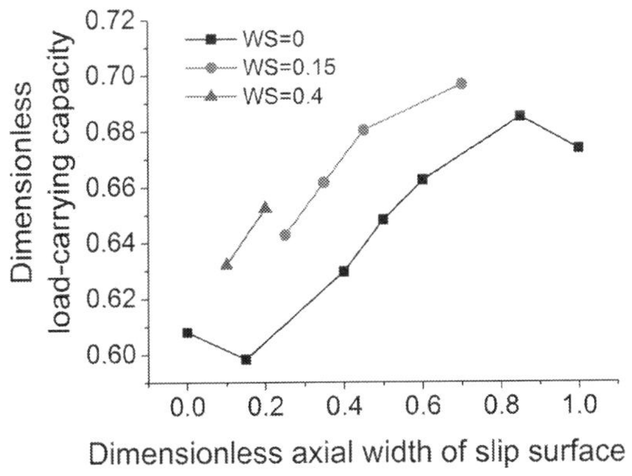

Figure 22. Load carrying-capacity for different slip surfaces.

The distribution curve of load-carrying capacity for WS = 0.4 lies above the curve for WS = 0.15, and the curve for WS = 0.15 lies above the curve for WS = 0, as shown in **Figure 22**. When WS = 0.4 and WL = 0.2, there is an increase of 3.6% in load carrying-capacity compared with this situation as WS = 0.0 and WL = 0.4; although the axial area of the former slip surface is smaller than the latter, the former slip surface is located much more closely to the bearing centre. This indicates that the enhanced impacts of slip surfaces on load carrying-capacity are much stronger if the slip surface is located near the bearing centre, under the same axial width.

It can be seen from **Figure 22**, the load-carrying capacity for the situation that WS = 0 and WL = 0.15 is smaller than that when WS = 0 and WL = 0. The load-carrying capacity for the situation WS = 0 and WL = 1.0 is also smaller than that when WS = 0 and WL = 0.85. When WS = 0, the starting position of slip surface is just located at the bearing axial outlet. When WS = 0 and WL =1.0, both the starting and the end positions of slip surface are just located at the bearing axial outlet. This situation, WS = 0 and WL = 0, means the size of slip surface is 0, namely there is no slip surface in bearing bush surface. Thus, it can be concluded that slip surface located at bearing axial outlet has an adverse effect on load-carrying capacity.

6. Conclusions

Boundary slip has complex influences on the tribological performances of journal bearings, including hydrodynamic and hybrid journal bearings. The impact laws of slip surfaces on journal bearings can be concluded as follows:

1. Boundary slip could produce a fluid hydrodynamic action in the downstream zone of the slip surface, and also could result in a negative fluid hydrodynamic action in the upstream zone of slip surface.

2. These two fluid hydrodynamic actions induced by boundary slip and convergence structure, respectively, may promote each other, when the slip surface is located in the pressure rising zone. In this moment, boundary slip has a beneficial influence and will improve load-carrying capacity, and the load-carrying capacity increases with the size/area of slip surface.

3. The negative fluid hydrodynamic action induced by boundary slip would damage the fluid hydrodynamic action produced by convergence structure, when the slip surface is located in the pressure drop zone. In this case, the boundary slip has an adverse influence, and the load-carrying capacity decreases with the size of slip surface.

4. Due to the pressure condition in cavitation zone, when the slip surface is located in cavitation zone, the boundary slip has no influence on the pressure distribution and load-carrying capacity of journal bearings.

5. The slip surface near the bearing centre in the axial direction has a much stronger influence on the enhancement of the load-carrying capacity.

6. The slip surface located at bearing outlet has an adverse influence on load-carrying capacity.

In summary, unreasonable design of slip surfaces not only cannot improve the tribological performances of journal bearings, including hydrodynamic journal bearings and hybrid journal bearings, but they also have adverse influences and would result in a decrease in load-carrying capacity. These results in this chapter can be a design criterion for the design of slip surfaces in journal bearings.

Author details

Qiyin Lin[1,2*], Baotong Li[2,3] and Hong Zhao[1,2,3]

*Address all correspondence to: qiyinlin88@gmail.com

1 Fuli School of Food Equipment Engineering and Science, Xi'an Jiaotong University, Xi'an, China

2 State Key Laboratory for Manufacturing Systems Engineering, Xi'an Jiaotong University, Xi'an, China

3 School of Mechanical Engineering, Xi'an Jiaotong University, Xi'an, China

References

[1] Jonathan P. Rothstein. Slip on superhydrophobic surfaces. Annual Review of Fluid Mechanics. 2010;42:89–109. DOI: 10.1146/annurev-fluid-121108-145558

[2] Bharat B, Yuliang W, Abdelhamid M. Boundary slip study on hydrophilic, hydrophobic, and superhydrophobic surfaces with dynamic atomic force microscopy. Langmuir. 2009;25(14):8117–8121. DOI: 10.1021/la900612s

[3] Hugh S, Steve G. Equation for slip of simple liquids at smooth solid surfaces. Langmuir. 2003;19(12):5065–5071. DOI: 10.1021/la034123j

[4] H A Spikes. The half-wetted bearing. Part 1: Extended Reynolds equation. Proceedings of the Institution of Mechanical Engineers, Part J: Journal of Engineering Tribology. 2003;217(1):1–14. DOI: 10.1243/135065003321164758

[5] Richard F. Salant, Alicia E. Fortier. Numerical analysis of a slider bearing with a heterogeneous slip/no-slip surface. Tribology Transactions. 2004;47(3):328–334. DOI: 10.1080/05698190490455348

[6] Cho-Yun Y, Cho-Yu Y, Cheng-Kuo S, Chih-Yung H. Design of slip boundary produced by a lotus structure applied to a hydrostatic bearing. Tribology Letters. 2014;55(1):55–64. DOI: 10.1007/s11249-014-0331-2

[7] Aurelian F, Patrick M, Mohamed H. Wall slip effects in (elasto) hydrodynamic journal bearings. Tribology International. 2011;44(7–8):868–877. DOI: 10.1016/j.triboint.2011.03.003

[8] Qiyin L, Zhengying W, Ning W, Wei C. Effects of large-area textured/slip surface on slider bearing. Journal of the Balkan Tribological Association. 2015;21(1):12–23.

[9] Qiyin L, Zhengying W, Ning W, Wei C. Effect of large-area texture/slip surface on journal bearing considering cavitation. Industrial Lubrication and Tribology. 2015;67(3):216–226. DOI: 10.1108/ILT-05-2013-0055

[10] Yongbin Z. A tilted pad thrust slider bearing improved by the boundary slippage. Meccanica. 2013;48:769–781. DOI: 10.1007/s11012-012-9630-6

[11] Qiyin L, Zhengying W, Yubin Z, Ning W. Effects of the slip surface on the tribological performances of high-speed hybrid journal bearings. Proceedings of the Institution of Mechanical Engineers, Part J: Journal of Engineering Tribology. Forthcoming. DOI: 10.1177/1350650116630202

[12] Qiyin L, Baotong L. Comparison of the influences of surface texture and boundary slip on tribological performances. Mathematical Problems in Engineering. 2015;2015:126824. DOI: 10.1155/2015/126824

Chapter 3

Wear Resistance of Thermal Spray WC-Co-VC Nanostructured Coatings

José Cabral Miramontes , Citlalli Gaona Tiburcio ,
Abraham Velasco Tellez , Carlos Poblano Salas and
Facundo Almeraya Calderón

Additional information is available at the end of the chapter

http://dx.doi.org/10.5772/64575

Abstract

Thermally sprayed WC-Co-VC coatings are widely used based on their resistance to abrasive wear. This chapter shows the fabrication procedure of bimodal WC-Co-VC coatings applied by a high-velocity oxy-fuel (HVOF) thermal spray process. We analyzed the effects of the mixture content of the nanostructure and microstructure phase on the mechanical properties and wear resistance of the coating. Additionally, VC was added to the bimodal mixture and it presented the best characteristics. The combination of VC additions and a bimodal WC particle size distribution in the WC-Co coatings proved successful in increasing their mechanical properties, which permitted the coatings processed in this work to show better mechanical properties than those reported in the literature for coatings having exclusively a bimodal WC particles size distribution or those only doped with VC additions. The effects of nanostructured phase contents on the microstructure and wear resistance of the coating are included.

Keywords: wear, thermal spray, nanostructured, coatings

1. Introduction

In industrial processes, the materials are exposed to corrosive and erosive environments that lead lifetime reduction for certain components and trigger, furthermore, high maintenance costs. In the different wear mechanisms, there is an intersection of phenomena and by this means a severe damage, for example, in a suction system can be found: cavitation and erosion; or heat (thermal fatigue) and erosion of a steam turbine blades or abrasion and corrosion, in a screw pulp pump by the presence of Cl^- ions. Once the wear mechanisms present on an equipment or component have been determined, there is a need for search for the alloy or coating, metallic, polymeric, ceramic or a mixture thereof that can prolong its service. The most critical area of structural and nonstructural components is usually the surface where the maximum loads are reached (e.g., mechanical and thermal stresses, corrosion, wear and their combinations thereof).

A material that works at high temperature normally is not operating in protective atmospheres. Usually the working environment is air and, in this case, the oxidizing agent is oxygen, but it is also likely that the atmosphere is composed of aggressive gases that damage the material, Carburization is the formation of metal carbides in a material as a result of exposure to a carbon containing atmosphere. In service, it can result in loss of mechanical properties over time in addition to wastage of the material. Sulfidation is a chemical reaction of a metal or an alloy, with sulfur in some form in its working environment. This produces compounds of sulfur that usually form on the solid surface or under the surface of the metal or alloy. Sulfidation most often causes serious deterioration of the solid surface and the vital functional properties of the affected substance. Nitridation is a process that results in the formation of nitrides in a material. It results from exposure to reducing, high temperature environments with high nitrogen activity. Since nitrides are inherently hard, brittle phases, nitridation can produce local or widespread loss of material strength and possible metal wastage or chemical attack in presence of salt deposits or ashes [1, 2]. In all environments that trigger corrosion in the above-mentioned problems, the oxygen activity is high enough to promote oxidative processes. Furthermore, the operating conditions of the materials forming part of the steam turbines or part of thermal plant components suffer corrosion processes at elevated temperature where the atmosphere contains a lot of water vapor, with this being more aggressive atmosphere than the air [3]. In reducing environments, the degradation of the material is controlled by the activity of species such as CO/CO_2 and H_2/H_2O, and they are usually more aggressive than oxidants, because the driving force for forming a protective oxide is less favorable [4].

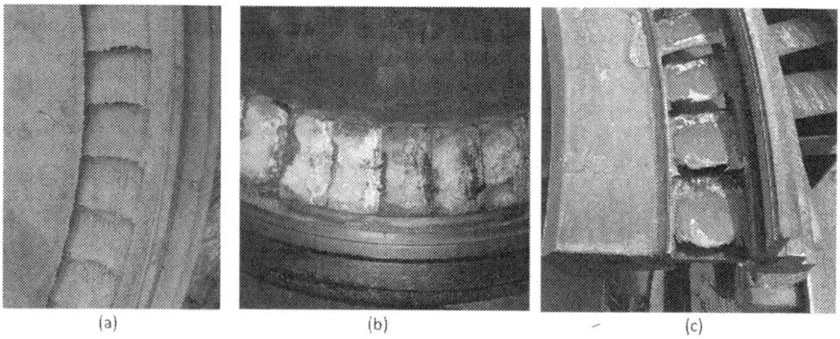

Figure 1. Steam turbine blades showing corrosion wear.

The protection of the metal components can be done with hard coatings to improve wear resistance. They are applied with different techniques, with the thermal spraying techniques being the most developed in recent years, where its increasing applications is to reduce the wear in industrial processes. In steam turbine, components like blades, body rotor, nozzle and auxiliary devices for cooling and lubrication must be developed to work properly under the required steam conditions for pressure and temperature. The most common vanes failures are associated with erosion, stress corrosion, corrosion fatigue and fatigue [5]. Some of these failures are shown in **Figure 1**.

The thermal spraying is one of the most versatile techniques ever used for coating materials application to protect components from wear by abrasion, adhesion, erosion, corrosion and fatigue [6]. This technique tends to extend the lifetime of the components, with the aim of obtaining high yields for longer and reducing the number of failures. The advanced technology of thermal spraying provides all types of techniques to the industries, being one of the most effective resources ever developed to combat premature wear by getting ahead of the failure or even when the equipment or item stops working in optimal conditions. Hence, it optimizes equipment availability, the maintenance costs are reduced and the lifetime of the equipment and machinery is maximized, increasing the reliability of critical systems. Through processes of high-velocity thermal spraying (HVOF), it is possible to obtain coatings with low porosity structures, low oxidation and high adhesion, thereby minimizing problems caused by the degradation processes on the components of steam turbines used for the electric power generation [7–10]. In this technique, oxygen and fuel gas at high pressures and flow rates are combined to produce a combustion which produces very high-speed particles and temperatures of about 2000°C, which can be considered low compared to other thermal spraying processes, such as vacuum or air plasma [11]. These flame characteristics are suitable for the application of corrosion- and wear-resistant coatings. **Figure 2** shows a schematic of high-velocity oxy-fuel (HVOF).

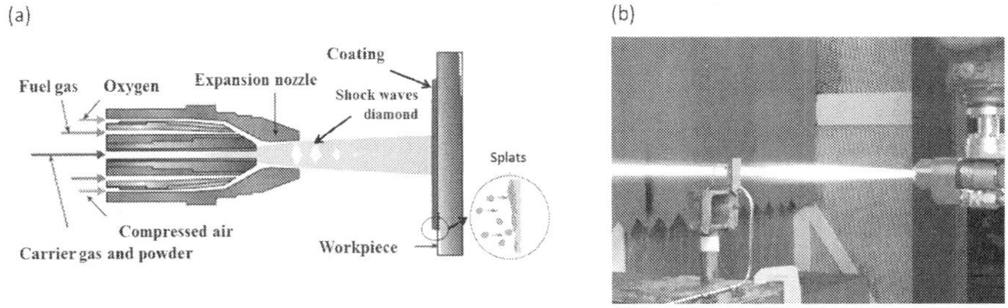

Figure 2. Schematic of an HVOF torch.

The materials that are often selected for applications at elevated temperatures must have the fundamental requirements to resist the corrosion and atmosphere deterioration where they are set. Usually, they are the austenitic stainless steels, nickel-, cobalt- and titanium-based alloys and besides the intermetallic materials [12, 13]. The selection of suitable material will depend on the operating temperature, the atmosphere and lifetime required for the working component in the equipment.

It is well known that Ni-based coatings show good corrosion resistance at high temperature and besides have good wear resistance after addition of W and Mo as alloying elements [15]. These alloys exhibit improved properties due to the formation of an amorphous phase that came from a rapid solidification and a suitable alloy composition [16, 17]. Some researchers have found that spraying the alloys Ni-Mo-Cr-B to make coatings using atmospheric plasma exhibited improved wear and corrosion resistance due to the formation of an amorphous phase coming from the rapid solidification and a suitable alloy composition [18]. An amorphous alloy phase in the protective coating is achieved by adding some appropriate refractory materials such as Mo and W and metalloids such as B and C in the base powder [19]. Those additions result in better resistance to abrasion and wear, while providing resistance to corrosion in aqueous and alkaline media [20, 21].

Protective coatings are widely used in the aircraft industry because they provide better surface properties of the structures where they are deposited. Besides, the characteristic that they must possess is the high structural performance due to the conditions of service that are subjected. Innovation in this area has focused on nanotechnology and gives special relevance in the world of materials. Regarding coatings, the extreme mechanical conditions that they must resist must be taken into account. Many components of an aircraft are subject to wear conditions; to address this problem, the pieces are coated with materials having good tribological properties. The components that are subject to wear and corrosion are generally those found in the undercarriage. The landing gear systems must withstand thousands of cycles of fatigue and shear stresses during the service life of the aircraft, whereby the material that forms the protective coating must have a high resistance to fatigue. Many components of the landinggear are coated with hard chrome and the inner surface of the cylinder sliding as well. However, environmental regulations require the replacement of those coatings where the use of hexavalent chromium [14] is required. Given this requirement, the coatings obtained by HVOF thermal spraying today are a viable alternative, enhancing the benefits obtained with traditional hard chromium coatings, both wear resistance and corrosion resistance.

The thermal spraying of WC-Co, WC-CoCr or Cr_3C_2-NiCr by HVOF, that is, applying a consistent coating made of metal matrix reinforced with high hardness ceramic inclusions [22] plus the addition of nanocrystalline WC particles, provides to the coatings an improvement in the hardness and tribological properties [23, 24]. The WC-Co coatings applied by HVOF are useful for operating temperatures up to 900°C, while the maximum pressure is up to 1200 MPa, its friction coefficient is only 0.01 (80 times lower than steel). The WC-Co coatings applied by HVOF containing nanoparticle sizes have better mechanical and tribological properties by providing an effective control of processing variables during thermal spraying application [25]. The wide applications as protective coatings permit to be an excellent choice on steam turbines in the industrial power generation sector and at the same time useful in the aeronautical industry.

This means that the corrosion behavior depends on the phases and microstructure of the applied coating.

Regarding the opposite corrosive behavior of the two main components of the alloy, Co and W, in respect to the influence of the medium [26], it is described that the cobalt shows stable passivity in alkaline solutions, while W is readily soluble; the situation observed in acid electrolytes is completely reverse. Therefore, these hard metals typically fail by corrosion of the less resistant phase, which is triggered by the testing environment [27]. On the other hand, it has to be taken into consideration the galvanic interactions between the existing phases, where typically the WC is nobler than Co [28].

It has been found, according to the literature reports, that VC effectively retards thickening of tungsten carbide grains after prolonged heat treatment, [29] and furthermore, promotes the formation of compounds (W, V) C, inhibiting, beside, the formation of brittle species [30]. The corrosion resistance of WC-Co alloys can be improved with the addition, in small quantities, of transition carbides. An example of this is the addition of Cr_3C_2, in amounts of only 0.5%, improving significantly the corrosion resistance [31]. However, the influence of the additions of VC was not so easy to assess as the case of Cr_3C_2.

Some studies have found that small additions of VC in WC-Co had a neutral effect on the corrosion resistance [31]. On the other hand, the tungsten carbide alloys reinforced with cobalt and chromium (cermets) and thermally sprayed by HVOF are currently good substitutes for hard chrome, typically present in some components such as aircraft landing gears, hydraulic rods, ball valves, printing rollers, etc. [14]. The use of nanocrystalline particles of WC is expected to improve properties in terms of hardness and tribological coatings [23, 24, 32, 33]; however, due to their high surface area-to-volume ratio, these suffer grain size growth and oxidation during their passage through the flame, which is detrimental to the coatings in good final properties such as hardness and wear resistance [34]. However, it has been reported, in different works [35, 25], a better mechanical and tribological properties using WC-Co coatings enriched with nanoparticles and deposited under HVOF technique, provided that an effective control of the processing variables was carried out during thermal spraying. The resistance against wear and erosion depends on the thickness of the coatings [36] and also on its hardness [37].

The processing and characterization of tungsten carbide coatings with a bimodal particle size distribution, that is, nanostructured and microstructured particle sizes of WC-Co, have been reported by several authors [35, 25]. The use of nano-sized particles in the compounds of cermet has resulted in a greater hardness than that found in their counterparts containing micron size particles. Furthermore, their combination, as bimodal particle size distribution, the under HVOF thermally applied WC-Co coating results in improved wear resistance, compared to those coatings produced solely from nano-sized powders. Excessive decarburization of nanoparticles by the torch flame and the formation of W_2C in the coating are considered to be responsible for the decrease in wear resistance of the mentioned coatings. Exposure to high temperature of WC nanoparticles in the torch flame during the HVOF process results in an undesirable increment of such particles. As a result, a decrease in hardness and wear resistance is observed. On the other hand, one effect of the mixed matrix hardening, done together by nanoparticles plus the strong arrangement of micrometer WC size grains, was proposed as possible reasons for the improved wear resistance of the bimodal coatings.

Different studies have reported the influence of VC additions in the kinetics of grain coarsening of WC-Co cermets [37, 38]. However, there is little information on the influence by additions of VC in the properties of WC-Co coatings applied with HVOF technique. Some authors [39, 40] have reported an increase in abrasion resistance of WC-Co-VC coatings compared to their commercial counterparts WC-Co coatings.

This research explored the influence by additions of VC powders in bimodal WC-Co coatings manufactured with HVOF thermal spraying, their mechanical properties and compared with those said properties, found in commercial counterparts.

2. Coatings fabrication

The nanostructured WC-Co powder was processed by mechanical milling in a Simoloyer CM01 high-energy ball mill for 20 h at 600 rpm under an argon atmosphere. The powder was then agglomerated employing a 2% methyl cellulose solution to form slurry which in turn was baked in a muffle at 80°C for 24 h under an inert nitrogen atmosphere. A solid block of material was then obtained and crushed into powder. The resulting powder was sieved through 450 and 635 meshes, +32 and +20 μm, respectively. Three powders were employed as feedstock in the present work. A commercial WC-12 Wt%Co powder, a bimodal mixture of commercial and nanostructured WC-Co powder having the same chemical composition (75 wt% commercial-25 wt% nanostructured), and a bimodal mixture doped with additions of 2 wt% vanadium carbide. The bimodal and bimodal doped with VC mixture were mixed mechanically in a double cone blender for 30 min at 20 rpm before being thermally deposited by HVOF.

All powder mixtures were sprayed on AISI 3014 stainless steel substrates employing a Sulzer Metco DJH 2700 hybrid HVOF gun, with propane being the fuel gas. A KUKA KRC robot arm

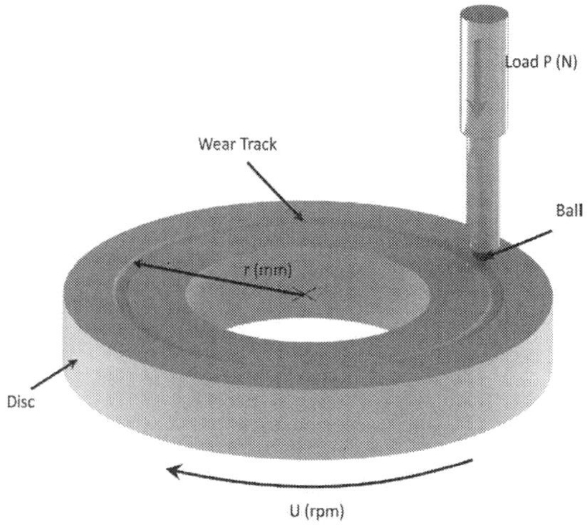

Figure 3. Schematically Pin-on-Disk test to ASTM G 99 [41].

was employed in order to control the gun transversal speed, at 1.5 m/s for all the experiments. Also, constant spray distance and powder feed rate (229 mm and 38 g/min), respectively, were employed.

The microstructure of the coatings was evaluated in the transversal section using a Philips XL 30 ESEM. Coating thickness, interface contamination, oxide contents and porosity percentage were determined from Backscattered electrons micrographs employing SigmaScan image analysis software. The Pin-on-Disk test is a model test for friction and wear determination of two solid surfaces being in sliding contact (pin or ball against coated disk). The test method and procedure are described in ASTM G99-94A [41] and shown in **Figure 3**. The normal loads applied to the ball are typically in the range of 1–10 N. As a result of rotational speed and radius of the sliding path, the relative velocity between ball and surface varies between $v = 0.19$ and 0.73 m/s. The relative humidity can be adjusted between 5 and 95% using different salt solutions.

Sliding wear measurements were performed in a CSM Instruments Pin-on-Disk 18-280 Tribometer employing a 10 N load. Samples were tested at ambient temperature using 6-mm-diameter 440C stainless steel balls as the counterpart. Testing time was 2000 min and the linear speed was 5 cm/s. Vickers microhardness of all coatings was measured transversally in an LECO M440H1 microhardness tester, using a 500-gf load and a 15-s dwell time.

3. Microstructured of the powders and coatings

The morphology of the initial powders used in coatings in this investigation is shown in **Figure 4**. The powders of WC-Co nanostructured show long and irregular conglomerates small particles of WC and Co, while the commercial powder consists of spherical shape particles, which is a typical characteristic of the sintered agglomerate powder mixtures. Furthermore, bimodal mixtures showed a combination of fine powder agglomerates of nanostructured WC-Co and spherical particles, near in shape to commercial powder. VC powder shows a smaller particle size in relation to the commercial and nanostructured powders.

The microstructure of the three-alloy powder obtained by thermal spraying coatings is shown in **Figure 5**. From the BSE micrographs, it can be observed that the microstructure of the three coatings are typical of WC-Co coatings with WC islands, having different size distribution are embedded in a cobalt matrix. As can be seen from **Figure 6**, the WC particles size distribution was similar to the three studied coatings. The characteristic feature of the observed coatings is the strong refinement of their microstructures. The coatings contained different phases, which can be distinguish by the diversification of the microstructure contrast. A good coatings adherence was observed. There are discontinuities between the substrate and coatings. The roughness of the substrate contributes to the better adhesion of deposited coatings and as it is necessary in the thermal coatings technologies. This result indicated that the deposited coatings have nanometric microstructure. Nanometric coatings were deposited by thermal spraying HVOF method, which confirms the possibility to apply nanostructured coatings by this technique [42]. The statistical analysis showed a slight influence of VC additions on the final size of WC grains within the coatings, as a lower mean WC size was measured in VC-doped coatings compared to that found in the bimodal WC-Co coating. On the other hand, the calculated mean WC particle size for bimodal mixture was slightly higher than that of the commercial coating.

Figure 4. Micrographs of the starting powders, (a) nanostructured WC-Co after agglomeration, (b) WC-Co commercial, (c) bimodal mixture and (d) VC powder.

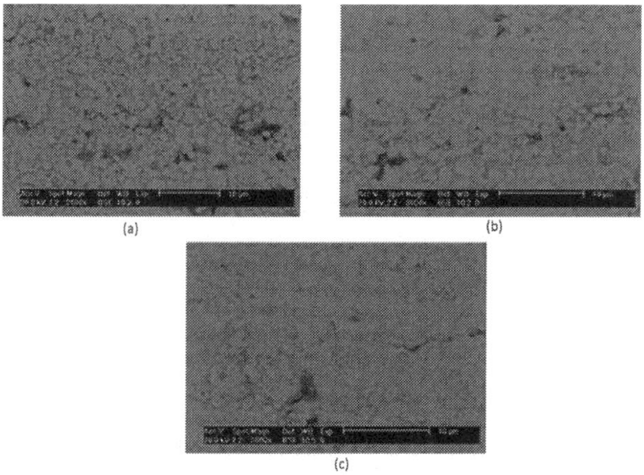

Figure 5. Backscattered electron micrographs in coatings WC-12Co, (a) commercial, (b) bimodal and (c) bimodal mixture + 2wt% VC.

http://dx.doi.org/10.5772/64575

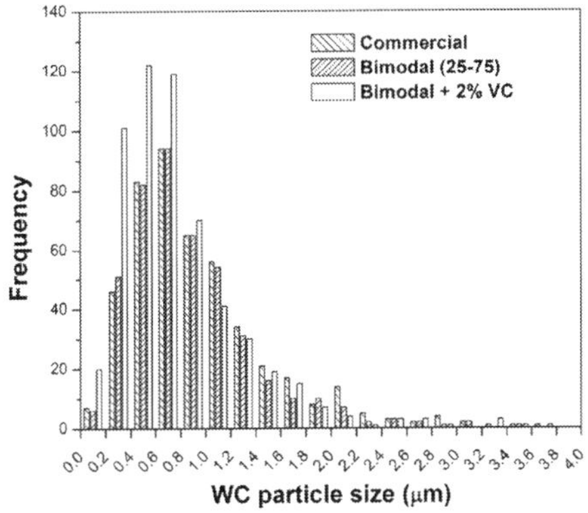

Figure 6. Particle size distribution of WC-Co coatings applied by HVOF.

The average porosity, oxide content, mean WC particle size, thickness and microhardness of the coatings are included in **Table 1**. The VC-doped and commercial coatings showed the highest levels of porosity, whereas the commercial alloy contained higher interface contamination and oxide contents than those found in bimodal mixture. Average microhardness levels of bimodal and VC-doped coatings were comparably higher than the microhardness measured in commercial samples. The thicknesses of all WC-Co coatings fell within allowable limits commonly accepted in the industry for the processing of OEM (Original Equipment Manufacturer) components and/or reconstruction of worn parts [43].

	Commercial alloy	Bimodal WC-12Co	Bimodal WC-12Co-(2% VC)
Thickness (μm)	154 ± 8	133 ± 6	99 ± 4
WC mean particle size (μm)*	1.1	1.5	1.2
Porosity (%)	4.4 ± 0.1	3.9 ± 0.1	4.6 ± 0.2
Oxide content (%)	0.27 ± 0.02	0.16 ± 0.02	0.19 ± 0.02
Interface contamination (%)	32.3 ± 1	21.1 ± 1	25.1 ± 4

All average values with standard error from 10 measurements.
* From statistical analysis shown in **Figure 5**.

Table 1. Summary of properties from microstructure evaluation of coatings.

The morphology of the WC grains in all HVOF coatings processed in this work is less angular than that found in sintered WC-Co cermet, as already mentioned by different authors [39–44]. Although the coatings were processed from powders having different particle size distributions, the average particle size of WC in the final coating was similar. The effects of processing conditions of thermal spraying on the final microstructure of the coatings were the initial microstructure and size of the precursor particles of the WC-Co powders as key factors. This can be attributed to the deposition conditions, which were the same for all processed samples. In all cases, a relationship 4.8 oxygen/propane was used in order to reach a temperature slightly lower than the maximum attainable (T_{max}), namely 2828°C, for a certain gas mixture using the relationship of 4.5 O_2/C_3H_8 [45]. Lower flame temperature that T_{max} is preferred when nano-structured materials are processed, in order to preserve a fraction of the nanostructures in the coating; however, very low flame temperatures should be avoided since they result in coatings that contain high porosity. The oxygen/fuel used here resulted in samples with levels of porosity within the range reported for WC-Co coatings applied by HVOF [35, 46]. Coatings with addition of 2% of VC showed higher porosity compared to commercial coatings; this same behavior has been reported by Luyckx and Machio [39] when processed WC-Co coatings with VC additions. The commercial WC-Co coating contains the highest amount of oxides; this kind of contamination appears to have an important effect on the hardness and wear performance as shown by the commercial coating with lower hardness and higher wear rate than the whole coatings studied here. An optimization of the oxygen/fuel ratio, distance and transverse velocity spray gun has to be performed in order to find the conditions that may lead to reduced content of oxides.

4. Microhardness

The microhardness of the three coatings obtained by thermal spraying is shown in **Figure 7**. The coating hardness is a key factor to optimize screening conditions and parameters, as well as comparisons between different deposited coatings. Bimodal WC-Co coatings and bimodal + 2% VC doped showed the highest hardness. These superior mechanical properties can be attributed to the bimodal structure of the tungsten carbides WC which provides a balanced amount of particles between micrometer sizes of WC and nano-sized particles of WC, where the latter being more prone to decarburization in the torch flame due to its high surface–to-volume ratio.

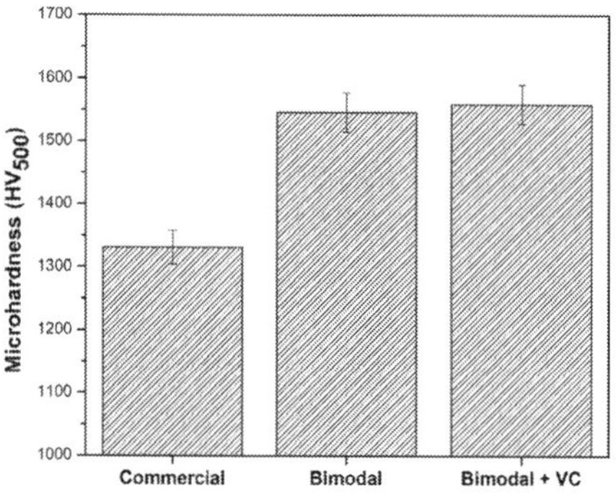

Figure 7. Microhardness of different coatings produced by HVOF.

5. Wear resistance

The variation of friction coefficient as a function of testing time from the Pin-on-Disc (POD) tests is included in **Figure 8**. When the steady state of wear conditions was reached by sliding in the POD testing set-up, the commercial coating showed greatest value ($\mu \approx 0.54$), while the bimodal mixture added with 2% of VC coatings showed comparable levels of wear slightly lower ($\mu \approx 0.48$) but still the lowest compared to the whole samples.

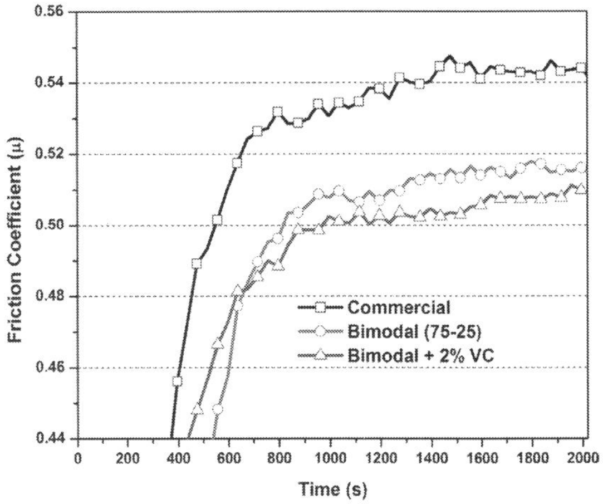

Figure 8. Friction coefficient as a function of testing time from POD test.

As final conclusion, the friction coefficient of bimodal WC-Co + 2% VC coatings was less than the commercial coatings. The addition of VC in bimodal coatings resulted in a slight increase in the mechanical and wear resistance properties. The same behavior is described by Guilemany et al. [25] for bimodal WC-Co coatings, who showed lower values of friction coefficient (μ) in nanostructured systems and in commercial, both with the same chemical composition. The wear resistance of WC-Co coatings can be explained by the limited amount of debris removed during the sliding wear tests, the nanostructured WC-Co coatings contain a lower volume fraction of cobalt than commercial coatings. Therefore, it has been shown in various studies that a bimodal particle size distribution of WC improves response to wear on WC-Co coatings use conditions without lubrication (Luyckx and Machio [39]). Therefore, this work shows that a combination of micro- and nanostructures is beneficial to improve the wear resistance of WC-Co coatings.

The major wear mechanism of the HVOF hard metal coating is connected with the gradual primary loss of the metal matrix from the areas between coatings hard particles, followed by the weakening of their particles and pulling them out of the coating surface. The design of the CSM Tribometer does not allow the wear debris to naturally fall off; they are trapped in the wear track and serve as an abrasive medium. Then the mechanism of wear changes from sliding to abrasive wear [47, 48].

The wear mechanism of the cermet coatings is described as the contact between the pin and coatings-carbides, which are slightly protruding from the matrix due to grinding. Such conditions lead only to a very slight wear [49].

The wear resistance strongly depends on the smoothness of the coating surface [50, 51], which becomes better if the size of grains becomes smaller. From this point of view, the refinement of microstructure coating by thermal spraying has strong influence on their wear properties.

The dependence of wear resistant to the microhardness level implicates for searching other reasons of resistance against the wear of the deposited coatings. Some data suggest that the wear mechanisms depend on the roughness of the coating face [50, 51].

These results indirectly indicate that the wear resistance is a complicated mechanism and depends not only on the hardness but also on the state of the coating surface and its internal microstructure. Specifically it was found that the refinement of microstructure to a nanometric dimensions favorably influences the wear resistance.

6. Conclusion

As general conclusion, it can say to that the vanadium carbide additions to bimodal WC-Co deposited by HVOF make coatings that slightly increase the mechanical properties of microhardness with 15% and tribological properties decrease in friction coefficient to 0.48. Then, the obtained hardness in bimodal WC-Co coatings is primarily influenced by the reinforcing WC particles.

Other conclusions are as follows:

- There is slight influence of the VC additions on the final WC grain size present in coating. Nevertheless, the mean WC grain size doped with VC was shorter than the grain size in bimodal WC-Co coatings. On the other hand, the WC particle size in the bimodal mixture was slightly larger than that in commercial alloy coatings.
- The commercial WC-Co coating contains the highest amount of oxides; this kind of contamination appears to have an important effect on the reduction of hardness and wear performance.
- In microhardness, the bimodal WC-Co coatings and bimodal + 2% VC doped showed the highest hardness. These superior mechanical properties can be attributed to the bimodal structure of the tungsten carbides WC which provides a balanced amount of particles between micrometer sizes of WC and nano-sized particles of WC
- In steady state, the friction coefficient for commercial coatings is larger ($\mu \approx 0.54$) than the bimodal mixtures + 2% of VC coating ($\mu \approx 0.48$), being slightly lower the wear in the last one.

Acknowledgements

We also thank the Academic Body UANL-CA-316 "Deterioration and integrity of composite materials" and the Advanced Technology Center (CIATEQ A. C.) for allowing the use of their facilities.

Author details

José Cabral Miramontes[1*], Citlalli Gaona Tiburcio[1], Abraham Velasco Tellez[1], Carlos Poblano Salas[2] and Facundo Almeraya Calderón[1]

*Address all correspondence to: jocamira@hotmail.com

1 Autonomous University of Nuevo León (UANL), Faculty of Mechanical & Electrical Engineering (FIME), Center for Research and Innovation in Aeronautical Engineering, Apodaca, Nuevo León, Mexico

2 Advanced Technology Center (CIATEQ), El Marqués, Querétaro, Mexico

References

[1] N. S. Stoloff. Mater. Sci. Eng. A. 1998;258:1–14.

[2] N. S. Stoloff, C. T. Liu, S. C. Deevi. Intermetallics. 2000;8:13–20.

[3] B. Szczucka Lasota, B. Formanek, A. Hernas. J. Material. Process. Technol. 2005;164–165:930–934.

[4] G. Ji, O. Elkedimc, T. Grosdidier. Surf. Coat. Technol. 2005;190:406–416.

[5] W. Z. Wang, F. Z. Xuan, K. L. Zhu, S. T. Tu. Eng. Fail. Anal. 2007;14:632–641.

[6] N. F. Ak, C. Tekmen, I. Ozdemir, H. S. Soykan, E. Celik. Surf. Coat. Technol. 2003;173–174:1070–1073.

[7] J. A. Cabral Miramontes, C. Gaona Tiburcio, F. Almeraya Calderón, F. H. Estupiñan López, G. K. Pedraza Basulto, C. Poblano Salas. Int. J. Corros. 2014;2014. Article ID 703806, 8 pages, http://dx.doi.org/10.1155/2014/703806

[8] J. G. Chacon Nava, A. Martinez Villafañe, F. Almeraya Calderón, J. A. Cabral Miramontes, M. M. Stack. Tribol. Int. 2010;43:1307–1317. doi:10.1016/j.triboint.2009.12.012

[9] T. Kinos. *Proc. NTPC*, 1994, 357.

[10] K.A. Khor, N.L. Leh: *Proc. NTSC*, 1993:613.

[11] J. R. Davis, editor. Handbook of Thermal Spray Technology. 1st ed. Prepared under the direction of the Thermal Spray Society Training Committee. 2004. 52, 54 and 263 p.

[12] M L. Lau, H. G. Jiang, W. Nuchter, E. J. Lavernia. Phys. Status Solidi. 1998;166:257.

[13] T. Grosdidier, A. Tidu, H. L. Liao. Scr. Mater. 2001;44:387.

[14] C. Bartuli, T. Valente, F. Cipri, E. Bemporad, M. J. Tului. J. Therm. Spray Technol. 2005;14:187195.

[15] J. C. Tan, L. Looney, M. S. J. Hashmi. J. Mater. Process. Technol. 1999;92:203.

[16] S. K. Das, E. M. Norin, R. L. Bye. Mater. Res. Soc. Symp. Proc. 1984;28:233.

[17] C. H. Lee, E. P. Yoon. Surf. Coat. Technol. 1998;99:203.

[18] A. H. Dent, A. J. Horlock, D. G. McCartney, S. J. Harris. Mater. Sci. Eng. 2000;A283:242–250.

[19] C. H. Lee, K. O. Min. Surf. Coat. Technol. 2000;132:49–57.

[20] A. H. Dent, A. J. Horlock, D. G. McCartney, S. J. Harris. Surf. Coat. Technol. 2001;139:244–250

[21] T. S. Sidhu, S. Prakash, R. D. Agrawal. Thin Solid Films. 2006;515:95–105.

[22] N. Espallargas, J. Berget, J. M. Guilemany, A. V. Benedetti, P. H. Suegama. Surf. Coat. Technol. 2008;202:1405–1417.

[23] B. R. Marple, J. Voyer, J. F. Bisson, C. J. Moreau. J. Material. Process. Technol 2001;117:418–423.

[24] J. He, M. Ice, S. Dallek E. J. Lavernia. J. Metall. Mater. Trans. A. 2000;31 A:555–564.

[25] J. M. Guilemany, S. Dosta, J. R. Miguel. Surf. Coat. Technol. 2006;201:1180–1190.

[26] M. Pourbaix. Atlas of electrochemical equilibria in aqueous solutions. National Association of Corrosion Engineers (NACE); 1974.

[27] S. Hochstrasser-Kurz, D. Reiss, T. Suter, C. Latkoczy, D. Günther, S. Virtanen, P. J. Uggowitzer, and P. Schmutz. Electrochem. Soc. 2008;155(8):415–426.

[28] G. Mori, H. Zitter, A. Lackner, M. Schretter. G. Kneringer, P. Rodhammer, H. Wildner, editors. 15th International Plansee Seminar; Reutte: Plansee Holding AG: 2001. p. 222–236.

[29] W. Kroemmer, P. Heinrich. C. Berndt, J. Heberlein, R. Tucker, C. Moreau, editors. Thermal Spray; Beijing, China: Global Coatings Solutions; 2007. p. 118–122.

[30] T. D. Xiao, Z. Zhang, D. M. Wang, R. W. Rigney, P. R. Strutt. Tungsten Hard Met. Refract. All. 2000;25–27:161–167.

[31] C. N. Machio, D. S. Konadu, J. H. Potgieter, S. Potgieter Vermaak, Van der Merwe. International of Corrosion. 2013(2013);:506759.

[32] A. I. Gusev. Effects of the nanocrysralline state in solids. 1st ed. Phys. Usp.; 1998. 49–76 p.

[33] F. gartner, t. K. Bormann, H. Kreye, N. Mitra. J. Metast. Nano Mater. 2000;8:933–940.

[34] P. H. Shipway, D. G. Mccartney, T. Sudapraset. Wear. 2005;259:820–827.

[35] Y. Qiao, T. E. Fischer, A. Dent. Surf. Coat. Technol. 2003;172:24–41.

[36] N. G. Hashe, J. H. Neethling, P. R. Berndt, H. O. Andrén, S. Norgren. Int. J. Refract. Met. Hard Mater. 2007;25:207–213.

[37] F. Arenas, I. B. de Arenas, J. Ochoa, S. A. Cho. Int. J. Refract. Met. Hard Mater. 1999;17:207–213.

[38] K. Choi, N. M. Hwang, D. Y. Kim. Powder Metall. 2000;43:168–172.

[39] S. Luyckx, C. N. Machio. Int. J. Refract. Met. Hard Mater. 2005;25:11–15.

[40] C. N. Machio, G. Akdogan, M. J. Witcomb, S. Luyckx. Wear. 2005;258(1–4):434–442.

[41] ASTM G99-94a. Standard test method for wear testing with a Pin-on-Disk apparatus (1995-00).

[42] J. Gang, J. P. Morniroli, T. Grosdidier. Scr. Mater. 2003;48:1599–1604.

[43] B. D. Sartwell, K. O. Legg, J. Schell, J. Sauer, P. Natishan, D. Dull, J. Falkowski, P. Bretz, J. Deverauz, C. Edwards, D. Parker. Naval Research Laboratory Report. 2004;:NRL/MR/6170-04-8762.

[44] H. De Villiers Lovelock, S. Luyckx. 1st ITSC Int. Thermal Spray Conference; Montreal, Canada. 2000. p. 647–656.

[45] W. Kroemmer, P. Heinrich. C. Berndt, J. Heberlein, R. Tucker, C. Moreau, editors. Thermal Spray; Global Coating Solutions; Beijing, China. 2007. p. 118–122.

[46] T. D. Xiao, Z. Zhang, D. M. Wang, Y. Wang, R. W. Rigney, P. R. Strutt. Tungsten Hard Met. Refract. Alloys 2000;25–27:161–167.

[47] S. Houdkova, M. Kasparova, F. Zahalka. Vrstvy a Poclaky; Trencin; 2007. p. 49.

[48] B. Bhushnan. Introduction to Tribology. 1st ed. NY, USA: John Wiley and Sons; 2002.

[49] G. Bolelli, V. Cannillo, L. Lusvarghi, T. Manfredini. Wear. 2006;261:1298–1315.

[50] H. jianhong, J. M. Schoenung. Mater. Sci. Eng. A. 2002;336:274–319.

[51] K. Holmberg, A. Matthews, editors. Coating tribology: properties, mechanisms, techniques and applications in surface engineering. 1st ed. B. V. Nederland: Tribology and Interface Engineering Series, Elsevier; 2009.

Chapter 4

Introducing New Coating Material Alloy with Potential Elements for High Corrosion Resistance for Oil and Gas Application

Mitra Akhtari-Zavareh and
Ahmed Aly Diaa Mohammed Sarhan

Additional information is available at the end of the chapter

http://dx.doi.org/10.5772/64483

Abstract

In petroleum and petrochemical industries, offshore and onshore systems have to function in an aggressive environment that exposes the production equipment components to thermal cycling and wear and corrosion. Although maintenance of material degradation in oil and gas is costly, internal and external parts of the equipment and pipelines must be well inspected and continually maintained. For this reason, highly advanced corrosion and wear–monitoring systems must be installed in the critical areas of the plant to protect pipes and equipment from seawater and crude oil. Therefore, researchers are in search of advanced materials and methods that could be applied in oil and gas pipelines and accessories for increasing their working time. The common manufacturing processing method for improving the surface of piping and accessories is overlay welding or cladding. This method has some limitations, such as its limitation for choosing materials. In addition, the high temperature of welding causes some defects on the final surface, such as thermal residual stress, cracking, and distortion in the substrate. The method is also time consuming and costly. However, the coating method provides a blend of unique properties with low cost. Thermal spray methods are cold spraying techniques that have a considerably less thermal stress, residual stress, and other defects. Among different thermal spray coating techniques, high-velocity oxygen fuel (HVOF) and plasma are the most commonly used thermal spraying coating processes to produce anti-wear and corrosion coatings with different types of materials such as metal, alloys, and ceramic composite . Furthermore, HVOF and plasma thermally sprayed coating processes induce microstructure heterogeneities, which increase the corrosion and wear resistance. In this research, one type of alloy with chemical composition NiCrCoAlY was chosen for increasing corrosion resistivity of carbon steel piping. A corrosion behavior of coated samples in seawater was investigated for 30 days. Potentiodynamic polarization and electrochemical impedance spectroscopy (EIS) results indicated that these types of alloys protected the surface of carbon steel piping from harsh environments. However, the corrosion protection of NiCoCrAlY deposited by HVOF technique is higher than a plasma coating technique.

Keywords: carbon steel piping surface, thermal spray coating techniques, alloy, corrosion

1. Literature review

Corrosion is generally defined as the destructive disintegration process through which a metal is gradually destroyed by electrochemical reactions with the environment [1]. There is no hard and fast rule to classify corrosion, but it can normally be divided into two main groups, uniform and localized corrosion [2]. In uniform corrosion, damage occurs on the entire surface at a uniform rate. In localized corrosion, damage occurs on localized spots due to heterogeneities in the material microstructure (galvanic, intergranular, and dealloyed) and the environment (crevices and hydrogen damage). Localized stress (stress corrosion cracking, corrosion fatigue, and fretting) and the geometry of closed chemical fluid flow system parts (erosion and cavitation) also cause localized corrosion damage [3, 4].

Corrosion damage depends on the extent of particular factors, e.g. aggressive ions or stress amplitude.

Corrosion has huge economic and environmental impact on infrastructures worldwide such as highways, bridges, oil and gas industries, chemical processing, and water and wastewater systems. Corrosion cannot be fully eliminated but its effects can be minimized using different protection and prevention methods [5]. The direct cost attributed to corrosion damage has been estimated in the order of 3–4% of industrialized countries' gross national product (GNP) [6].

According to the National Association of Corrosion Engineers (NACE), the global corrosion cost, through direct and indirect losses, was $US 552 billion in 2001, which increased to $US 1.3 trillion in 2009 [7]. Direct and indirect damage to the environment is massive if corrosion-related problems are overlooked in sensitive oil/gas industries and nuclear power plants. For instance, radiation and poisonous gas leakage due to severe pitting in nuclear plants can put workers' health at risk [8].

Corrosion measurement and prevention cover a large field of technical activities such as measuring corrosion rate; controlling physical parameters like temperature, pH, and pressure stress; protecting against corrosion like cathodic and anodic protection [9]; chemical dosing; and prevention by material selection or organic/inorganic coatings. Measuring corrosion rate by weight loss is one of the corrosion testing methods commonly used since the past till now [10]. A material is exposed to an environment in which it needs to be in service for a prolonged time (90 days). The corrosion rate is measured from the net weight loss per unit time, i.e. difference in weight before and after the exposure divided by exposure time. In this way, corrosion damage can be assessed for the future by extrapolating the weight loss results. With advancements in electrochemical sensor technology, different sorts of resistance probes, linear polarization resistance probes, and H_2 evolution probes are used for monitoring corrosion systems [11, 12].

Metallic and nonmetallic coatings along with cladding or surface modifications are also common corrosion prevention methods. Their effective use depends on carefully selecting and regular service monitoring of corrosion. Corrosion prevention, monitoring, and testing can save billions of dollars besides minimize hazards [5, 6].

In oil and gas, many components, such as valve, wellhead, Xmass tree, exchanger, and so on, exposed to the harsh environments and cause the long life of equipment significantly decreased. For this reason, choosing a suitable alloy that has a specific properties depend on environment is essential for increasing corrosion resistivity of equipment [13].

The selection of corrosion resistant alloys is an important step that is directly affected on the long life system. Because, any mistakes for choosing the materials it make some issues for system [5, 13]. Companies with high research facilities first simulate a certain part of the field environment under study. Then, a group of alloys selected some different alloys based on the available information. Then, test all the candidate alloys at the same time and finally choose the best alloys for a certain condition. This method can easily take 1–3 years to accomplish at significant cost [6, 14].

Another selection method is literature review on corrosion data that generally applied to the expected fields. It is quickest and cheapest technique because it neglected some group of alloys that is not suitable for specific condition. Then, the selected alloys are tested under specific condition [13, 15]. However, the chance of error in this method is greater than the previous method, e.g., introducing potential for Corrosion Resistance Alloys (CRA) failure or using a more expensive alloy than required [16].

Other resources for material selection are using available standards, such as the 2003 ISO 15156 publications derived from the previous NACE 0175 publication for "sour service." ANSI/NACE MR0175/ISO 15156 gives general information for service in oil and gas productions and in natural gas sweetening plants in hydrogen sulfide (H_2S)-containing environments, and it also recommended some materials in the appropriate design codes, standards, and regulations. Generally, it can be applied to help avoid costly corrosion damage to the equipment itself [13, 17].

Finally, before accepting the final Alloys for specific condition, it is necessary to simulate the field of environment [14, 16].

Temperature, chloride ion concentration, CO_2 and H_2S, environment pH, and the presence or absence of sulfur (S) are some parameters that directly affect the corrosion properties of alloys [18].

These parameters that cause the risk of stress corrosion cracking (SCC), rate of metal solution from pits, and other things are considerably increased [18].

Generally, three items directly affect the corrosion rate of sample, temperature, pressure, and the chloride content in sodium chloride. In order to make the scale for these parameters is universal, centigrade (°C), pounds per square inch, psi and gram/liter, respectively [19].

Furthermore, the absence of oxygen is critical to the application of these alloys under the conditions shown. If the environment has an oxygen (typically greater than 10 parts per billion), the other alloys should be consider [20, 21].

According to the information presented above, nickel alloys have a long life for oil and gas process fluids. Also it has a high performance and relatively low cost, and at the end of the structure's lifecycle, nickel alloys are completely recyclable. For this reason, this alloy is offered for both the environments [9, 22].

The Nickel Cobalt Chromium Aluminum Yttrium family of gas-atomized powders is designed to produce thermal sprayed coatings with excellent high-temperature oxidation and high-temperature corrosion resistance. The presence of cobalt improves coating ductility and high-temperature corrosion resistance. The presence of chromium and yttrium improves oxidation resistance by increasing the activity of aluminum and improving the spallation resistance of the oxide scale. The function of chromium and aluminum is to provide a reservoir that continually replenishes the oxide scale. Maintaining the chromium and aluminum ratio is critical for avoiding coating embrittlement [23, 24].

Adding tantalum to the chemical composition exhibits superior oxidation resistance at high temperatures. This chemical composition can serve as overlay coating on moving and rotating instruments to improve their performance and service life, even under harsh environmental conditions. Gas atomization ensures excellent chemical homogeneity and high purity, which results in consistent coating results. **Figure 1** represents a SEM photomicrograph of Nickel Cobalt Chromium Aluminum Yttrium powder. The SEM photomicrograph shows gas-atomized morphology that is typical of these materials [25].

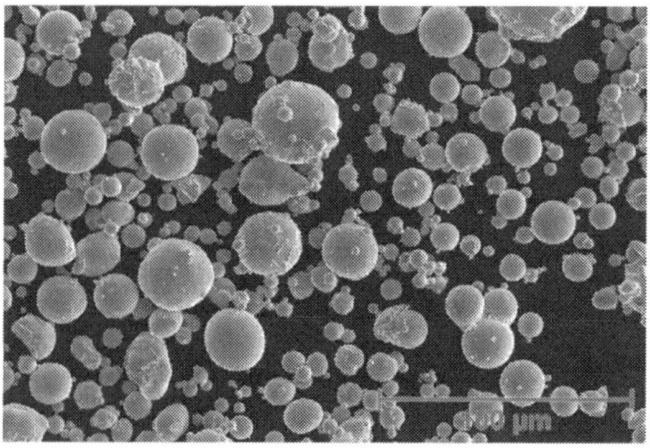

Figure 1. SEM photomicrograph of Nickel Cobalt Chromium Aluminum Yttrium powder morphology.

This chemical composition is a premium grade NiCoCrAlY gas-atomized powder that produces thermal spray coatings with excellent surface resistance against oxidation and corrosion at high temperatures [26].

High-temperature corrosion resistant bond coats act as thermal barrier and oxide-based abradable coatings for hot section components. They also produce superior low oxide coatings that machine well and closely resemble wrought alloys in terms of characteristics [27].

A noteworthy point of preference of thermal spray procedures is to a great degree wide assortment of materials that can be utilized to make coatings. For all intents and purposes, any material that melts without breaking down can be utilized. Another real point of interest is the capacity in most warm splash procedures to apply a covering to a substrate without significantly warming it. In this manner, materials with high melting points can be Joint to the substrate without changing the characteristics or thermal distortions of the parts. A third advantage is the ability, in most cases, to strip and recoat worn or damaged coatings without changing the part's properties or dimensions. Also, applying thin coatings on low-cost substrates results in increased efficiency and cost savings [28–30].

There are five different types of thermal spray coatings: flame spraying, wire arc splashing, detonation gun deposition, plasma spray, and high speed oxy-fuel.

Coating produced by HVOF and plasma processes is characterized by lamellar structure embedded with solid particles, oxide, and inclusions (surface residue from shot blasting or surface cleaning) [14]. Molten or semi-molten particles deform upon hitting the substrate. Deformed particles in the coating are called splats and are approximately 1–20 μm thick. The high striking speed of particles produces compact coating, but some voids/pores still form at the surface and interlamellar particle boundaries [31]. The impaction of solid particles at high speed also adds strength due to the peening effect (relieving tensile stress or adding compression by low level mechanical stresses). The high temperature and presence of oxygen in the environment also cause the formation of oxides. The adhesion between substrate and coating is predominantly by mechanical interlocking [24].

The utilizations of thermal spray coatings are to a great degree changed, yet the vital use classes include upgrading surface wear and/or erosion resistance. Different applications incorporate their utilization for dimensional reclamation, as thermal obstructions, thermal conduits, and electrical channels or resistance; for electromagnetic protection; and improving or retarding radiation. Thermal spray coatings are utilized for all intents and purposes in each industry, including aviation, farming, car, essential metals, mining, paper, oil and gas creation, chemicals and plastics, and biomedicine [25].

2. Methodology

The substrate material used is carbon steel (S45) because it is one of the most popular materials used in oil piping production in both upstream and downstream domains. The substrate material was supplied by Kelvin Steel, Glasgow, with the chemical composition in percentage as follows: C, 0.42–0.50; Mn, 0.5–0.80; Si, 0.17–0.37; Cr (max%), 0.25; Cu (max%), 0.25; Ni (max %), 0.25; S (max%), 0.035; P (max%), 0.035; and Fe balance.

For the layer deposited onto the carbon steel surface, gas-atomized and spheroidal NiCoCrAlTaY powders with nominal size ranges of 45+11 and 37+15 μm, respectively, were used for plasma and HVOF methods.

For increasing the adhesion of deposited layer to substrate, the surface preparation is a very important step. Roughness of sample directly affects the adhesion of deposited layer, which can be controlled by different parameters such as blasting pressure, angel, and so on. [32].

Grit blasting was carried out with a high efficiency sand blaster with Alumina grit (size 10–20 mesh), 8/10 mm nozzle, and operating at a blasting pressure of >0.5 MPa. The distance between substrate and nozzle was 150 mm with a 30° angle. The grit blasting time was dependent on obtaining the required surface roughness. Upon grit blasting completion, powders with different chemical compositions were sprayed using HVOF and/or plasma gun systems.

The coating thickness selected for all samples in this research after trial and error was approximately 400 μm. Generally, increasing some initial parameters, such as arc current, power feed rate, and so on, helps increase the thermal spray coating thickness.

For the deposition of this powder by HVOF machine, a Model 9MP machine was employed. **Table 1** shows the parameters for HVOF coating. For plasma coating, the same specific setup was used as for HVOF coating in this material powder group. The setup of powder coated by plasma machine is tabulated in **Table 2**.

Model: 9MP	**DJ2700 hybrid**		
Nozzle	Standard		
Powder port			
Type		DJ2702	
Injector		#9	
Angle		90°	
Suction and spreader		L/L	
Gases	Pressure (psi)	Flow (FMR)	SCFH
Oxygen	155	282.2	645
Natural gas (CH_4)	110	140.2	320
Carrier gas (N_2)[1]	105	350	799
Spray distance	254 mm (10″)		
Spray rate	38 g/min (5 lb./h)		

[1]Using nitrogen as a shroud gas will reduce coating oxide content.

Table 1. Parameters of NiCrCoAlY powder for HVOF coating.

Gun	**F4**		
Nozzle	Standard		
Powder port			
Type	#2 (1.8 mm/0.071")		
Gauge	#6		
Angle	90°		
Disc rpm[1]	23		
Suction and spreader	L/L		
Gases	Pressure (psi)	NLPM	SCFH
Primary (air)	75	65	1218
Secondary (H_2)	50	14	32
Carrier (air)	100	2.3	5.2
Amps	600		
Voltage	66		
Spray distance	114–140 mm (4.5–5.5")		
Spray rate	40–57 g/min (5.3–7.5 lb./h)		

[1]As a starting point, adjust to indicate spray rate.

Table 2. Parameters of NiCrCoAlY powder for plasma coating.

The coating morphology was observed through a high resolution FEI Quanta 200F field emission scanning electron microscope (FESEM). Using an EDX-System (Hitachi SU8000) instrument attached to a FESEM instrument, energy dispersive X-ray (EDAX) analysis was performed to investigate the elemental composition of the samples.

The surface of the substrate and coated samples before and after corrosion was also studied by a relatively destructive method. The X-ray diffraction (XRD) equipment is an Empyrean X-ray diffractometer with Cu Ka radiation (k = 1.54178 A operating at 45 kV, 30 mA, 0.026° step size, and scanning rate of 0.1 s 1 in a 2 h range from 10° to 90°).

For measuring the corrosion rate of coated sample, electrochemical techniques, such as DC (polarization) and AC (impedance), were used and applied using a potentiostat/galvanostat model AutoLab PGSTAT30 from Ecochemie (Netherlands) [33]. Polarization experiments were performed using a three-electrode cell, where the samples were the working electrode (WE) while a platinum wire and saturated calomel electrode (SCE) was the counter and reference electrodes, respectively. The electrolyte was 3.5% NaCl. In this test, the linear scan voltammetry, corrosion current (I_{corr}), and corrosion potential (E_{corr}) were calculated from the Tafel plots with a scan rate of 0.001 Vs^{-1}.

A frequency range of 10^5–10^{-2} Hz with amplitude potential of 5 mV vs SCE was used to conduct electrochemical impedance spectroscopy (EIS). To obtain the charge transfer resistance values across the electrode-electrolyte interface that could relate to parameters from the polarization output, simulations with analog circuits were performed with the experimental data. Having been installed in a computer interfaced with a potentiostat, general purpose electrochemical software (GPES) and frequency response analyzer (FRA) were used to run the AC and DC techniques, respectively.

3. Results, analysis and discussion

3.1. Microstructural analysis of surface

Figure 2 displays FESEM micrographs of plasma and HVOF-coated samples before corrosion testing. The top view image indicates a relatively dense, uniform, and crack-free coating deposited on the substrate. However, the HVOF-coated samples have a semi-uniform surface compared to the plasma-coated samples.

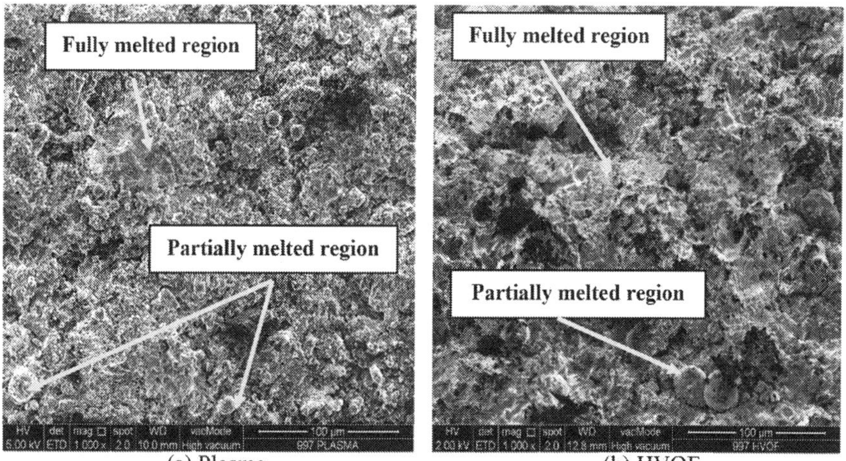

Figure 2. FESEM of NiCoCrAlY-coated samples at different magnifications: (a) plasma and (b) HVOF.

A small amount of semi-molten and nonmolten particles is visible in the fully melted region (Ni) area. The presence of semi-molten and nonmolten particles provides the coating with high bonding strength and good wear resistance [34, 35]. A few pores appear in black in the micrographs of both types of coating. The porosity size and Average number of porosity with the HVOF method are greater than with plasma.

The coatings deposited by plasma spray (**Figure 2a**) exhibit the most visible defects. Shrinkage of the molten droplets, as they splat onto the substrate and rapidly solidify, is one of the main causes of porosity and visible defects in plasma spray coatings. Coatings deposited by HVOF

(**Figure 2b**) feature the most desirable structure with minimal porosity, absence of cracks, and a clean interface with the substrate. The finer pores within HVOF coating demonstrate the superior compaction and deposition of particles during spraying. Also, HVOF coating has a rough surface due to the overlap of different splats and the roughness of each single splat [36, 37].

X-ray diffraction patterns and EDX for NiCoCrAlY plasma and HVOF-coated samples are shown in **Figure 3**. According to the coated samples in **Figure 3a** and **b**, Only three phases were detected γ (Co, Ni, Cr), β(Co, Ni, Y)Al, and Al_2O_3. The morphology of the coatings consists of single-phase splats separated by Al_2O_3veins as a result of the oxidation occurring in the deposition process.

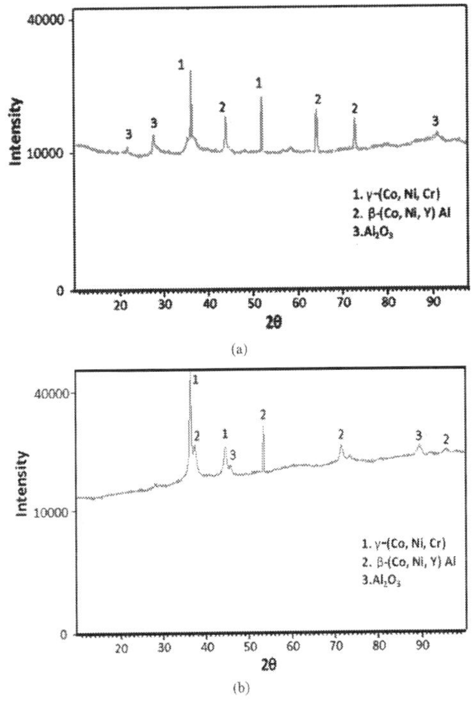

Figure 3. X-ray diffraction patterns for NiCoCrAlY-coated samples with: (a) plasma and (b) HVOF.

Previous studied showed the samples that coated with NiCoCrAlY by vacuum plasma coating(VPS) have high performance. Because it can provide good adhesion to the metallic substrate with high density, while this method is more expensive rather than other techniques of thermal spray coatings. For this reason, high-velocity oxygen fuel (HVOF) has a good option for disposition of NiCoCrAlY alloys instead of VPS. However, existing the free oxygen in the combustion gas causes increasing homogeneity in the melted powders and remove Oxidation particles, because they need high temperature. The aluminum and yttrium elements in the metallic powder have high affinity with oxygen and are thus easily oxidized during thermal spraying; consequently, the coatings have high oxide content [38–42].

3.2. Electrochemical corrosion analysis

The open circuit potential (OCP) of coated samples for day 3 and day 30 is shown in **Figure 4**. The plasma-coated sample (blue) shows the potential moved from −0.421 on day 3 toward noble regions to −0.341 at the end of that period, and the OCP of the HVOF-coated sample (red) was more negative compared to the plasma-coated sample, with a change from −0.305 on day 3 to −0.413 on day 30.

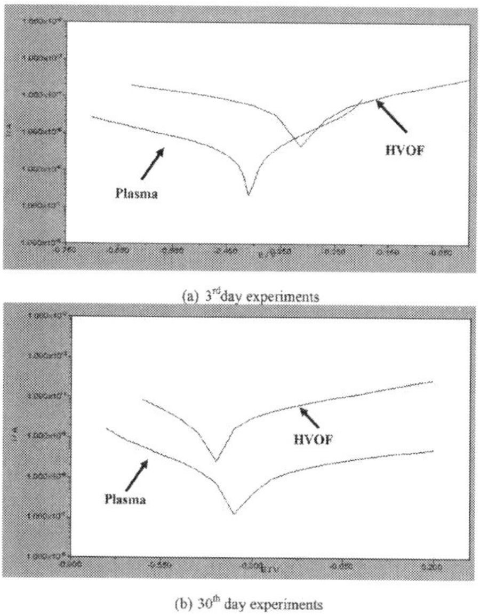

Figure 4. Polarization curves (log I vs E) of NiCoCrAlY with plasma and HVOF methods for (a) 3rd day and (b) 30th day experiment.

The charge transfer resistance R_{ct} was consistent with the OCP values from polarization measurements and I_{corr} illustrated in **Table 3**. It can be seen that the R_{ct} for plasma-coated samples changed from 11.735 to 10.098 kΩ from day 3 to day 30, and for the HVOF-coated samples, it changed from 12.193 to 8.242 kΩ. The higher R_{ct} (EIS data) is due to the greater resistance to charge transfer across the electrode-electrolyte interface, which is consistent with the positive OCP result values belonging to the plasma-coated sample. From the computer simulations for the plasma-coated sample, the R_s (QR) circuit diagram accurately matches the experimental data. Only one semicircle is observed in the Nyquist plot (**Figure 5a** and **c**).

Day	Plasma-coated sample				HVOF-coated sample			
	OCP		I_{Corr}/A (×10^{-5})	R_{CT}/kOhm	OCP		I_{Corr}/A (×10^{-5})	R_{CT}/kOhm
	Mean	SD			Mean	SD		
3	-0.421	0.00061	5.683	11.735	-0.305	0.00054	5.267	12.193
6	-0.413	0.00052	5.701	11.380	-0.319	0.00050	5.326	11.871
9	-0.404	0.00072	5.784	11.034	-0.328	0.00071	5.635	10.714
12	-0.389	0.00088	5.832	10.784	-0.289	0.00074	4.897	11.341
15	-0.379	0.00069	5.975	10.514	-0.340	0.00069	5.593	10.891
18	-0.367	0.0010	6.030	10.321	-0.367	0.00081	5.772	10.462
21	-0.360	0.00092	5.436	10.989	-0.378	0.00073	5.829	9.843
24	-0.354	0.00085	5.591	10.679	-0.382	0.0005	6.084	9.245
27	-0.348	0.00074	6.096	10.364	-0.395	0.0011	6.203	8.884
30	-0.341	0.00092	6.231	10.098	-0.413	0.00089	6.289	8.242

Table 3. OCP, I_{corr}, and R_{CT}/kOhm of NiCoCrAlY samples with plasma and HVOF coating methods in 3.5% NaCl solution for 30 days.

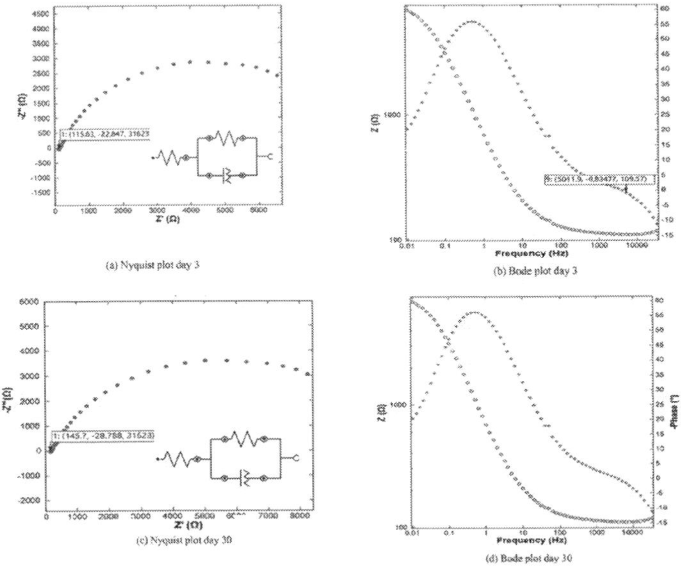

Figure 5. EIS of NiCoCrAlY plasma-coated samples. (a) Nyquist plot day 3, (b) Bode plot day 3, (c) Nyquist plot day 30, and (d) Bode plot day 30.

The Bode phase diagrams of the plasma-coated samples (**Figure 5b** and **d**) show one-phase maxima, consistent with the presence of one semicircle in the Nyquist plots from **Figure 5a** and **c** with R_s (Q_1R_1) circuit. The R_s (QR) circuit diagram accurately matches the start day of the plasma-coated samples. Only one semicircle and one maximum phase were observed in the Nyquist and Bode plots (**Figure 5a** and **b**). **Figure 6a** and **b** shows that the R_s (QR) circuit diagram accurately matches the experimental data at the beginning of measurement. Whereas for the HVOF-coated samples, at the end of the period, the equivalent circuit R_s (Q_1 [R_1(Q_2R_2)] accurately fits the experimental data for HVOF coating, because two semicircles and maxima phases were observed in the Nyquist and Bode plots (**Figure 6c** and **d**). The resistance between RE and WE is the solution resistance R_s, which is in series with two time constants (parallel arrangement of R and Q) that are also in series.

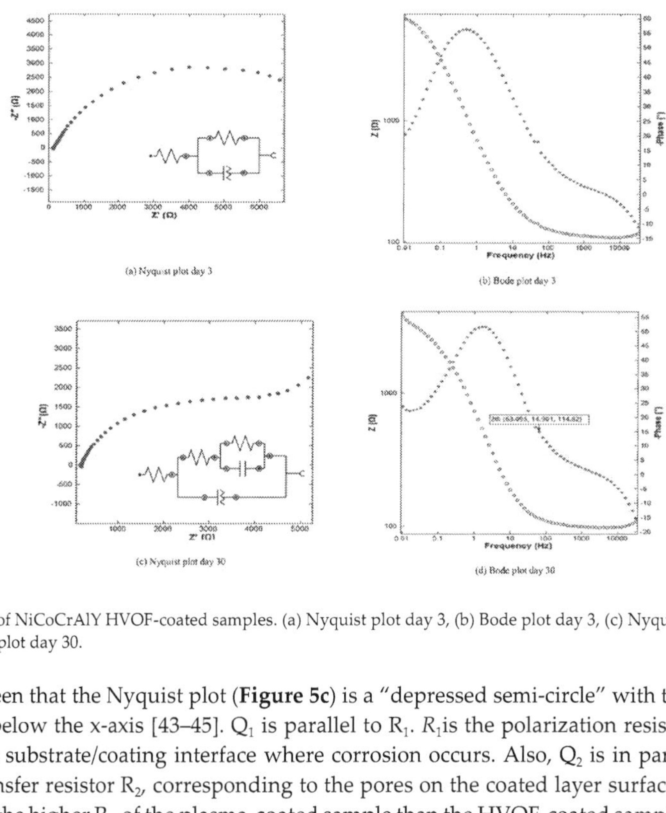

Figure 6. EIS of NiCoCrAlY HVOF-coated samples. (a) Nyquist plot day 3, (b) Bode plot day 3, (c) Nyquist plot day 30, and (d) Bode plot day 30.

It can be seen that the Nyquist plot (**Figure 5c**) is a "depressed semi-circle" with the center of the circle below the x-axis [43–45]. Q_1 is parallel to R_1. $R_1$is the polarization resistance of the area at the substrate/coating interface where corrosion occurs. Also, Q_2 is in parallel with a charge transfer resistor R_2, corresponding to the pores on the coated layer surface. The main reason for the higher R_{ct} of the plasma-coated sample than the HVOF-coated sample at the end of the period is the presence of a bigger oxide layer on top of the plasma-coated samples that protects the deposited layers from electrolyte penetration to the interface.

In these simulations, the plasma coating had higher charge transfer resistances, R_{ct}, than the HVOF coating due to the higher resistance against corrosion. The charge transfer resistance, R_{ct}, values for plasma and HVOF-coated samples are presented in **Table 3**.

According to **Table 3**, the corrosion current density of plasma-coated samples increased from 5.683×10^{-5} to 6.030×10^{-5} from day 3 to day 18, and then suddenly dropped to 5.436×10^{-5}A on day 21. After that, it rose again up to 6.231×10^{-5}A on day 30. The corrosion current density for HVOF-coated samples moved from 5.267×10^{-5} to 5.635×10^{-5} on day 3 to day 9, but suddenly dropped to 4.897×10^{-5}A on day 12. From the middle of the period until the end, the current density increased from 5.593×10^{-5} to 6.289×10^{-5}A. The decreasing amount of corrosion current density related to the oxide layer created on top of the deposited layer protected the samples from the corrosive environment. Therefore, the amount of R_{ct} over this period increased.

3.3. Microstructural analysis of corrosion

FESEM micrographs of both coated sample types after 30 days of immersion in crude oil and seawater are shown in **Figures 7** and **8**. According to **Figures 7a** and **8a**, when samples were exposed to the oil environment, the corrosion rate on the coated samples' surface was lower than in NaCl electrolyte (seawater) solution. According to **Figure 7**, the amount of pitting corrosion on the plasma samples was less than the HVOF samples, because there were fewer fine holes produced than on the HVOF samples.

(a) Crude oil (b) Seawater

Figure 7. FESEM micrographs of NiCoCrAlY plasma-coated sample surfaces after 30 days: (a) in crude oil and (b) in seawater.

In **Figure 8**, the HVOF-coated samples corroded significantly after 30 days and large holes appeared on their surface. This coating displayed evident crevices and pitting corrosion in the microstructure containing intersplat porosity, that was detected on the top surface of HVOF-coated samples by previous studies [46, 47]. Also, the size and amount of corroded area inside NaCl solution (seawater) is obviously greater than in the crude oil.

Regarding **Figures 7** and **8**, there is a big difference between the corroded and uncorroded samples with the HVOF coating method in both environments, whereas in plasma-coated samples, the difference between corroded and uncorroded samples is considerably less than the HVOF method. However, both methods protected the substrate from corrosion.

Figure 8. FESEM micrographs of NiCoCrAlY HVOF-coated sample surfaces after 30 days: (a) in crude oil and (b) in seawater.

Generally, there are several factors affecting corrosion resistance, including adhesion force between coating layer and substrate, cooling rate, generation of oxides and compounds, and porosity [48, 49]. The NiCoCrAlY alloy has an excellent corrosion resistance and oxides generated affect chemical stability, bond strength, and durability. As it has high adhesion to the base metal and creates a dense oxide layer, it yields high durability [50].

4. Conclusion

Morphological, microstructural, and compositional analyses were performed for both coating methods in order to characterize the samples before and after the corrosion tests. The characterization for seawater (3.5% NaCl) and crude oil after 30 days showed that both methods can protect the substrate from corrosion, but the corrosion rate with HVOF is higher than with the plasma coating method. This result suggests that the potential impact of the plasma thermal spraying process against a corrosive environment is better than that of HVOF spraying systems. Therefore, from this chemical composition group, the plasma samples were chosen for corrosion resistivity.

Author details

Mitra Akhtari-Zavareh* and Ahmed Aly Diaa Mohammed Sarhan

*Address all correspondence to: akhtari.mitra@yahoo.com

Department of Mechanical Engineering, Faculty of Engineering Building, University of Malaya, Kuala Lumpur, Malaysia

References

[1] Fantana, F, Goldoni, D & Grandi, G. (1992). Visualization of ribosomal gene activity in oogenesis of Reticulitermes lucifugus (Isoptera: Rhinotermitidae) revealed by silver staining. Cytologia, 57(2), 223-226.

[2] Jones, DA. (1996). Principles and Prevention of Corrosion (2nd ed.): Prentice Hall,The University of Michigan, USA

[3] Perez, N. (2004). Electrochemistry and Corrosion Science: Kluwer Academic. pp. 3. Department of Mechanical Engineering, University of Puerto Rico, USA

[4] Akhtari Zavareh, M, Sarhan, A, Razak, B, & Basirun, WJ. (2015). The tribological and electrochemical behavior of HVOF-sprayed Cr 3 C 2–NiCr ceramic coating on carbon steel, Ceramics International 41, 5387-5396.

[5] Cramer, SD.&Covino, BS. (2003). Corrosion: Fundamentals, Testing, and Protection (Vol. 13A): Materials Park, OH, USA

[6] Cramer, SD&Covino, BS. (2006). Corrosion: Environments and Industries (Vol. 13): ASM International, Materials Park, OH, USA

[7] Setayesh, H, Moradi, H, & Alasty, A. (2015). A comparison between the minimum-order & full-order observers in robust control of the air handling units in the presence of uncertainty. Energy and Buildings, 91, 115-130.

[8] Villalobos, ME, Mizuno, A, Dahl, BC, Kemmotsu, N, & Müller, R-A. (2005). Reduced functional connectivity between V1 and inferior frontal cortex associated with visuomotor performance in autism. Neuroimage, 25(3), 916-925.

[9] Parker, ME,&Peattie, EG. (1984). Pipe Line Corrosion and Cathodic Protection: A Practical Manual for Corrosion Engineers, Technicians, and Field Personnel: Gulf Professional Publishing, Burlington, USA

[10] Younis, AA, Ensinger, W, El-Sabbah, MMB, & Holze, R. (2013). Corrosion protection of pure aluminium and aluminium alloy (AA7075) in salt solution with silane-based sol–gel coatings. Materials and Corrosion, 64(4), 276-283.

[11] Akhtari Zavareh, M, Sarhan AADM, Zavareh, PA, & Basirun, WJ. (2015). Electrochemical corrosion behavior of carbon steel pipes coated with a protective ceramic layer using plasma and HVOF thermal spray techniques for oil and gas, Ceramics International, 42, 3397-3406.

[12] Sampath, S, Jiang, XY, Matejicek, J, Prchlik, L, Kulkarni, A, & Vaidya, A. (2004). Role of thermal spray processing method on the microstructure, residual stress and properties of coatings: an integrated study for Ni–5 wt.% Al bond coats. Materials Science and Engineering: A, 364(1), 216-231.

[13] Craig, BD,&Smith, L. (2011). Corrosion resistant alloys (CRAs) in the oil and gas industry. Nickel Institute Technical Series No. 10073.

[14] Eadie, RL, Szklarz, KE, & Sutherby, RL. (2005). Corrosion fatigue and near-neutral pH stress corrosion cracking of pipeline steel and the effect of hydrogen sulfide. Corrosion, 61(2), 167-173.

[15] Akhtari Zavareh, M, Sarhan, AAD, Bushroa, AR, & Basirun, WJ. (2014). Plasma thermal spray of ceramic oxide coating layer to enhance wear and corrosion resistance of carbon steel for oil and gas application, Ceramics International. 40, 14267-14277.

[16] Javaherdashti, R. (2000). How corrosion affects industry and life. Anti-corrosion Methods and Materials, 47(1), 30-34.

[17] Nešić, D, Tan, Y, Moase, WH, & Manzie, C. (2010). A Unifying Approach to Extremum Seeking: Adaptive Schemes Based on Estimation of Derivatives. 49th IEEE Conference on Decision and Control (CDC),15-17 December, Hilton Atlanta Hotel, Atlanta, GA, USA

[18] Lucci, A, Demofonti, G, & Spinelli, CM. (2011). CO_2 Anthropogenic Pipeline Transportation. Paper presented at the Proceedings of the Twenty-first International Offshore and Polar Engineering Conference, Maui, Hawaii, USA.

[19] Kirk-Othmer (2007), Encyclopedia of Chemical Technology, Volume 25, 5th Edition, Wiley USA.

[20] Lucio-Garcia, MA, Gonzalez-Rodriguez, JG, Casales, M, Martinez, L, Chacon-Nava, JG, Neri-Flores, MA, & Martinez-Villafañe, A. (2009). Effect of heat treatment on H2S corrosion of a micro-alloyed C–Mn steel. Corrosion Science, 51(10), 2380-2386.

[21] Selman, C. (2011). Condensation Corrosion Modelling, Reality and Design in Deep Water Wet Gas Pipelines. Paper presented at the The Twenty-first International Offshore and Polar Engineering Conference.

[22] M. Akhtari Zavareh, A. Sarhan, B. Razak and WJ. Basirun, (2014). Electrochemical Characterizations of Different Ceramic Composite Coatings on Carbon Steel Piping Using High Velocity Oxy-Fuel Spray, Proceedings of the World Congress on Engineering and Computer Science 2014, San Francisco, USA, 22-24 October 2014, Vol. 2, 664-668.

[23] Islak, S, Buytoz, S, Ersöz, E, Orhan, N, Stokes, J, Saleem Hashmı, M, Tosun, N. (2013). Effect on microstructure of TiO_2 rate in Al_2O_3-TiO_2 composite coating produced using plasma spray method, Optoelectronics And Advanced Materials – Rapid Communications, Vol6, No 9-10, p 844-849.

[24] Marple, Basil Richard, Lima, Rogerio S, Moreau, Christian, Kruger, Silvio E, Xie, L, & Dorfman, MR. (2007). Yttria-stabilized zirconia thermal barriers sprayed using N2-H2

and Ar-H2 plasmas: influence of processing and heat treatment on coating properties. Journal of Thermal Spray Technology, 16(5-6), 791-797.

[25] Tuominen, J, Vuoristo, P, Mäntylä, T, Kylmälahti, M, Vihinen, J, & Andersson, PH. (2000). Improving corrosion properties of high-velocity oxy-fuel sprayed inconel 625 by using a high-power continuous wave neodymium-doped yttrium aluminum garnet laser. Journal of Thermal Spray Technology, 9(4), 513-519.

[26] Mumtaz, KA, Erasenthiran, P, & Hopkinson, N. (2008). High density selective laser melting of Waspaloy®. Journal of Materials Processing Technology, 195(1), 77-87.

[27] Torabian, S, Haddad, E, Rajaram, S, Banta, J, & Sabate, J. (2009). Acute effect of nut consumption on plasma total polyphenols, antioxidant capacity and lipid peroxidation. Journal of Human Nutrition and Dietetics, 22(1), 64-71.

[28] Berndt, ML, & Berndt, CC. (2003). Thermal spray coatings. Corrosion: Fundamentals, Testing, and Protection, 13, 803-813.

[29] Berndt, CC. (2001). Thermal Spray 2001: New Surfaces for a New Millennium: Proceedings of the International Thermal Spray Conference: ASM International, Materials Park, Ohio, USA

[30] Akhtari Zavareh, M, Sarhan, AADM, Akhtari Zavareh, P, Razak, BBA, Basirun, WJ, Ismail, MBC. (2016). Development and protection evaluation of two new, advanced ceramic composite thermal spray coatings, Al 2 O 3–40TiO 2 and Cr 3 C 2–20NiCr on carbon steel petroleum oil piping, Ceramics International, 42, 5203-5210.

[31] Al-Fadhli, HY, Stokes, J, Hashmi, MSJ, & Yilbas, BS. (2006). The erosion–corrosion behaviour of high velocity oxy-fuel (HVOF) thermally sprayed inconel-625 coatings on different metallic surfaces. Surface and Coatings Technology, 200(20), 5782-5788.

[32] Mann, BS, & Arya, Vivek. (2001). Abrasive and erosive wear characteristics of plasma nitriding and HVOF coatings: their application in hydro turbines. Wear, 249(5), 354-360.

[33] Hartfield-Wünsch, SE, & Tung, SC. (1994). The Effect of Microstructure on the Wear Behavior of Thermal Spray Coatings: ASM International, Materials Park, OH (United States).

[34] Somasundaram, B, Kadoli, R, & Ramesh, MR. (2014). Evaluation of cyclic oxidation and hot corrosion behavior of HVOF-sprayed WC-Co/NiCrAlY coating. Journal of Thermal Spray Technology, 23(6), 1000-1008.

[35] Wu, YS, Zeng, DC, Liu, ZW, Qiu, WQ, Zhong, XC, Yu, HY, & Li, SZ. (2011). Microstructure and sliding wear behavior of nanostructured Ni60–TiB_2 composite coating sprayed by HVOF technique. Surface and Coatings Technology, 206(6), 1102-1108.

[36] Richer, P, Yandouzi, M, Beauvais, L, & Jodoin, B. (2010). Oxidation behaviour of CoNiCrAlY bond coats produced by plasma, HVOF and cold gas dynamic spraying. Surface and Coatings Technology, 204(24), 3962-3974.

[37] Zhang, Q, Li, C-J, Li, C-X, Yang, G-J, & Lui, S-C. (2008). Study of oxidation behavior of nanostructured NiCrAlY bond coatings deposited by cold spraying. Surface and Coatings Technology, 202(14), 3378-3384.

[38] Yuan, FH, Chen, ZX, Huang, ZW, Wang, ZG, & Zhu, SJ. (2008). Oxidation behavior of thermal barrier coatings with HVOF and detonation-sprayed NiCrAlY bondcoats. Corrosion Science, 50(6), 1608-1617.

[39] Jiang, J, Zhao, H, Zhou, X, Tao, S, & Ding, C. (2012). The effect of ion implantation on the oxidation resistance of vacuum plasma sprayed CoNiCrAlY coatings. Applied Surface Science, 261, 422-430.

[40] Nijdam, TJ, Jeurgens, LPH, & Sloof, WG. (2005). Promoting exclusive α-Al_2O_3 growth upon high-temperature oxidation of NiCrAl alloys: experiment versus model predictions. Acta Materialia, 53(6), 1643-1653.

[41] Zavareh, MA, Hamdi, M, Ghahnavyeh, RR, Roudan, MA, Shafieirad, M. (2013). Fabrication of TiB2-TiC composites optimized by different amount of carbon in the initial Ti-BC powder mixture, Applied Mechanics and Materials, 315, 720-723.

[42] Miguel, JM, Guilemany, JM, & Vizcaino, S. (2003). Tribological study of NiCrBSi coating obtained by different processes. Tribology International, 36(3), 181-187.

[43] Zavareh, MA, Sarhan, AAD, Roudan, MA, Zavareh, PA. (2014) TiC–TiB_2 composites: A review of processing, properties and applications, International Journal of Innovative Research in Science & Engineering, 4, 2347-3207.

[44] Mochizuki, H, Yokota, M, & Hattori, S. (2007). Effects of materials and solution temperatures on cavitation erosion of pure titanium and titanium alloy in seawater. Wear, 262(5), 522-528.

[45] Woo, Y-B, Lee, S-J, Jeong, J-Y, & Kim, S-J. (2014). Evaluation on cavitation characteristics of CoNiCrAlY/ZrO 2–Y2O3 coating layer by atmospheric pressure plasma coating process. Materials Research Bulletin, 58, 78-82.

[46] Zavareh, MA, Sarhan, AAD, Roudan, MA, Zavareh, PA. (2014). A new approach to fabricate TiB_2-TiC composite utilizing self-propagation high temperature via pressure less sintering, International Journal of Engineering & Technology Sciences, 2, 193-203.

[47] Mansfeld, F. (1990). Electrochemical impedance spectroscopy (EIS) as a new tool for investigating methods of corrosion protection. Electrochimica Acta, 35(10), 1533-1544.

[48] Dent, AH, Horlock, AJ, McCartney, DG, & Harris, SJ. (2000). Microstructure formation in high velocity oxy-fuel thermally sprayed Ni–Cr–Mo–B alloys. Materials Science and Engineering: A, 283(1), 242-250.

[49] Zavareh, MA, Sarhan, AADM, Zavareh, PA. (2015). Comparison and wear behavior evaluation of Cr_3C_2-25%NiCr composite coated on carbon steel by two different thermal spray techniques, International Journal of Material Science Innovations, 3, 87-96.

[50] Wei-jie, L, Yong, L, Yan, W, Chao, H,&Hui-ping, T.(2011), Hot corrosion behavior of Ni–16Cr–xAl based alloys in mixture of Na2SO4–NaCl at 600 °C, Transactions of Nonferrous Metals Society of China,Vol.21, No12,p. 2617-2625

Section 2

Friction and Wear Mechanism

Chapter 5

Tribocorrosion

Rana Afif Majed Anaee and

Majid Hameed Abdulmajeed

Additional information is available at the end of the chapter

http://dx.doi.org/10.5772/63657

Abstract

This chapter includes some of the definitions of corrosion as important problem in many industries. The main forms of corrosion have been discussed in this chapter which depended on the appearance of corroded surface with some details of their mechanisms and how can prevent them. This chapter reviews tribocorrosion and explains the simplest method for tribocorrosion test by certain set-up using electrochemical test with three electrodes. Tribocorrosion test is needs to special arrangement of cell to be predicted which also has been mentioned in this chapter.

Keywords: tribocorrosion, corrosion forms, electrochemical test, wear rate

1. Introduction

Many materials such as metals and alloys, plastics, rubber, ceramics, composites, wood, etc. have a wide range of applications as a constructional material, and the selection of an appropriate material for a given application is the important. Generally, there are no general rules that govern the choice of a material for a specific purpose, and a logical decision involves a consideration of the relevant properties, ease of fabrication, availability, relative costs, etc. of a variety of materials; frequently, the ultimate decision is determined by economics rather than by properties.

There are many properties that may be limited by the selection of the proper material, which includes mechanical, physical and chemical properties. Mechanical and physical properties can be expressed in terms of constants, while the chemical property is dependent on the presented environmental conditions during service.

The term corrosion of non-metallic materials refers to their deterioration by chemical media, but a similar concept is not necessarily applicable to metals. Many authorities [1] consider that the term metallic corrosion includes all interactions of a metal or an alloy with its environment. Other definition of corrosion refers to the undesirable deterioration of a metal or an alloy. This definition is also applicable to non-metallic materials such as glass, concrete, etc. [2].

On the other hand, Fontana and Staehle defined the term [3] 'Corrosion' as the reaction of metals, glasses, ionic solids, polymeric solids and composites with its environments (such as liquid metals, gases, non-aqueous electrolytes and other non-aqueous solutions) [4]. Vermilyea defined corrosion as remove atoms or molecules from the material [5]. Finally, Evans defined the corrosion as a chemical thermodynamic and kinetic process, causing the introduction of electrochemical cells which deteriorate the lattice structure of a material [6].

2. Forms of corrosion

Corrosion forms can be classified according to the appearance of the corroded surface. This classification identifies the corrosion failure as a corrosion form by visual inspection, either by the naked eye or possibly by a magnifying glass or microscope. Since each form of corrosion has its characteristic causes, important steps to a complete diagnosis of failure can often be taken after a simple visual inspection [7].

On this basis, the following corrosion forms can be defined as follows:

1. Uniform (general) corrosion.
2. Galvanic (two-metal) corrosion.
3. Thermogalvanic corrosion.
4. Crevice corrosion (including deposit corrosion).
5. Pitting, pitting corrosion.
6. Intergranular corrosion (including exfoliation).
7. Selective attack, selective leaching (de-alloying).
8. Erosion corrosion.
9. Cavitation corrosion.
10. Fretting corrosion.
11. Stress corrosion cracking.
12. Corrosion fatigue.

A simple illustration of the various forms of corrosion is shown in **Figure 1**.

http://dx.doi.org/10.5772/63657

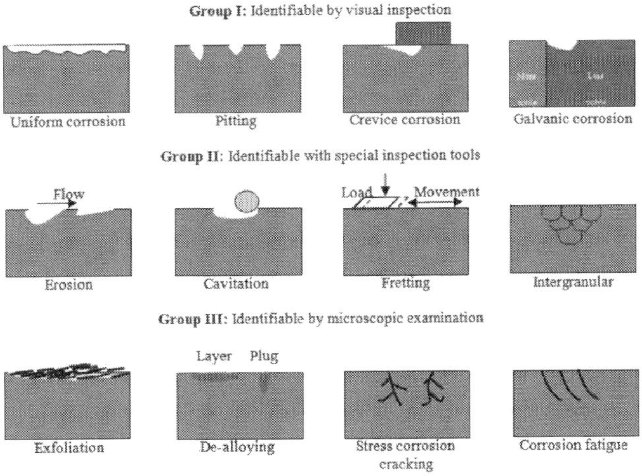

Figure 1. Main forms of corrosion grouped by their ease of recognition.

2.1. Uniform (general) corrosion

By definition, attacks of this type are quite evenly distributed over the surface, and consequently they lead to a relatively uniform thickness reduction. The necessary conditions for uniform corrosion have already been presented. Homogeneous materials without a significant passivation tendency in the actual environment are liable to this form of corrosion. The most common form of corrosion is uniform corrosion, but it is not a dangerous type of attack due to the following:

a. By simple tests, we can predict the reduction in thickness and then we can add the corrosion allowance.

b. Reducing the corrosion rate to acceptable level may be achieved by changing the environment or material or applying coatings and cathodic protection.

2.2. Galvanic corrosion

This form of corrosion occurs with contact between a more noble metal and a less noble one in the presence of an electrolyte, where the latter is oxidized and the other is reduced.

Galvanic corrosion can be prevented or made harmless in several ways:

1. Select the proper material.
2. Using weld consumable is more noble than the base metal.
3. Avoiding the combination of a large area of cathode with a small area of anode.
4. Using insulation materials to insulate the parts in galvanic couples.

5. Applying metallic coating on one of the materials similar to the other material.
6. Applying paints.
7. Using cathodic protection.
8. Adding corrosion inhibitors.
9. Preventing the access of humidity to be present between the contacted materials.
10. Avoiding electrolytic deposition of a more noble material on a less noble one.

2.3. Thermogalvanic corrosion

Thermogalvanic corrosion takes place when a material is subject to a temperature gradient in a corrosive environment. The hot surface is the anode and the cold one the cathode (see **Figure 2**) because of the anodic properties that are dependent on the temperature, as well as the variation in temperature affects the cathodic reaction in corrosion process.

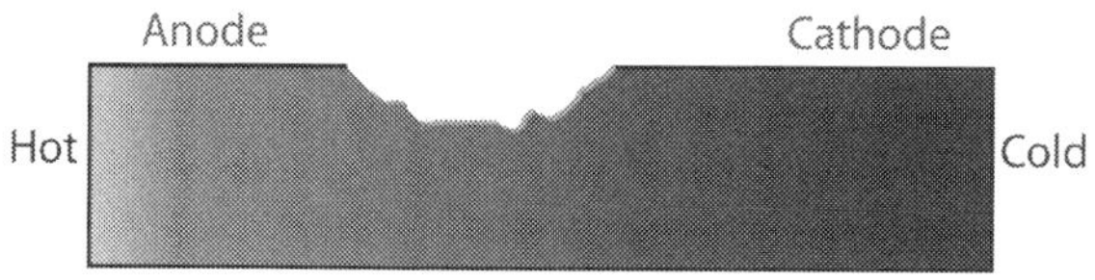

Figure 2. Thermogalvanic corrosion.

This form of corrosion requires little information about the service conditions to distinguish between thermogalvanic and other forms [7]. Thermogalvanic corrosion can be prevented by several ways:

1. Selecting the proper design.
2. Avoiding uneven heating/cooling.
3. Checking the continuous insulation for heat-insulated pipes.
4. Applying cathodic protection or coatings.

2.4. Crevice corrosion

This form of corrosion concentrates in crevices that are sufficiently wide for liquid to penetrate into it and sufficiently narrow for the liquid to be stagnant. This attack can be seen beneath flange gaskets, paint-coating edges, nail and screw heads, tube plates in heat exchangers, in overlap joints and between tubes, etc. Most materials undergo crevice corrosion that is passive beforehand, or can easily be passivated such as stainless steels, aluminium, unalloyed or low alloy steels in corrosive media containing chlorides and can also occur in other salt solutions. This form of corrosion occurs in stagnant or slowly flowing seawater and even at higher velocities too [7].

http://dx.doi.org/10.5772/63657

A special form of crevice corrosion that can develop on steel, aluminium and magnesium beneath a protecting film of lacquer, enamel, phosphate or metal is the so-called filiform corrosion, which leads to a characteristic stripe pattern. It has been observed most frequently in cans exposed to the atmosphere. Crevice corrosion is affected by several factors, of a metallurgical, environmental, electrochemical, surface physical, and last but not least a geometrical nature. One of the most important factors is the crevice gap.

Crevice corrosion can be prevented or reduced by several ways:

1. Selecting a proper material.
2. Avoiding crevices and deposition as far as possible.
3. Applying cathodic protection.
4. Measures for preventing deposition are as follows:
   a. Inspection and cleaning during and between service periods.
   b. Separation of solid material from flowing media in process plants.
   c. Gravel filling around buried piping and structures.

2.5. Pitting corrosion

Pitting corrosion occurs on more or less passivated metals and alloys in environments containing halogen ions. The characterization of pitting shows narrow pits with a radius of the same order of magnitude as, or less than, the depth, and with different shapes of pits as illustrated in **Figure 3**. This form of corrosion is dangerous because penetration can occur without a clear warning in addition to difficult prediction of the pit growth.

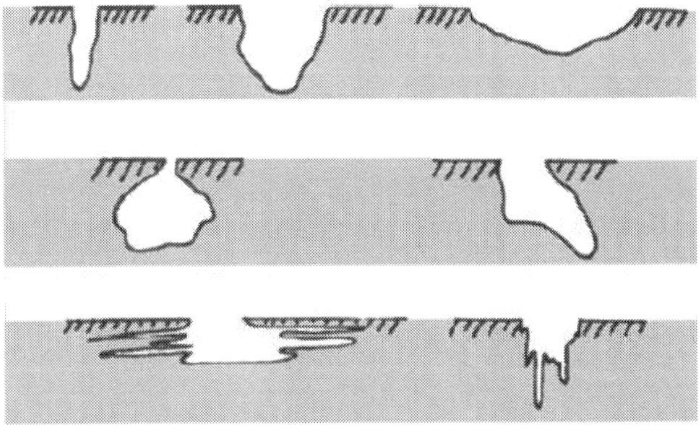

Figure 3. Different shapes of corrosion pits.

Determining pitting corrosion in laboratory is difficult because it needs longer time for the pit to be grown, for example, many months before the pits are visible in addition to the critical size of the pit, that is, the maximum pit depth, which increases with increasing surface area.

Most materials liable to pitting are aluminium in environments containing chloride, stainless steels in strongly oxidizing chloride solutions, and copper and its alloys in fresh water at pH <7.4 and temperatures >60°C, and when the ratio of sulphate to carbonate is relatively high.

The mechanism of pitting consists of the following stages:

1. Local breakdown of passivity (pit nucleation).

2. Early pit growth.

3. Late or stable pit growth.

4. Repassivation (if possible).

There are many theories to interpret the mechanism of pitting: some authors suggested that pitting is initiated by the adsorption of halide ions that penetrate the passive film at weak points of the oxide film, for example, at irregularities in the oxide structure due to grain boundaries or inclusions in the metal. This absorption leads to strong increase of the ion conductivity in the oxide film [7].

Other authors suggested that the initial adsorption of aggressive anions at the oxide surface catalytically enhances the transfer of metal cations from the oxide to the electrolyte and then local thinning of the oxide film can occur. A third theory suggests that the attacks may be start at defects in the passive layer and then introduce.

The next stage proceeds essentially by the anodic dissolution inside the pit (at the bottom) and the cathodic reaction outside the pit. Metal chlorides form at the bottom and then hydrolyze to produce an acidic solution, which also may allow an additional cathodic reaction represented by the reduction of hydrogen ions. Metal cations from the dissolution reaction migrate and diffuse toward the mouth of the pit where they react with OH^- ions to form metal hydroxide deposits that may cover the pit.

There are many factors affecting pitting as follows:

a. pH and chloride concentration: the pitting potential and pitting resistance normally increase with increasing pH and decreasing chloride concentration. The negative effect of increased Cl^- concentration on the pitting potential E_p is of great significance. For an 18-8 CrNi steel, E_p can be expressed by

$$E_p = 0.168 - 0.0881 log a_{Cl^-} \tag{1}$$

where a_{Cl^-} *is the activity of chloride ions.*

While for aluminium

http://dx.doi.org/10.5772/63657

$$E_p = -0.504 - 0.124 \log a_{Cl^-} \quad (2)$$

b. Flow velocity: this factor has a complex effect on corrosion rate. At high velocity, washing away the corrosive environment in the pit can occur, but it is leads to increase the transport of oxygen to the active area, so the pit may be passivated before it gets the chance to grow. Then, the cover of pits washes away to a higher extent, and pit initiation in the neighbourhood of an active pit occurs more easily. The result of this process is smaller but more numerous pits, which makes pitting corrosion less serious.

c. The gravity force: heavy pitting occurs on horizontal top surfaces, while hard pitting occurs on underside surfaces. Vertical surfaces are intermediate as to the extent of pitting due to the higher density of corrosive environment in the pits.

d. Cu^{2+} and Fe^{3+} ions have more noble potentials that cause initiation of pit and acceleration of pit growth.

e. Metallurgical properties: some impurities, inclusions and secondary phase participate in the localization of pits on materials because they act as efficient cathodes.

f. The insulating ability of the oxide: the oxide layer plays a good role in slowing the growth of pits if it has efficient insulation.

g. Surface roughness: the smooth surfaces get few, large pits while rough surfaces get numerous smaller pits.

h. Temperature: pitting potential decreases with increasing temperature.

i. Galvanic contact: the tendency of pitting increases with contact to more noble material.

There are many methods to prevent pitting corrosion:

1. Selecting a proper material such as aluminium alloys with magnesium (e.g. AlMg 4.5 Mn), stainless steels with sufficient content of Cr, Ni and Mo, some Ni alloys and titanium.
2. Applying cathodic protection using sacrificial anodes of Zn or Al alloys.
3. Changing the environment.

2.6. Intergranular corrosion

Intergranular corrosion occurs on or at grain boundaries and it is a dangerous form of corrosion due to decreasing of the cohesive forces between the grains and it cannot withstand tensile stresses, followed by reducing the toughness of the material, and fracture can occur without warning. Finally, the grains may fall out to form pits or grooves on the surface.

This form of corrosion is presented in the presence of galvanic elements in or at the grain boundaries, which may be the impurities that segregate at the grain boundaries, larger amount of a dissolved alloying element at the grain boundaries or smaller amount of a dissolved alloying element at the grain boundaries.

In most cases, the zone of less noble material in/at the grain boundaries acts as an anode and the other parts of the surface form the cathode. But in some cases, precipitates at the grain boundaries may be more noble than the bulk material and then they act as efficient local cathodes [7].

2.7. Selective corrosion (selective leaching)

This form of corrosion occurs in alloys that contain clearly less noble metal than the other. In this attack, the less noble element is removed from the material leaving a porous material with very low strength and ductility. The selectively corroded areas are sometimes covered with corrosion products or other deposits and it is difficult to be discovered. Serious material failure may therefore occur without warning. The most common example of selective corrosion is dezincification of brass (**Figure 4**), which occurs by removing the zinc from the alloy and retaining the copper. The Zn-depleted regions have a characteristically red copper colour in contrast to the original yellow brass. Dezincification occurs in two forms:

a. Uniform dezincification which occurs in the presence of high Zn content and acidic solutions.

b. Localized dezincification which occurs in the presence of lower Zn content and in neutral, alkaline and slightly acidic solutions.

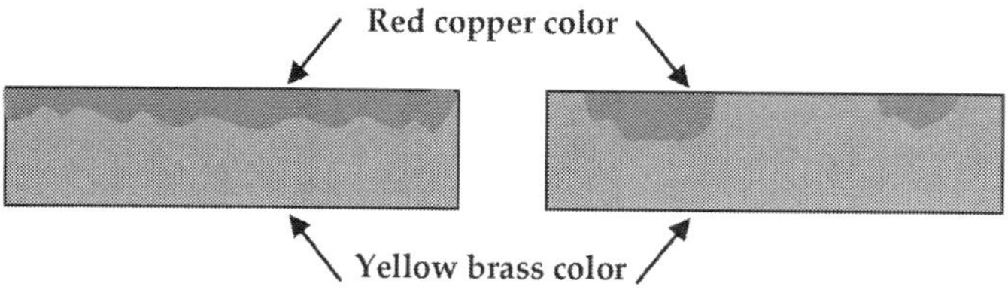

Figure 4. (a) Uniform dezincification and (b) Localized dezincification of brass.

The dezincification increases with increasing Zn content and temperature and is promoted by a stagnant solution, porous surface layers and oxygen in the corrosive medium.

2.8. Erosion and abrasion corrosion

Erosion corrosion occurs in metallic material immersed in corrosive fluid with a relative movement between the material and medium by wearing of the corrosion products, deposits or salt precipitates, which lead to the metallic surface being metallically clean and more active. Erosion corrosion accompanies with pure mechanical erosion, by tearing out the solid particles in the fluid and causing plastic deformation and leaving grooves or pits with a pattern determined by the flow direction and the local flow conditions. The corrosive fluid in erosion may be two-phase or multiphase flow, that is, liquid-gas and liquid-solid particle flow. This form of corrosion takes place in pumps, turbine parts, propellers, valves, heat exchanger tubes, bends, nozzles and equipment exposed to liquid sputter or jets [8].

http://dx.doi.org/10.5772/63657

The mechanism of erosion and abrasion corrosion can be divided into three types:

a. Impingement corrosion which occurs in two-phase or multiphase flow as shown in **Figure 5a**.

b. Turbulence corrosion which occurs in the regions of strong turbulence such as the inlet end of heat exchanger tubes as shown in **Figure 5b**.

c. Removal of corrosion products leads to increase in corrosion by wearing the particles moving along the corroding surface, or by wearing between components in moving contacts with each other.

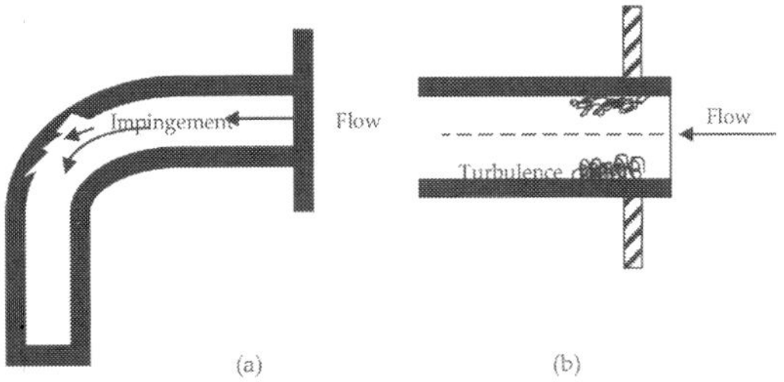

Figure 5. Type of erosion: (a) Impingement and (b) turbulence corrosion (nest).

In some cases, deposits may also be removed by high shear stresses. The shear stresses may vary considerably as a consequence of flow fluctuations or repeated impacts. The pure erosion rate can be expressed by the following formula:

$$W\ (\mathrm{mm/y}) = K_{\mathrm{mat}} \times K_{\mathrm{env}} \times c \times v^{n} \times f(\beta) \tag{3}$$

where $\mathbf{K}_{\mathrm{mat}}$ is a material factor depending on a complex manner on (among other properties) hardness and ductility of the substrate, $\mathbf{K}_{\mathrm{env}}$ is an environmental factor that includes the effects of size, shape (sharpness), density and hardness of the particles, **c** is the concentration of particles, **n** is the so-called velocity exponent, **v** is the particle velocity and **β** the impact angle. This equation is very useful in connection with erosion testing in the laboratory [7].

2.9. Cavitation corrosion

This corrosion form is similar to erosion corrosion and its appearance is deep pits grown perpendicularly to the surface. These pits localize close to each other or grow together over smaller or larger areas to produce a rough and spongy surface.

Cavitation corrosion occurs at high-flow velocities and fluid dynamic conditions causing large pressure variations, as often is the case for water turbines, pump rotors, propellers and the external surface of wet cylinder linings in diesel engines. The mechanism starts by rapidly collapsing vapour bubbles formed in low-pressure zones when they suddenly enter a high-pressure zone. This collapse causes a concentrated and intense impact against the metal, which induces the high local stress and then local plastic deformation of the material can occur. Repeating these steps leads to microscopic fatigue and crack formation, and subsequent removal of particles from the material itself.

2.10. Fretting corrosion (fretting oxidation)

Fretting corrosion occurs in fitting components in the presence of slightly relative motion (slip) between them. This motion may be ranged from less than a nanometre to several micrometres. This form of corrosion takes place in joints, press fits, and other components which subjected to load.

The fretting corrosion involves mechanical contribution including elements of adhesive wear and microscopic fatigue crack development, and then delamination in metal lattice can occur. The formed debris adheres to the fretting surfaces and tappers between these, and may partly escape [7].

The mechanism of fretting corrosion proposes that the relative motion between the parts may promote oxidation of the surface. This formed oxide film is partly worn off, and the fresh metal surface is highly active and oxidizes again. The repeating of this circular process leads to fretting damage. Another explanation is that, firstly, metal particles are released from the metal surface by adhesive wear, and then the particles oxidize forming debris, which moves out from the interfacial area between the components. Repeating of this process leads to deterioration at a high rate.

Fretting corrosion may be reduced or prevented by the following:

1. Use of lubricants, gaskets or sealing materials, which hinder the access of oxygen and then decrease adhesive wear.
2. Using proper and hard materials in components.
3. Appling coating on soft material.

2.11. Stress corrosion cracking

Stress corrosion cracking occurs by crack formation caused by static tensile stresses and corrosion. This stresses may be formed by external or internal stresses. The initial crack is discoloured and brittle, dull or darkened by oxide layers, and the crack may be discovered as a pit being narrow, and filled or covered by corrosion products. The factors affecting stress corrosion cracking are as follows:

1. Environmental and electrochemical factors.
2. Metallurgical factors.
3. Mechanical stress and strain.

Stress corrosion cracking can be prevented by affecting one or more of the above factors as follows:

http://dx.doi.org/10.5772/63657

a. Selecting the right material.

b. Achieve annealing to remove residual stresses in materials to below the threshold values.

c. Changing the environment to be less corrosive.

d. Appling cathodic protection by sacrificial anodes.

e. Adding corrosion inhibitors.

2.12. Corrosion fatigue

Corrosion fatigue is crack formation due to varying stresses combined with corrosion. The crack development can be divided into four stages:

1. Initiation.
2. Crack growth stage I, with crack growth direction about 45° relative to the tensile stress direction.
3. Crack growth stage II.
4. Fast fracture as shown in **Figure 6**.

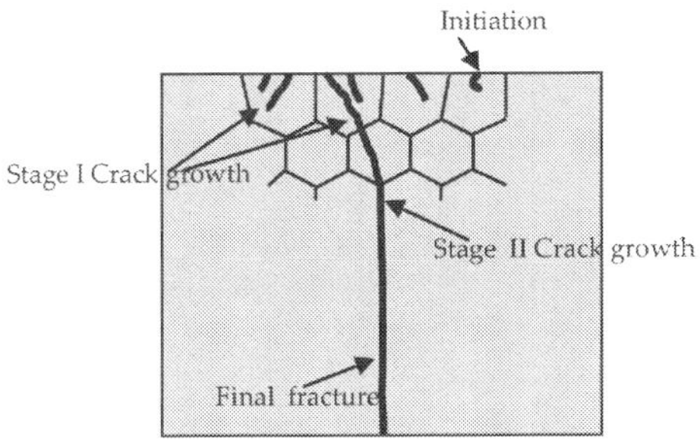

Figure 6. The stages of development of a fatigue fracture.

The mechanism of corrosion fatigue follows the following steps:

i. Stress concentrations in certain site lead to form corrosion pit and groove.

ii. Mutual acceleration of corrosion and mechanical deterioration.

iii. Absorption and adsorption of species from the environment, including absorption of hydrogen and resulting embrittlement.

3. Tribocorrosion

Tribocorrosion is a material degradation resulting from simultaneous action of wear and corrosion. Under these conditions, the material selection is a challenge since the material has to effectively withstand wear, corrosion and their combined effect. The importance of tribology and tribocorrosion is to know the equipment failure due to wear and corrosion. Potential financial savings in the industry by reduction in maintenance, downtime, breakdowns and replacement of equipment are considerable [8–10].

Application of good practices and known tribological principles can save industry and society from this cost. Research within tribology contributes thereby to an increase in efficiency and availability of production processes, longer lifetime of equipment and safer operation [9–11].

Wear and friction together with lubrication form the scientific discipline called 'tribology'. A major area of tribology is to design surfaces sliding and rolling against each other in such a way that friction and wear are minimized. By reducing friction and wear, many benefits, both economical and environmental, can be achieved in several technological fields of application. When developing technical equipment, the choice of material for a certain detail is often a compromise. Usually, it is not possible to optimize all material properties (mechanical, tribological, electrical and chemical) at the same time. The best material choice with respect to one property might not be possible to use due to poor performance with respect to another property [9, 10].

Sometimes the application of any material needs to knowledge about the surface of the component. For instance, low wear and low friction are mainly controlled by the surface properties. An elegant way to achieve this can be to use a material with poor tribological performance but otherwise optimal properties, for example, high toughness, low weight or low price, as base material and coat it with a thin layer of a wear resistant material with low friction coefficient.

Wear is very complex phenomenon and usually involved progressive loss of material. Wear processes can be categorized into five main forms:

Abrasion Wear: The definition of abrasion wear is produced by a hard/sharp particle or protuberance imposed on and moving on a softer surface.

Adhesion Wear: It is defined as localized bonding between contacting solids leading to material transfer or loss from either contacting surface.

Corrosion Wear: Corrosive wear occurs when the sliding of surfaces is in corrosive environment. This wear action generally takes place in two stages. First, there is corrosive attack on the surface. Then, the sliding action wears off the corroded surface film.

Surface fatigue: Contacts between asperities with very high local stress that are repeated a large number of times during sliding or rolling, with or without lubrication. High plastic deformation causes crack initiation, crack growth and fracture.

Erosion Wear: As illustrated in previous section, the erosion is referred to the continuous loss of a solid surface due to mechanical interaction between the surface and a fluid, which may be a multiphase fluid or impinging solid or liquid particles.

Accurate prediction of the rate of erosion requires detailed investigation of the solid particle motion before and after impact. The difficulty arises mainly from the fact that most flows

http://dx.doi.org/10.5772/63657

occurring in industrial processes are turbulent, which makes the particle path and impact characteristics difficult to predict [12].

4. Measurement of corrosion under normal conditions

The exposure of material to corrosive medium leads to reduction in weight or thickness of material. This reduction in weight or thickness refers to corrosion rate measurement. The common expression of this loss is mils per year (mpy), inches per year (ipy) and milligrams per square decimeter per day (mdd). Conversion of ipy to mdd or vice versa requires knowledge of the metal density.

To determine the corrosion rate, a specimen with known surface area is exposed to the corrosive environment for a period of time and then the loss in weight has been determined by the weighting of corroded specimen. The corrosion rate (C.R.) of the metal can be calculated using the formula:

$$\text{C.R.} = \frac{\Delta \text{W}}{\text{area . time}} (\text{mdd, gmd...}) \tag{4}$$

$$\text{C.R.(mpy)} = \frac{22.237\,\Delta\,\text{W}}{\text{density . area . time}} \tag{5}$$

where ΔW is in grams, density in g.cm^{-3}, area in inch^2 and time in days, and then weight loss to penetration loss can be converted by the following relation:

$$C.R._{(mpy)} = \frac{1.44}{density} C.R._{(mdd)} \tag{6}$$

According to Faraday's law:

$$Q = \frac{n\text{F}W}{M} \tag{7}$$

where Q is the charge passed (Coulombs), n is the number of electrons involved in the electrochemical reaction, F is Faraday constant, W is the weight of electroactive species (g) and M is the molecular weight (g).

$$W = \frac{QM}{nF} \tag{8}$$

$$W = \frac{QE.W.}{F} = \frac{it(E.W.)}{F} \tag{9}$$

Or

$$\frac{W}{t} = \text{Corrosion Rate}(C.R.)(\frac{g}{s}) \tag{10}$$

$$C.R.\left(\frac{cm}{s}\right) = \frac{i(E.W.)}{dFA} \tag{11}$$

$$C.R.(mpy) = \frac{i(E.W.)\times 31.6\times 6\times 10^{6}\times 10^{3}}{dFA\times 2.5\times 10^{6}} \tag{12}$$

$$C.R.(mpy) = \frac{0.13\times i_{corr}(E.W.)}{d} \tag{13}$$

where mpy is milliinches per year, i_{corr} is the corrosion current density ($\mu A.cm^{-2}$), E.W. is the equivalent weight of the corroding species (g) and d is the density of the corroding species ($g.cm^{-3}$).

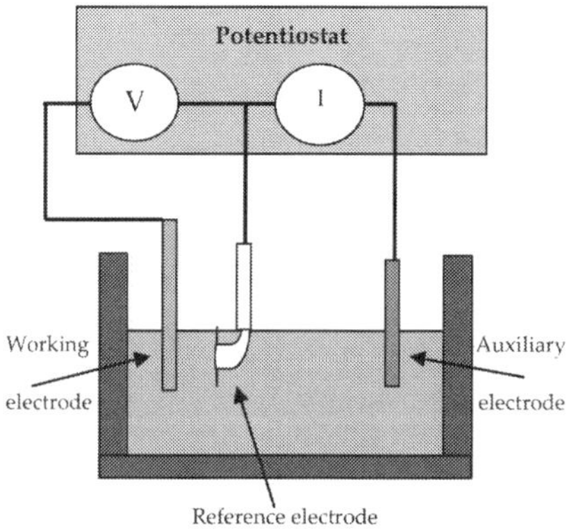

Figure 7. Arrangement of electrodes in corrosion test.

http://dx.doi.org/10.5772/63657

The most exact method to determine corrosion rate is achieved using Potentiostat. This device controls the voltage difference between a working electrode and a reference electrode. Three electrodes used in this method include an auxiliary electrode (such as Pt and Graphite) to inject the current into the cell, and then the current flow will be recorded by Potentiostat, another electrode is a reference electrode (such as Calomel and Ag/AgCl) to measure the potential of a specimen, which represents the third electrode as working electrode. These three electrodes are arranged as in **Figure 7**, which are connected to Potentiostat. This arrangement is obtained using electrochemical cell as in **Figure 8**.

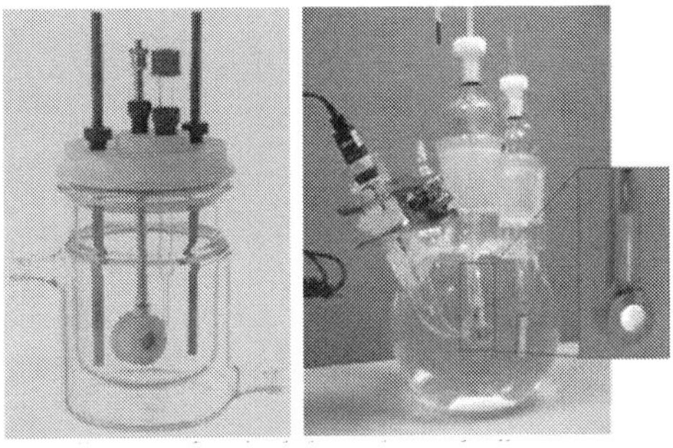

Figure 8. Standard electrochemical cells.

4.1. Measurement of corrosion potential (E_{corr})

The potential of the specimen (working electrode) is measured with respect to a reference electrode, which is kept in a separate container and connected electrically with the working electrode placed in a container in contact with the electrolyte via a salt bridge (Luggin). The potential of the working electrode is measured by a voltmeter, which is connected between the working and the reference electrode through the negative terminal of the voltmeter connected to the working electrode and the positive to the reference electrode. The corrosion potential is also referred to as the open-circuit potential (OCP) as the metal surface corrodes freely and it is called mixed corrosion potential. After applying voltages with ±200 mV around OCP, we can get polarization behaviour of working electrode in environment, which is immersed in it. The obtained polarization is the behaviour of the material in active region only, and by drawing the tangents to the anodic and cathodic curves, both the corrosion current density (i_{corr}) and corrosion potential (E_{corr}) can be determined (which is extrapolated to the point of intersection) using the four-point method [13].

4.2. Measurement of corrosion current density (i_{corr})

When applying the voltages on working electrode through Potentiostat, the current of working electrode can be recorded by auxiliary electrode through ammeter, which is connected between the working and auxiliary (counter) electrode.

The anodic polarization is tended in the noble direction starting from the OCP and the overpotential is positive, since the polarized potential is more positive than the equilibrium potential. The cathodic polarization has a small overpotential and is tended in the active direction; however, at a sufficiently large overpotential, only one reaction, either anodic or cathodic, takes place depending on which direction the potential is impressed [13].

When a metal is corroded in its environment, the anodic process (represented by oxidation reaction) intersects the cathodic process (represented by reduction reaction) at the corrosion potential (E_{corr}). The current corresponding to i_{corr} is the corrosion current density (i_{corr}), which can be determined from the intersection with X-axis.

4.3. Determination of corrosion rate by electrochemical measurements

4.3.1. Tafel extrapolation method

When the polarization curves for the anodic and cathodic reactions are obtained by applying potentials about ±200 mV around OCP, the corrosion potential and THE corresponding current will be recorded. Plotting the logarithms of current (log I) versus potential and extrapolating the currents in the two Tafel regions gives the corrosion potential and the corrosion current density i_{corr} and this is the principle of Tafel technique as illustrated in **Figure 9**. By knowing i_{corr}, the rate of corrosion can be calculated in desired units by using Faraday's law [13].

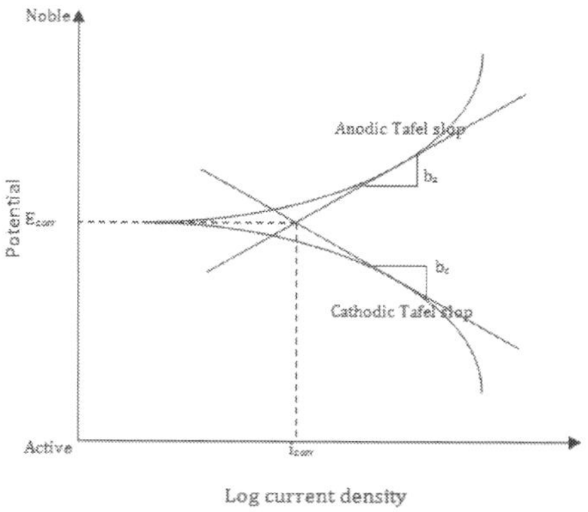

Figure 9. A hypothetical Tafel plot.

4.3.2. Polarization resistance (linear polarization) method

In this technique, the potential-current density plot gives approximately linear in a region of within ±10 mV of the corrosion potential with slope represented by (ΔE)/(ΔI). When both anodic and cathodic reactions proceed in the vicinity of corrosion potential (E_{corr}), they are exponentially dependent upon the applied potential. Over a small potential range (20 mV),

http://dx.doi.org/10.5772/63657

the exponents are linearized and an approximate linear potential-current relationship is obtained [13].

Polarization resistance (R_p) of a corroding metal is defined using Ohm's law as the slope of a potential (E) versus current density (log i) plot at the corrosion potential (E_{corr}). Here, R_p = (ΔE)/(ΔI) at ΔE = 0. By measuring this slope, the rate of corrosion can be measured. The correlation between i_{corr} and slope (dI)/(dE) is given by (**Figure 10**):

$$\frac{\Delta E}{\Delta I} = \frac{b_a b_c}{2.3 i_{corr} (b_a + b_c)} \tag{14}$$

where ba and bc are Tafel slopes.

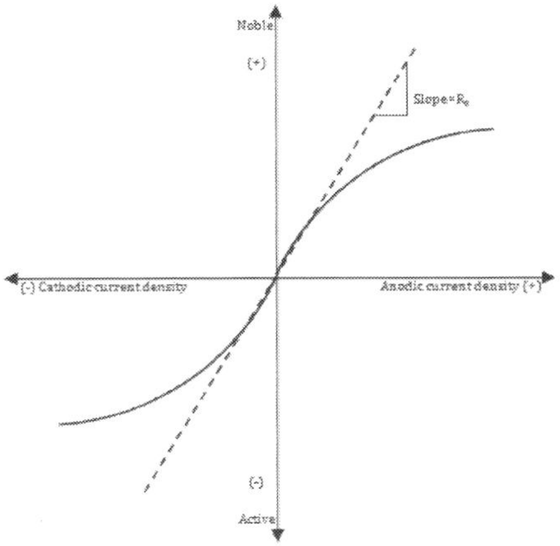

Figure 10. Hypothetical linear polarization plot.

5. Tribocorrosion test principle

Tribocorrosion test can be achieved depending on the basic principles of electrochemical test using Potentiostat. This technique requires three electrodes: counter, reference and working electrode. The working electrode (experimental specimen) is mounted as O-ring with a known surface area, which is exposed to the electrolyte (see **Figures 11** and **12**).

Tribocorrosion test is achieved by applying a known force on a pin, or a ball, in contact with the surface of specimen to create the wear, and then OCP, linear polarization and cyclic polarization can be recorded in situ. The wear rate of the material calculates from the volume lost.

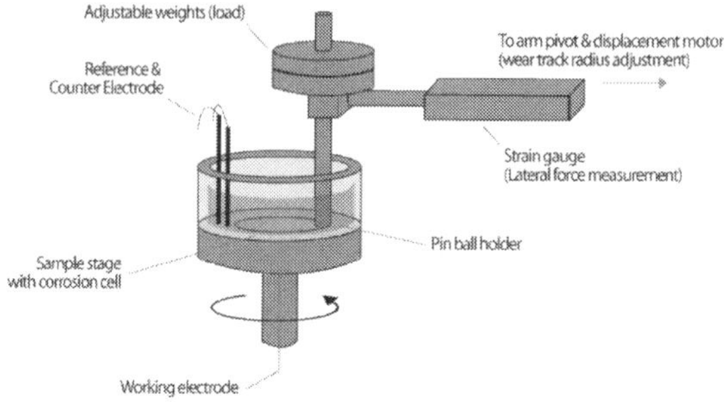

Figure 11. Schematic of the tribocorrosion test.

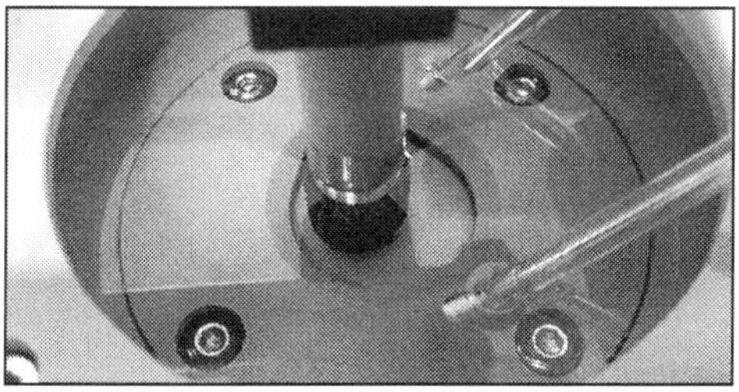

Figure 12. Tribocorrosion module of the pin-on-disk tribometer.

5.1. Tribocorrosion test at open-circuit potential

Using three-electrode cell as illustrated in **Figure 11**, a platinum wire, a Ag/AgCl, NaCl (sat) and mounted specimen with Tribometer are used. After applying the force on material surface, OCP can be recorded beyond maintaining the specimen in the electrolyte for 1 h to obtain a stable OCP, followed by the tribocorrosion test for 10 min (1000 cycles). After the wear process, the sample was kept in the solution for another 10 min to allow surface repassivation. Dry wear tests must be performed at the same condition for comparison.

5.2. Tribocorrosion test at controlled potentials

Cathodic and anodic polarization conditions are carried out after OCP test by changing the applied potentials ±200 mV around OCP to show the acceleration of corrosion process by wear. The evolution of the current was monitored before, during and after the wear process. The wear rate, K, was evaluated using the formula:

$$K = \frac{V}{F \times s} = \frac{A}{F \times n} \tag{15}$$

where V is the worn volume, F is the normal load, s is the sliding distance, A is the cross-section area of the wear track and n is the number of revolution.

The set-up of simultaneous measurements of wear and electrochemical parameters (open-circuit potential) is shown in **Figure 13**.

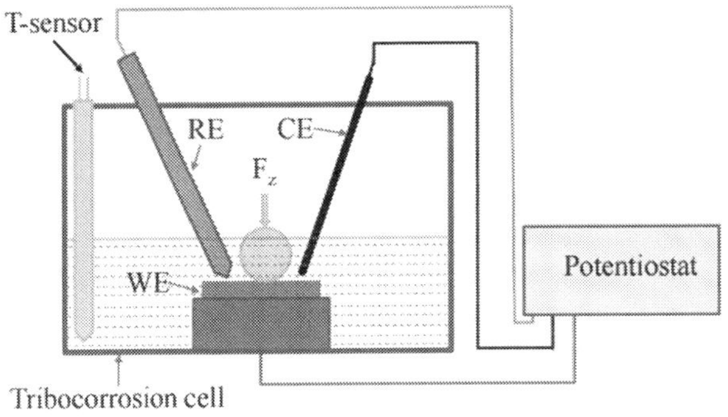

Figure 13. Schematic tribocorrosion test, WE: working electrode, RE: reference electrode and CE: auxiliary (counter) electrode.

5.3. Tribocorrosion tests at anodic/cathodic potential

In order to further emphasize and demonstrate the contribution of corrosion in tribocorrosion, the tribocorrosion tests are performed at anodic/cathodic potentials. The synergistic effect of severe corrosion and wear leads to the formation of a substantially larger wear track. By contrast, the tribocorrosion test that performs at a cathodic potential exhibits a lower wear rate showing the importance of corrosion suppression for tribocorrosion protection.

The tribocorrosion studies take a wide range of researches due to wide applications in a variety of industries, such as mining, food, oil, automotive, offshore marine, nuclear and biomedical. Tribocorrosion is a surface degradation process resulting from simultaneous tribological and electrochemical actions in a corrosive environment (*Wood, Landolt, Stack and Mischler*). Mathew et al. highlighted the study tribocorrosion behaviour of Ti_6Al_4V coated with a bioabsorbable polymer for biomedical applications to predict OCP of Ti_6Al_4V immersed in artificial saliva solution at 37°C in contact with Al_2O_3 in addition to evaluate the coefficient of friction [14]. Mathew et al. in another study highlighted tribocorrosion in the hip joint [15]. These studies confirm the importance of tribocorrosion tests.

These studies cannot be predicted from the knowledge of wear and corrosion of the material because tribocorrosion can accelerate the mass loss in synergistic effects of wear and corrosion. The degradation of materials can cause tribocorrosion and affect the friction, wear and lubrication behaviour of the materials [16]. According to ASTM G 119, wear and corrosion form a synergistic couple:

$$T = W_o + C_o + S \tag{16}$$

where T is the rate of total material loss due to tribocorrosion, W_o is the rate of mechanical wear in the absence of corrosive agent, C_o is the rate of corrosion in the absence of mechanical agent and S is the change in the rate of material loss due to wear-corrosion synergy.

To obtain synergy (S), the following tests are performed: *Test 1:* electrochemical polarization to find corrosion potential (E_{corr}) and corrosion current (i_{corr}) to calculate C_o. *Test 2:* wear test in corrosive environment, no external potential or electrical current to obtain *T*. *Test 3:* repeat Test 2 to obtain electrochemical corrosion rate in the presence of wear (C_W) and *Test 4:* repeat Test 2 with a potential of 1 V cathodic to E_{corr} to obtain W_o.

$$S = \Delta C_W + \Delta W_C \tag{17}$$

$$\Delta C_W = C_W - C_o \tag{18}$$

$$\Delta W_C = T - (W_o + C_W) \tag{19}$$

$$\text{Total synergism factor} = \frac{T}{(T - S)}$$

$$\text{Corrosion augmentation factor} = \frac{C_W}{C_o}$$

$$\text{Wear augmentation factor} = \frac{W_o + \Delta W_C}{W_o}$$

6. Conclusion

- Corrosion is an important problem in many industries, so it is necessary to know how one can discover this problem before the failure can occur.
- There are some difficulties to discover many types of corrosion because the appearance of metallic surface may be covered by corrosion products; therefore, it is necessary to know some details about service conditions.
- Tribocorrosion is a dangerous form of corrosion because it leads to reduction in material surface and then the failure can occur without clear warning.
- The prediction of tribocorrosion rate needs some requirements to be accurate and then using some methods to reduce it or continuously monitor it.

http://dx.doi.org/10.5772/63657

Author details

Rana Afif Majed Anaee[1*] and Majid Hameed Abdulmajeed[2*]

*Address all correspondence to: Dr.rana_afif@yahoo.com and majed_hamed60@yahoo.com

1 University of Technology, Department of Materials Engineering, Corrosion Science (University of Baghdad), Baghdad, Iraq

2 University of Technology, Department of Materials Engineering, Applied Mechanic (University of Belgrade), Baghdad, Iraq

References

[1] Hoar T. P.: Electrochemical principles of the corrosion and protection of metals. Journal of Applied Chemistry. 1961; 11:121–130. DOI: 10.1002/jctb.5010110401.

[2] Winston Revie R. and Uhlig H. H. Corrosion and Corrosion Control: An Introduction to Corrosion Science and Engineering. 4th ed. Wiley, New York; 2008.

[3] Fontana M. G. and Staehle R. W. Advances in Corrosion Science and Technology. Plenum Press, New York; 1990.

[4] Fontana M. G. Corrosion Engineering. 3rd ed. Mc Graw-Hill Book Co, Singapore; 1987.

[5] Vermilyea D. A. Proc. 1st International Congress on Metallic Corrosion, London, 1961, Butterworths, London, 62; 1962.

[6] Evans U. R. The Corrosion and Oxidation of Metals. Arnold, London; 1960.

[7] Bardal E. Corrosion and Protection. Springer, London; 2003.

[8] Hon S., Woydt M., Rymuza Z., Kajdas C., Furey M., Viz¡intin J., Tysoe W., Kotvis P., and Buyanovsky I. Surface Modification and Mechanisms Friction Stress and Reaction Engineering. Ch. 13: Corrosion and Its Impact on Wear Processes. 2004. ISBN: 978-0-203-02154-5; DOI: 10.1201/9780203021545.ch3

[9] Stack M. M.: Mapping tribo-corrosion processes in dry and in aqueous conditions: some new directions for the new millennium. Tribology International. 2002; 35:681–689.

[10] Mischler S., Triboelectrochemical techniques and interpretation methods in tribocorrosion: A comparative evaluation. Tribology International. 2008;41:573–583.

[11] Wood R. J. K.: Tribo-corrosion of coatings: a review. Journal of Physics D-Applied Physics. 2007; 40:5502–5521.

[12] Badr H. M., Habib M. A., Ben-Mansour R., and Said S. A. M.: Numerical investigation of erosion threshold velocity in a pipe with sudden contraction. Computers and Fluids. 2005; 34(6):721–742.

[13] Ahmad Z. Principles of Corrosion Engineering and Corrosion Control. Springer; 2006.

[14] Mathew M. T., Souza J. C. M., Tajiri H. A., Morsch C. S., Buciumeanu M., Silva F. S., and Henriques B.: Tribocorrosion behavior of Ti6Al4V coated with a bio-absorbable polymer for biomedical applications. Journal of Bio- and Tribo- Corrosion. 2015; 1(27). DOI: 10.1007/s40735-015-0029-5

[15] Mathew M. T., Uth T., Hallab N.J., Pourzal R., Fischer A., and Wimmer M.A.: Construction of a tribocorrosion test apparatus for the hip joint: validation, test methodology and analysis. Wear. 2011; 271: 2651–2659. DOI: 10.1016/j.wear.2011.01.085

[16] Landolt D., Mischler S., Stemp M.: Electrochemical methods in tribocorrosion: a critical appraisal. Electrochimica Acta. 2001; 1(46): 3913–3929.

Chapter 6

Biotribology of Artificial Hip Joints

Binnur Sagbas

Additional information is available at the end of the chapter

http://dx.doi.org/10.5772/64488

Abstract

Tribology is the science of interacting surfaces; when these surfaces are in a biological system, it is called as biotribology. With the increasing rate of joint replacement operations and need for artificial prosthesis, biotribology is becoming a very important and rapidly growing branch of tribology. Based on this fact, in this chapter, basic tribological concepts are presented in terms of friction, lubrication, and wear; then with these fundamental backgrounds the biotribological behavior of natural and artificial hip joints are discussed in detail. Moreover, material pairs that are used in artificial joint replacements and the application of surface modification for the enhancement of the tribological properties of these materials are handled. Furthermore, the determination of tribological behavior of joint materials such as wear, coefficient of friction, friction torque, and frictional heating by using conventional techniques and hip joint simulator are discussed. Finally, the measurement and analysis of wear in both retrieved prosthesis and experimental studies are discussed referring the latest research articles.

Keywords: hip joints, friction, wear, biotribology

1. Introduction

The term tribology, which is derived from Greek words *tribos* meaning rubbing and friction and *logos* meaning science, was first used in the Jost Report in 1966 [1]. It is the science of interacting surfaces and includes the subjects such as friction, lubrication, and wear. Tribology is multidisciplinary science that encompasses mechanical engineering, materials science, surface engineering (surface coating, surface modification, and surface topography analysis), lubricants, and additives chemistry [1]. *Nanotribology* is the area in which friction and wear are studied at the micro- and nanoscale levels. The term *biotribolgy* is used when the interacting surfaces are a part of a human body or an animal such as total hip and knee joints.

It was first used by Dowson in 1970 and was explained as tribology in biological systems such as skin, hair, eyes, synovial joints, spine, and oral tribology [2].

Artificial prosthesis is replaced with natural parts of the hip joints when these natural parts lose their functionalities because of osteoarthritic and traumatic situations. Different material pairs such as metal-metal, ceramic-ceramic, ceramic-polymer, and metal-polymer have been used for artificial femoral head and acetabular insert. Service life of these joint parts is restricted because of the tribological deficiency in the joint materials. Lubrication regimes, wear debris, friction coefficient, and frictional heating are the primary factors that affect the implant durability. Wear debris leads to adverse tissue reactions that cause aseptic loosening of the components and finally leads to implant loss [2]. Frictional heating may cause property changes in sliding materials, tissue damage, and changes in lubricant properties such as protein precipitation especially in the biotribological zone.

The measurement, characterization, and analysis of the biotribological properties of artificial hip joints are vitally important for developing long durable joint replacements. Radiographic, gravimetric, volumetric, and optical techniques are the current methods for measuring and evaluating wear in the total joint replacement components. For surface characterization, optical and laser profilometry, white light interferometry, and digital microscopy are the prominent methods that can be applied both macroscopically and microscopically for characterizing damage modes such as burnishing, abrasion, scratching, pitting, plastic deformation, fracture, fatigue damage, and embedded debris.

Although the researches have been conducted for determining the wear behavior of the total joint replacements in order to improve design, material and manufacturing quality, and the service life of these joints, the phenomena of how to design wear-resistant artificial joint parts and select ideal material pairs for these kinds of replacements are still unknown. Therefore, the examination of friction and wear characteristics of prosthesis material pairs, both for in vivo clinical applications and in vitro laboratory simulations, is still one of the most important topics for researchers [3–7]. This chapter describes how to handle biotribological behavior of natural and artificial hip joints in terms of friction, lubrication, wear, frictional heating, and wear evaluation referring the basic tribological rules and the latest research articles.

2. Basic concepts of tribology

Handbooks define tribology as the science and technology of interacting surfaces in relative motion with all practices. Basically, tribology deals with the study of friction, lubrication, and wear phenomenon that can be explained with a series of engineering subjects such as mechanics, solid mechanics, fluid mechanics, physics, applied mathematics, rheology, machine design lubricant chemistry, material science, thermodynamics, and heat transfer [8–12].

Tribology is vitally important for any national and global economy. It causes large amount of energy loss, material waste, time loss, labor loss, etc. [13]. For instance, early wear of an acetabular insert of artificial hip joint means revision surgery that takes time of surgeons and causes waste of materials. Furthermore, new surgery means pain for patients who need rest after the surgery. Therefore, it reduces life quality and causes labor loss. Tribological problems, both in industrial and biological areas, cause enormous cost to global economy and waste of natural resources, as well as affect the social system. Therefore, understanding tribology in terms of friction, lubrication, and wear is crucial for the development of wear-resistant materials and designs.

In this section, the basic concepts of tribology such as friction, lubrication, and wear are summarized for better understanding of biotribology in hip joints.

2.1. Friction

Friction is generally explained as the resistance to relative motion between articulating surfaces [13, 14]. It is the main cause of wear and energy loss [13]. An energy input is provided for the motion of sliding surfaces and maintaining the motion. This energy is dissipated into the system, mainly as frictional heat that causes property changes in sliding materials [14], tissue damage, and changes in lubricant properties such as protein precipitation especially in the biotribological zone.

A deep understanding of friction and wear processes first requires the investigation of the influence of numerous effects accompanying the friction process, i.e., mechanical, electrical, hydroacoustic, physicochemical, and other effects, and their influence on physicochemical properties, structure of working surfaces, etc. [15]. Friction is generally classified as static friction, sliding friction, and rolling friction (see **Figure 1**).

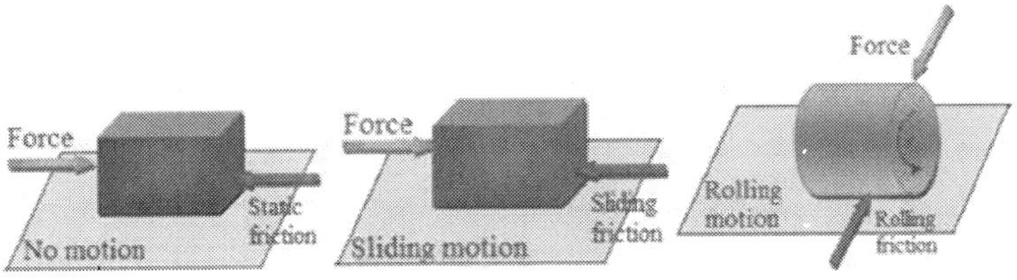

Figure 1. Scheme of friction force and motion.

In static friction, there is no motion [14, 16]. To slide a contacting body over another, a tangential force must be overcome which is named as *friction force* (F). It acts on the sliding surface plane and is usually proportional to the normal force (N) [14]:

$$F = \mu \cdot N \tag{1}$$

The proportionality constant termed as the coefficient of friction (μ) is used for quantifying sliding or kinetic friction, and it is defined as the ratio between the friction force F and the normal load N:

$$\mu = \frac{F}{N} \tag{2}$$

The coefficient of friction generally ranges from 0.03 for well-lubricated bearing to 1 for dry sliding [14, 16]. These values change according to operating parameters, such as sliding speed, applied load or contact pressure, temperature, presence of lubricant, and the properties of materials in contact such as surface roughness of sliding pairs [17].

For comfortable walking, the coefficient of friction must be 0.2–0.3; on ice walking, the μ-value between shoe and ice pair nearly becomes 0.05. In a synovial joint, with the very efficient natural lubrication, the coefficient of friction is 0.02 [14].

Surfaces are not perfect at the microscopic level. Peaks, valleys, asperities, and depressions can be seen at high magnification even on the best polished surface (see **Figure 2**) [16]. When these two surfaces are brought together, they touch from the tips of the surface asperities. At that point, adhesion or cold welding, which generally refers to resistance to separating bodies from each other, may occur, and plastic deformation may take place on a very local scale. To start the sliding motion, these formations must be broken by the friction force. The main contribution to friction action is extended by adhesion and deformation, but additional contributions may occur, such as wear debris, presence of oxides, or adsorbed films [8, 14].

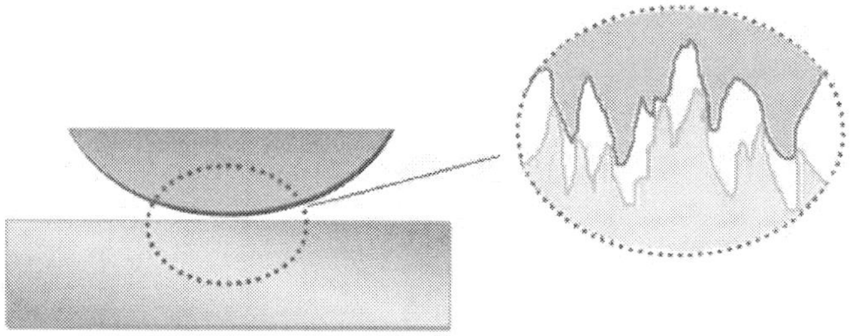

Figure 2. Microscopic detail of a real surface contact.

2.2. Lubrication

The main purpose of using lubricant is reducing the effect of normal and shear stresses on the solid surface contact [17]. Lubrication is one of the most effective ways of minimizing friction and delaying wear. Unfortunately, it is not the exact solution of the wear because wear occurs even with lubrication. Especially abrasive wear and delamination problem may occur under lubricated conditions [14].

Different types of lubrication regimes may arise between the sliding surfaces. In all lubricating modes, the surfaces are separated by a solid, a semisolid, a pressurized liquid, or the gaseous form of a lubricating film. *Dry-film (solid-film) lubrication* is a system in which a coating of solid-state lubricant separates the sliding surfaces and the lubricant itself wears away. *Boundary lubrication* is the regime in which the interacting surfaces react with the lubricant components. Each surface is covered by a chemically bonded fluid or a semisolid film that may or may not separate opposing surfaces. In *thin-film lubrication,* the lubricant usually is not bonded to the surfaces and it does not separate sliding surfaces. Moreover, lubricant viscosity affects friction and wear. In *fluid-film lubrication,* the sliding surfaces are separated by a fluid film and the physical properties of the lubricant such as viscosity and pressure viscosity designate the performance of the lubricated surfaces. This lubrication regime can be divided into two subcategories such as hydrodynamic lubrication and elastohydrodynamic lubrication. *Hydrodynamic lubrication* is a regime in which the formation of the fluid film depends on the shape and relative motion of the sliding surfaces with sufficient pressure for separating the surfaces [8, 14]. For the mathematical explanation of hydrodynamic lubrication, an equation derived by Reynolds known as the "Reynolds equation" is used [13]. In elastohydrodynamic lubrication, friction and film thickness between the sliding surfaces are defined by the elastic properties of the contacting bodies. Although fluid-film lubrication is a desired regime, the boundary lubrication cannot be avoided. Boundary lubrication occurs during starting up and stopping stages of the motion [12]. In *mixed lubrication,* boundary and the fluid film lubricated regions are considered simultaneously [18].

To minimize the friction and wear of the sliding surfaces, understanding and determining lubrication mechanism are very important tools for the optimization of the bearing materials and geometries, both in engineering system and artificial joints. For the theoretical prediction of the lubrication regime, some classical engineering methods can be applied to artificial joints [11]. At that point, it is necessary to provide some basic information about geometrical and surface features of hip joint. Joints in a human body may be classified anatomically and physiologically such as plane, ball-in-socket, ellipsoid, hinge, condylar, pivot, and saddle [19]. Hip joint is considered as ball-in-socket geometry where contacting surfaces fit together. In theoretical calculations sometimes ball-on-plane equivalent configuration may be used for simplifying the geometry. In ball-in-socket types of geometries, contacting bodies have same diameters but with a clearance between the elements for suitable fit of the bodies and tribological reasons. R_{head}, R_{cup}, and $c = R_{cup} - R_{head}$ represent the femoral head radii, acetabular cup radii, and radial clearance, respectively. Another important parameter of sliding bodies is the average surface roughness R_a of the frictional surfaces. It is relevant for the determination of the lubrication regime [3], Schematic drawing of relation between the surface roughness and the film thickness can be seen in Figure (**Figure 3**).

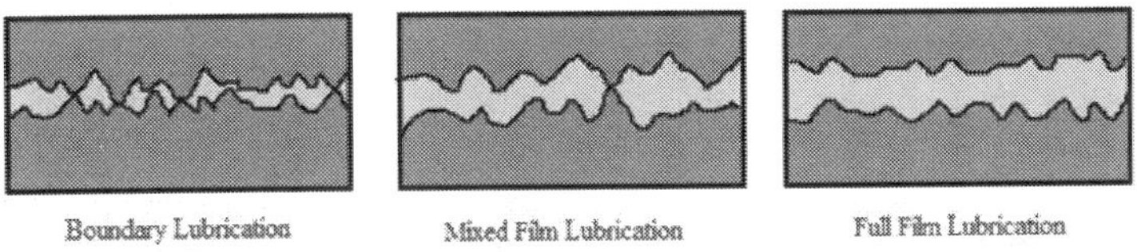

Figure 3. Schematic drawing of relation between the surface roughness and the film thickness.

The parameter λ, which is the ratio between the minimum film thickness $h_{\min}$ and the composite roughness of the sliding surfaces R_a, is generally used for determining the distance between the sliding surface asperities [16]:

$$\lambda = \frac{h_{\min}}{R_a} = \frac{h_{\min}}{\left[(R_{a\,\text{head}})^2 + (R_{a\,\text{cup}})^2\right]^{1/2}} \tag{3}$$

With the evaluation of λ, the lubrication regime can be identified in the following ranges [3]:

$0.1 < \lambda < 1$: boundary lubrication,

$1 \leq \lambda \leq 3$: mixed lubrication, and

$\lambda > 3$: full film lubrication.

The precision measurement of surface roughness for both femoral head and acetabular cup is important for the accurate determination of λ.

For the determination of film thickness, the following equation formulated for engineering can be used [11]:

$$\frac{h_{\min}}{R'} = 2.8\left(\frac{\eta u}{E'R'}\right)^{0.65}\left(\frac{W}{E'R'^2}\right)^{-0.21} \tag{4}$$

where R' is the equivalent radius that depends on the femoral head radius R_{head} and the radial clearance c.

$$\frac{1}{R'} = \frac{1}{R_{\text{head}}} - \frac{1}{R_{\text{cup}}} = \frac{c}{R_{\text{head}}(R_{\text{head}} + c)} \tag{5}$$

In the ball-on-plane equivalent configuration, the entraining velocity *(u)* can be calculated from the angular velocity of the femoral head (ω):

$$u = \frac{\omega R_{\text{head}}}{2} \tag{6}$$

The equivalent elastic modulus (E') can be determined by following equation:

$$\frac{1}{E'} = \frac{1}{2}\left(\frac{1-\nu_{\text{head}}^2}{E_{\text{head}}} + \frac{1-\nu_{\text{cup}}^2}{E_{\text{cup}}}\right) \tag{7}$$

where E_{head}, ν_{head} and E_{cup}, ν_{cup} are the Young's modulus and Poisson ratio of the femoral head and acetabular cup material, respectively [3].

For the prediction of lubrication between the femoral head and acetabular cup bearing surfaces and its effect on friction, generated during articulation, the Stribeck diagram can be used (see **Figure 4**). In this diagram, the relation between the lubrication and friction is commonly illustrated [9, 11, 14, 20]

$$z = \frac{\eta u r}{W} \tag{8}$$

where z is the Sommerfeld number, η is the lubricant viscosity, u is the sliding speed, r is the radius, and W is the load.

Traditionally, the Stribeck curve is divided into three regions. Boundary lubrication is seen when the thickness of the lubricating film is less than or equal to the average surface roughness of the articulating surfaces. When the thickness of the lubrication film increases, a transition stage, called mixed lubrication, is generated. The articulating surfaces are separated from each other while in contact with some asperities, and the combination of the fluid film and boundary lubrication can be seen in this regime. Full fluid film lubrication occurs with the continuous decrease in the friction coefficient, and the articulating surfaces are completely separated [9, 16].

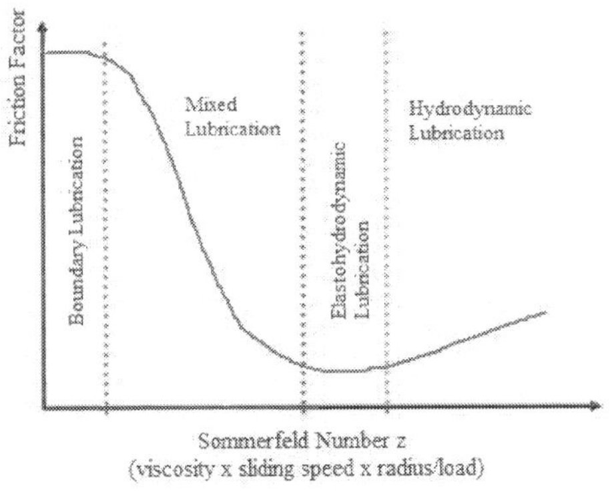

Figure 4. The Stribeck curve.

2.3. Wear

Wear is defined as the progressive loss of material from the surface of a body [18]. It is a complex phenomenon that involves multifarious events in a wildly unpredictable manner [10]. Numerous types of wear have already been defined in different studies related to tribology.

Although it is difficult to classify wear types without considering mating materials, for a general classification wear can be divided into five different groups, such as abrasive, adhesive, surface fatigue, erosive, and corrosive [11].

Abrasive wear is generally categorized according to contact types of the surfaces such as two-body and three-body wear. When the asperities of the harder surface abrade the softer one, it is called as two-body abrasion. If there are hard particles trapped between surfaces and abrade one or both of the surfaces, it is known as three-body abrasion [14]. *Adhesive wear* occurs due to high local friction that leads to tearing and fracture. This type of wear is generally defined as transfer of material from one surface to another during relative motion. The particles, broken away from one surface, may attach to another surface and act as an abrasive. This kind of wear may be seen between ceramic, polymers, and metallic material pairs or their combinations [9, 14, 17]. *Corrosive wear* is the wear that contains both mechanical wear and chemical reaction in which metal ions are released [9]. Besides the chemical and electrochemical reactions, environmental conditions govern the oxidative wear [11]. *Fatigue wear* is the displacement of the particles from the microscopic contact area of material surface by cyclic loading. It may lead to the generation of debris from the surface or cracks' propagation into the bulk material [11, 17]. *Erosive wear* is the loss of surface layer of the material caused by hard particles attacking to the surface. The attacking particles may be in solid, liquid, or slurry form. This type of wear may involve plastic deformation and brittle fracture. Erosive wear is similar to abrasive wear and generally be confused with it, but there is a definite distinction between erosive and abrasive wear. In erosive wear, the force is transferred to the surface by the particles due to their slowing down, while in abrasive wear, the force is externally applied. This kind of wear is not common in hip joints [8, 14].

3. Biotribology in hip joints

Biotribology is one of the newest and popular term dealing with friction, wear, and lubrication of interacting surfaces in a human or an animal body, while tribology is the study about these phenomena in engineering systems. Synovial joints, spine, skin, eyes, hair, and oral tribology are some examples of biotribological systems [21]. Because biotribology contains all tribological activities in these complex and natural biological systems, it is interdisciplinary like tribology. Moreover, besides the disciplines related to tribology, biotribology is also associated with biomechanics, biology, biochemistry, physiology, clinical medicine, and pathology. From the engineering point of view, for the development of long-lasting artificial interacting surface parts, it has to be well understood how these natural systems and artificial joints behave under biotribological conditions.

3.1. Natural synovial joints

There are over 300 joints in a human body. These joints have been categorized into two main groups such as synarthroses and diarthroses. Diarthroses joints are generally named as synovial joints because they have a cavity between the interfaces of the bones containing synovial fluid. Moreover, joints may be classified according to their anatomical and physiological properties such as plane type, ball-in-socket, ellipsoid, hinge, condylar, pivot, and saddle-type joints. Hip joint is a ball-in-socket type configuration that has three degrees of freedom such as flexion-extension plane, abduction-adduction, and lateral-medial rotation [19]. Articular cartilage, bone, synovial fluid, ligaments, tendons, tissues, and soft tissue capsule are the main components of a natural synovial joint [8].

Natural hip joint is surrounded by a synovial membrane that builds synovial capsule. Synovial fluid separates the articulating bone surfaces by filling the joint cavity, and it performs under both fluid film and boundary lubrication regime. The articulating surfaces are coated by articular cartilage, which has very low coefficient of friction such as 0.02 [22]. Synovial membrane provides fresh synovial fluid into the cavity. The volume of this fluid is about 2 ml. Hyaluronic acid, lubricin, and globulin are the main biological components that provide unique lubricant properties to synovial fluid and make it the best lubricant for synovial joint [23, 24]. A healthy synovial fluid is a dialysate of blood plasma, which is comprised mainly of water (85%), hyaluronic acid, and protein [25]. All of these components make different contributions to the function of synovial fluid. For instance, hyaluronic acid serves to increase the viscosity of synovial fluid [26], whereas lubricin decreases the shear strength at the asperity contact interfaces [24, 27] and lipid layer on synovial membrane plays an effective role in the boundary lubrication mechanism of articular cartilage [28]. The synovial fluid composition changes from a healthy person to an osteoarthritis patient [29]. Therefore, the functionality and lubrication property of the synovial fluid may vary.

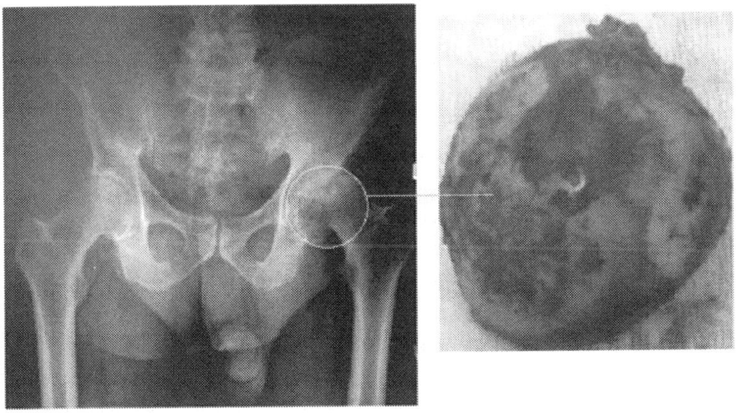

Figure 5. Radiographic image of diseased femoral head and its photography.

Ball-in-socket configuration, synovial fluid, and synovial capsule build a joint system that can transmit seven to eight times body weight during walking, climbing up stairs, jumping, or some other vigorous activities [3]. These systems provide approx. 70 years or more lifetime to the synovial joint with an overall wear factor (k) 10^{-9} mm^3/Nm [8, 22]. However, the lifetime of this special system sometimes may end earlier than the expected time because of various clinical factors such as osteoarthritis, rheumatoid arthritis, necrosis, and trauma [22]. At that point, artificial joint elements are replaced with the fractured natural joint parts. For understanding the failure mechanism of the natural joint, biotribological studies have been conducted widely over the past 50 years. These researches are generally focused on the measurement of friction in synovial joints, the mechanisms of joint lubrication, measurement and analysis of cartilage wear and damage, and the lubrication properties of synovial fluid and its ingredients [8]. Studies in this area have made great contribution to understand normal joint function; however, the entire tribological system cannot be clarified yet because of the complexity of the natural synovial joint. A radiographic image of diseased femoral head before total joint replacement surgery and its photography after surgery can be seen in **Figure 5**.

3.2. Artificial hip joints

Artificial joint replacements have been one of the best solutions for patients affected by the clinical factors specified above. Although applications of total hip joint arthroplasty are very successful, the revision rate of these artificial joints is still unexpectedly high. For instance, in England 11 and 12% of all total hip replacements failed in 2011 and 2012, respectively [24]. The expected service life of these replacements is about 15–20 years. This duration is not enough for patients younger than 60 years old. Sometimes the joints may fail prematurely before the expected service life. This causes a lot of problems for both patients and surgeons because premature failure brings pain with revision surgery, need of extra money, and spending extra time. Biotribological behavior of the artificial joints is the primary failure factor that limits the service life of the prosthesis.

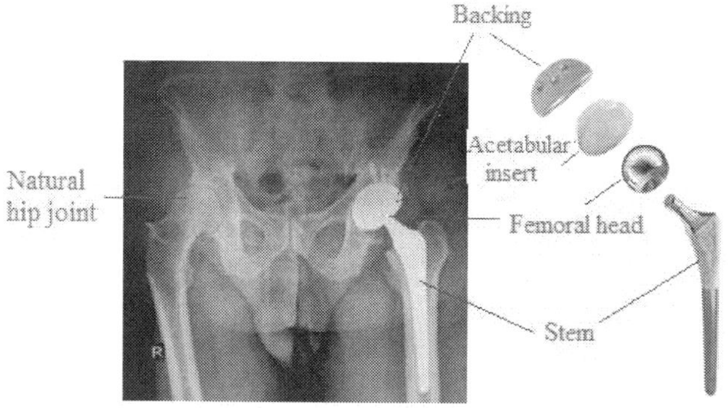

Figure 6. Components of artificial total hip joint.

Different material combinations such as ceramic-on-ceramic (CoC), metal-on-metal (MoM), ceramic-on-metal (CoM), metal-on-polymer (MoP), and ceramic-on-polymer (CoP) are used in total hip replacements. The most common combination is metal femoral head ultrahigh-molecular-weight polyethylene (UHMWPE) cup [30, 31]. UHMWPE has been the most preferred acetabular cup material for the past four decades with excellent biocompatibility, chemical stability, impact load damping properties, and low friction coefficient [32, 33]. However, the wear debris of UHMWPE induces adverse tissue reactions and third-body wear damages that cause aseptic loosening and implant failure. Therefore, in order to extend the implant life, especially for young and more active patients, improvement in the UHMWPE properties such as low friction coefficient, third-body wear resistance, generation of small amounts of wear debris, and low cellular reactions to such wear debris became a need [31, 34, 35]. Components of the total hip joint can be seen in **Figure 6**.

Radiation-induced cross-linking has been a very effective way for the modification of UHMWPE microstructure, so the first-generation cross-linked UHMWPE has been developed [35–37]. This cross-linked UHMWPE has shown higher wear resistance than the conventional UHMWPE, but mechanical properties, oxidation, and delamination resistance have been decreased because of the postheat treatment operation [33, 34, 38]. With the aim of reducing the restriction of cross-linked UHMWPE, the second-generation UHMWPE has been developed by the addition of α-tocopherol or vitamin E as a natural antioxidant [36]. Tribological studies reported that the addition of vitamin E enhanced the oxidation and delamination resistance of the conventional and cross-linked UHMWPE while maintaining the mechanical properties by stabilizing the residual-free radical without postirradiation melting [31, 39–41].

As an alternative to the hard-on polymer implant pair, hard-on-hard artificial joint materials such as MoM, CoC, or CoM have been used for increasing wear resistance and so the service life of the artificial joints. These types of bearing pairs have shown better wear resistance and given less wear debris than the conventional MoP pairs. CoC bearings are recognized as the most wear-resistant pairs because of their very hard, smooth surface, and effective lubrication, but squeaking is a restrictive problem that causes distress to the patients as reported in clinical studies [42–44]. A photographic image of the surgical application of ceramic femoral head can be seen in **Figure 7**.

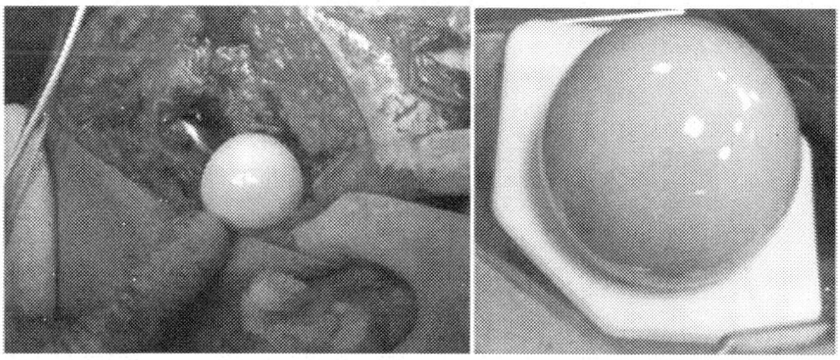

Figure 7. Surgical application of ceramic femoral head.

MoM joints have been reported as low volumetric wear rate bearing besides high stability with large heads [45]. Metal ion release, metallic wear debris, and tribochemical reactions are the main problems for these artificial joints [46, 47]. Moreover, during the gait cycle MoM total hip joints are exposed to higher impacts, which may cause severe pain in patients than hard-on-polymer joints. For combined advantages of CoC and MoM pairs, CoM artificial hip joints improved recently. Simulator studies showed a reduced wear of CoM joints than MoM pairs. In addition, limited chromium release has been reported for CoM total hip joints in short-term clinical results [48].

For understanding the tribological behavior such as friction, lubrication, and wear of the artificial joints, advanced in vitro simulator tests have been performed for simulating an artificial joint with real articulating components under dynamic loading, multidirectional sliding, and lubrication conditions. Extremely useful outcomes are obtained from these studies and the results have been used for the improvement of design and properties of artificial joint materials. However, tribological behavior of the artificial joints does not only depend on implant material and design but also on patient-related factors such as lifestyle, body weight, age, gender, and synovial fluid that are poorly understood and cannot be simulated adequately in vitro test studies [24]. Biotribological behaviors of different artificial joint materials have been discussed in the following section referring the literature studies.

3.3. Friction, lubrication, wear

As mentioned above, friction forces are caused by the adhesion and cohesion forces. The proper definition and understanding of friction in synovial joints serve to predict and determine quantity of wear in artificial hip joints [49]. The coefficient of friction (COF) in artificial hip joint surfaces varies according to joint material pairs, geometry of joint parts, lubrication condition, and loading [16]. In order to make tribological assessments such as the determination of wear behavior, friction factor or COF of artificial materials, conventional tribological methods such as pin-on-disc, ball-on disk, and pin-on-plate can be used (see **Figure 8**). Furthermore, for the simulation of real joint conditions such as working with real joint parts and applying multidirectional loads, more complicated joint simulators have been used. The pressure distribution between the contact areas of the hip prosthesis cannot be known precisely. Because of this, it is difficult to determine COF accurately. At that point, a dimensionless parameter called friction factor (*f*) is used instead of COF. In experimental studies, the frictional torque (*T*) is measured first and then the friction factor is calculated by using the following equation:

$$f = \frac{T}{RL} \tag{9}$$

where *R* is the radius of the femoral head and *L* is the normal load [11, 20, 50, 51].

For simulator studies, the test conditions are standardized in ISO 14242-1. In these studies, the tests have been performed with the anatomical position of the components, e.g., in flexion-extension, adduction-abduction, and internal-external rotation planes with 1 Hz frequency under the dynamic loading condition and 25% bovine calf serum lubrication. In some friction simulator studies, this configuration was simplified, for example, the prosthesis was inverted with respect to anatomical position and the tests were performed just in the flexion-extension plane ± 25° [11, 50, 52, 53]. These kinds of friction simulator friction factors are used in the lubrication of bovine serum, for CoC bearing it is reported as between 0.04 and 0.07, for MoP bearing 0.06–0.08, for CoP 0.05–0.08, and for MoM as 0.12–0.27 [11, 16, 50].

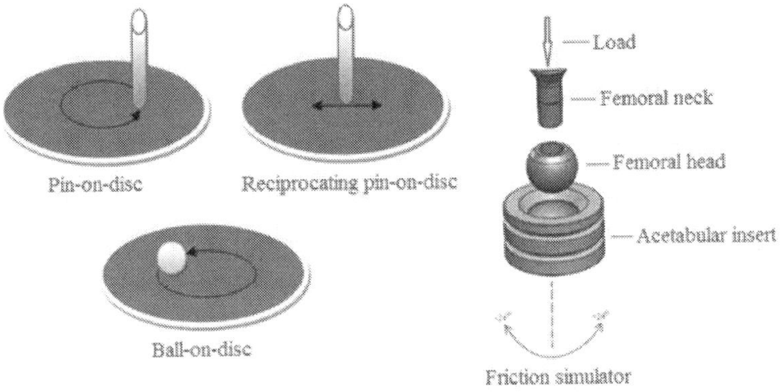

Figure 8. Schematic drawings of the conventional wear testing devices (pin-on-disk, reciprocating pin-on-disk, and ball-on-disk) and friction simulator.

Measuring, modeling, and predicting of friction on the total hip prosthesis have played an important role in developing artificial hip joints material and geometry. Moreover, the determination of friction can be used for the estimation of the lubrication regime between articulating surface of artificial hip joints as mentioned in the previous section.

Lubrication is one of the most complex and important factors that affect the implant friction and wear [31, 38]. However, the lubrication mechanism of natural joint cannot be fully understood, it is definitely known that there is a perfect lubrication system in a healthy natural joint. Different lubrication mechanisms, such as boundary lubrication, hydrodynamic lubrication [54] elastohydrodynamic lubrication, fluid film lubrication, and mixed lubrication, have been reported for lubrication in natural joints [16, 55]. However, in artificial joints the alternatives are limited because of joint material properties, surface qualities, and implant geometries. As a result of this, boundary lubrication and mixed lubrication regimes are leading lubrication mechanisms in artificial hip joints [19].

In MoP artificial joints, boundary lubrication regime is a dominant mechanism with λ values in the range 0.1–1. The surface asperity contact cannot be avoided because of the soft and rough surface of UHMWPE and thereby boundary lubrication occurs. In MoM bearing, the mixed lubrication regime can be seen generally with λ in the range of 0.6–2.9. CoC bearing surfaces are so hard and they have superior manufacturing tolerances and minimum surface roughness as compared with metal or polyethylene. The fluid film lubrication regime is predicted for these pairs with λ values between 5.7 and 28.3 [16, 55–57].

For improving surface quality and enhancing the lubrication properties of artificial joints, surface modification techniques have been an effective solution for artificial joint materials. Ion implantation, electron beam or gamma radiation, plasma surface treatment, and surface texturing are the most common surface modification techniques that are applied to improve the wear resistance of articulating surfaces without changing bulk material [58].

Surface texturing has been a well-known way for many years especially in machine bearings but it has started to be excessively popular in the last decade for artificial joints [5, 31]. In some previous studies, surface patterning was studied on artificial joint materials to reduce friction and wear. Young et al. [59] measured 43% lower friction coefficient of surface patterned samples than unpatterned disk samples by using pin-on-disk testing apparatus. Ito et al. [60] formed concave dimples on the metal femoral head surface, and they reported that the dimples served in reducing polyethylene abrasive wear by providing better lubrication and capturing wear particles. In reference [61], it is reported that lower friction coefficient and wear factor were obtained with surface-dimpled disk samples (see in **Figure 9**). Besides reducing wear rate and friction coefficient, surface dimples also serve in decreasing frictional heating of acetabular insert and femoral head [31, 62–65]. By acting as a reservoir of lubricant and capturing wear debris and bone cement particles, surface dimples provide better lubrication and thereby better tribological behavior [66].

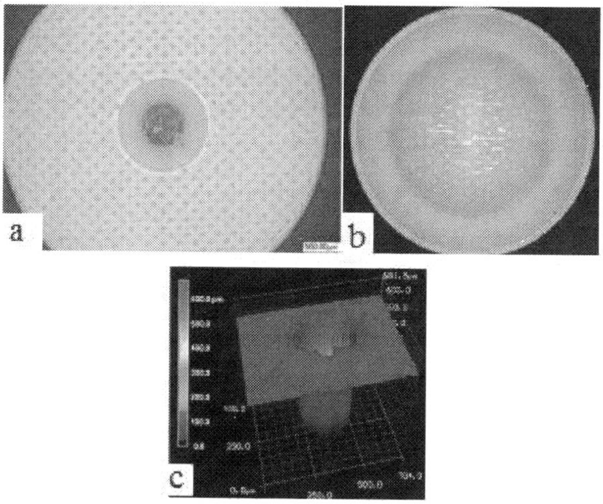

Figure 9. Microscopic images of surface dimples. (a) Dimples on UHMWPE disk sample before wear test, (b) dimples on the inner surface of UHMWPE acetabular insert after wear test with PMMA third-body particles, and (c) laser scanner microscope image of a dimple for characterization of its dimensions [31, 61].

There are many factors that affect the lifetime of a total hip prosthesis. Short-term failures are generally of biological origin, whereas long-term failures are related to material properties with a biological response [19]. Wear has been considered as the primary factor that limits service life of hip implants. The wear debris generated during articulation of joint materials such as UHMWPE, metallic, or ceramic counter face could cause adverse tissue reactions, aseptic loosening, osteolysis, and finally implant loss [22, 31, 67, 68]. Microscopic image of worn surfaces of retrieved prosthesis can be seen in **Figure 10**.

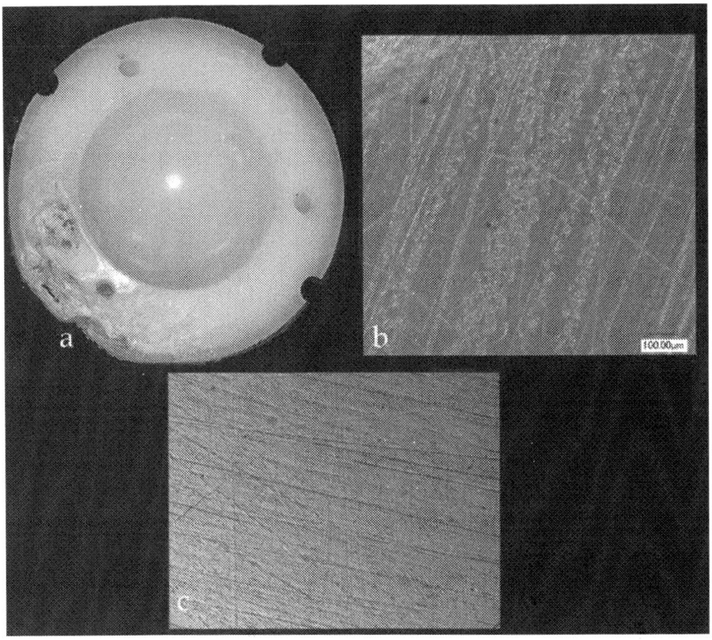

Figure 10. Worn surfaces of retrieved hip prosthesis. (a and b) UHMWPE acetabular insert and (c) CoCrMo femoral head.

Sliding surfaces in hip joint are conformal surfaces that fit together [16]. This makes geometry, size, and manufacturing tolerance of the joint components vitally important because initial wear occurs if there is any mismatch during bedding-in of prosthesis. This is a problem especially for MoM bearings because MoP pairs compensate this problem by polyethylene creep [56]. Wear of CoC pairs is negligible along normal walking because the wear volume is too low [69]. A well-functioning polyethylene acetabular insert exposes clinical wear such as 50–100 μm of penetration per year that means nearly 80 mm^3 volume loss per year [11, 16, 70]. In experimental hip joint simulator studies, MoP bearing wear rate was 40 mm^3/million cycle, whereas CoP bearing wear rate was 25 mm^3/million cycle. For MoM bearing, the wear rate was reported to be 1.0 mm^3/million cycle and for CoC artificial pairs the wear rate was recorded to be 0.1 mm^3/million cycle [11].

Three main wear mechanisms have been reported for UHMWPE such as fatigue, adhesion, and abrasion wear. *Fatigue wear* occurs when the surface of the material is weakened by cyclic loading. *Adhesive wear* is generated by the transfer of material from one surface to another when these two surfaces are articulating against each other under load. The transferred particles could break off and fuse together and then may act as third-body particles causing abrasive wear. Adhesion and fatigue wear generally work together. *Abrasive wear* occurs when hard asperities on the sliding surfaces or third-body particles are trapped between these surfaces. These asperities cause loss of the material from softer surface of the articulating pairs [10]. These wear mechanisms primarily occurs at the microscopic scale, whereas fatigue wear may occur at the macroscopic scale in the form of delamination [56]. The oxidation of polyethylene causes degradation and nearly 80% decrease in fatigue strength [71]. It is clear that the wear resistance of polymers is directly related to its mechanical properties and physical morphology [68].

Literature works about retrieved prosthesis show that third-body abrasive wear is a very important parameter affecting the service life of artificial joints [72]. By scratching the metal femoral head, third-body particles promote the wear rate of UHMWPE acetabular cup. PMMA bone cement particles are believed to be the main cause of third-body particles [73]. In addition, bone particles, metal beads or fibers from porous coatings and hydroxyapatite coatings, and corrosion products from the metal tapers and metal fragments from other fixation devices may be the source of third-body particles [74–77]. Microscopic images of PMMA particles can be seen in **Figure 11**.

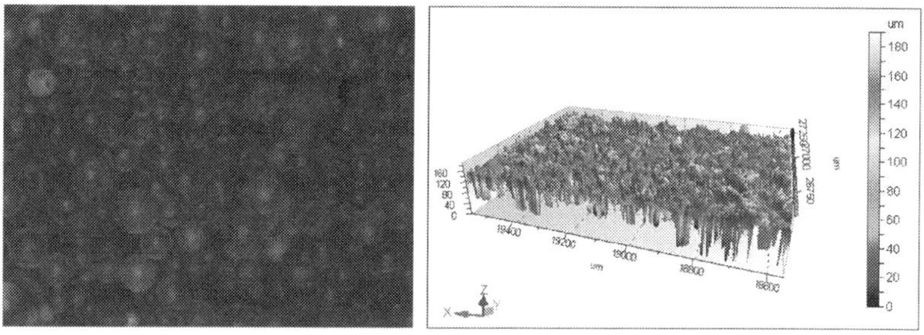

Figure 11. (a) Microscopic image and (b) surface texture of the PMMA particles used as bone cement for fixation of the artificial joints [78].

Bragdon et al. [72] reported three possible interacting mechanisms of third-body abrasive particles with acetabular cup and femoral head sliding surfaces after being trapped at the interface. In the first mechanism, particles may embed in polyethylene surface that causes to reduce the contact area between the head and the cup. In the second mechanism, third-body particles may adhere to the femoral head under pressure and finally some particles may roll freely between the surfaces. In a pin-on-disk configuration, the embedded particles may cause pitting, and the free particles that roll between surfaces may cause scratches on the UHMWPE disk surface as can be seen in **Figure 12**.

http://dx.doi.org/10.5772/64488

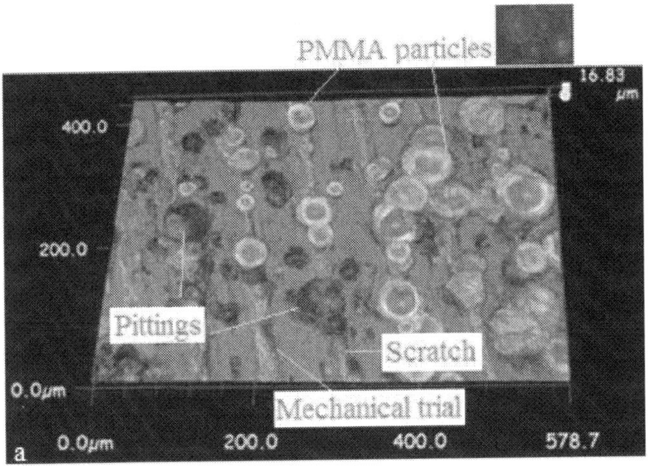

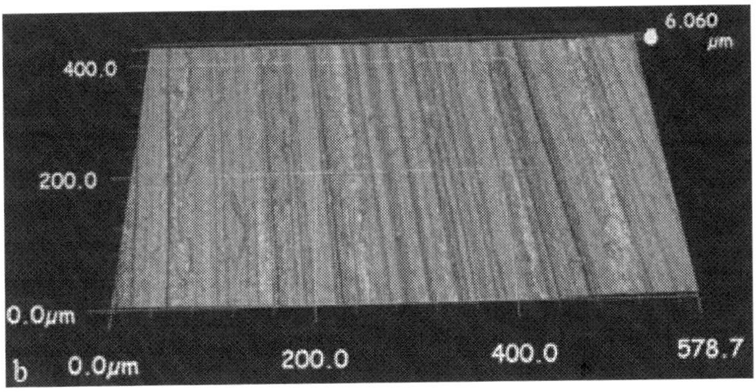

Figure 12. Worn surface of UHMWPE (a) with PMMA third-body particles, (b) without-third-body particles in 25% bovine calf serum lubrication [78].

3.4. Frictional heating of articulating surfaces

Most portion of frictional work between articulating surfaces is converted into heat during the wear process. This heat causes temperature rise in artificial joint parts, and the temperature rise may influence the properties of lubricant, the rate of wear, fatigue, creep, and oxidative degradation of bearing materials [14, 79–81]. Moreover, temperature rise may contribute cup loosening by causing bone necrosis and surrounding tissue damages [52]. In literature studies, frictional temperature rise has been measured with different ways by various researchers. These are experimental methods such as in vitro and in vivo measurements, theoretical calculations, and computer simulations [52, 79–84]. An example of experimental setup for the measurement of frictional heating between UHMWPE acetabular insert and CoCrMo femoral head can be seen in **Figure 13**.

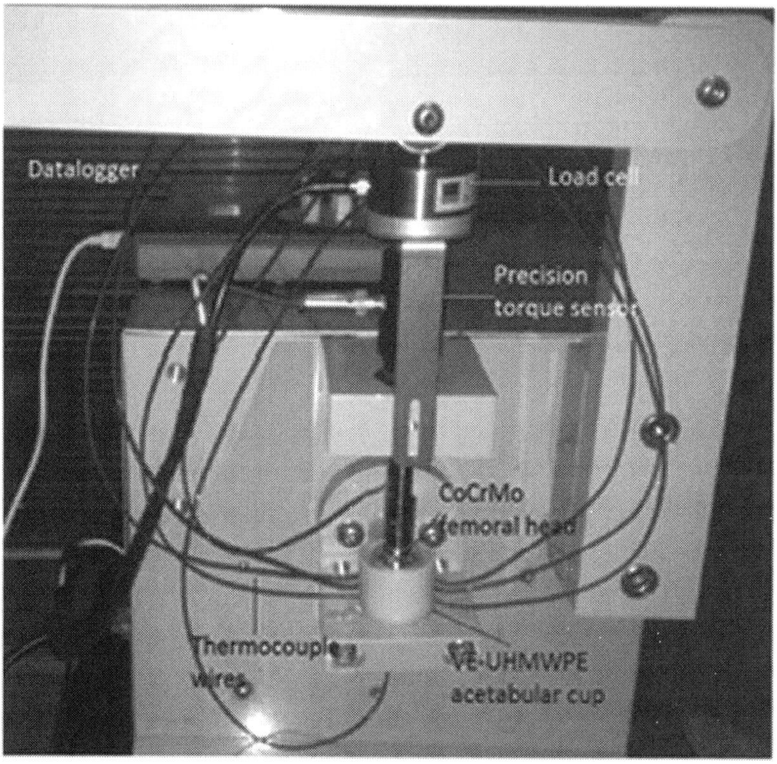

Figure 13. Friction simulator for measurement frictional heating between artificial hip joint components [31].

In an in vitro study, Lu and McKellop [52] reported that frictional heating may promote the protein precipitation from the lubricant and as a result property change of the lubricant. Exhaustion of proteins may cause incomplete boundary lubrication mechanism and so these changes may accelerate adhesive wear rate. In another study, Bergmann et al. [82] measured the temperature rise of hip prosthesis in patients' bodies after 1 hour walking, and they reported that the maximum temperature value was 43.1°C in vivo. In a computer simulation study [83], synovial fluid's temperature was found as 46°C by 2D and 3D finite element analysis. By using a thermomechanical finite element model of the ball-cup interaction, the peak temperature of the contact surface was predicted as 51°C [85]. In an experimental study [86], frictional temperature between zirconia femoral head/UHMWPE measured as 39°C at the contact point of femoral head whereas it was measured as 36°C between zirconia femoral head/ vitamin E blended UHMWPE.

Frictional temperature rise is related with lubrication condition and material properties such as thermal conductivity and elastic modulus. For example, while two materials with higher elastic modulus used as frictional pair, the contact area would be smaller and a larger area would be exposed to the lubricant during the cycle. Therefore, with better lubrication and cooling, frictional temperature would be lower [31, 52]. In reference [31], the surface dimples' effects on frictional temperature rise of artificial joints was studied. It is reported that frictional temperature rise (ΔT) of surface-dimpled UHMWPE/CoCrMo sample was 8.39°C, whereas undimpled samples temperature rise was recorded as 11.22°C. Surface dimples acted as a reservoir for lubricant and provided continuous lubrication for cooling the surfaces, so by increasing surface lubrication quality, lower temperature values were recorded [31, 59, 87, 88] The biological defects occur at about 40, and 6°C temperature rise between articulating surfaces may cause fibrous tissue formation and possibly prosthetic loosening. Therefore, it is clear that service life of artificial joints can be negatively affected by frictional heating [80, 83].

3.5. Measurement of wear in artificial joints

The measurement, evaluation, and analysis of wear are vitally important for understanding the wear mechanism of artificial joints and improvement of new materials with new designs. Wear measurement of artificial joint materials may be applied in vivo and in vitro conditions with different measurement techniques. In vivo evaluation of the implant provides in situ monitoring of the patient more closely and for considering revision surgery by determining wear amount. Moreover, the wear rate of the implants and histological changes can be determined with periodical measurements. In vitro measurement studies include the examination of retrieved prosthesis, which provides data about wear mechanisms and wear rate of the materials, and the evaluation of lubricant, wear debris and debris distribution. The objectives of all these evaluations are to determine the wear rate and lifetime of implant and to understand wear mechanisms, tissue reactions, and other related events that occur in artificial joint system. Therefore, it would be possible to develop new materials and designs [19, 89].

Radiographic, gravimetric, volumetric, and optical techniques are current methods for measuring and evaluating the wear in the total joint replacement components [90]. X-ray techniques, magnetic resonance imaging (MRI), microcomputed tomography, and biochemical markers are methods for the measurement of in vivo cartilage wear. It is possible to see bone deterioration on joint by an X-ray technique where the MRI scans provide a very detailed view of the damaged tissue [25].

Two- or three-dimensional techniques, radiostereometric analysis, and manual or computer-assisted plain radiography techniques are the best known radiographic methods used for measuring polyethylene wear. Radiographic techniques allow estimation about femoral head migration into the cup [89–92].

The most common used techniques to evaluate wear are the measurement of weight change called as gravimetric method and the measurement of dimensional changes known as volumetric wear [8]. Gravimetric method is the standardized method with ASTM F2025-06, F1714-96, and ISO 14242-2 used for the measurement of wear in the total joint prosthesis. In this method, the weight difference between the initial weight and the weight after wear test were determined. According to ISO 14242-2:2000, a balance with the accuracy of 0.1 mg must be used for weight measurements. Because of the fluid absorbing property of polymer component, the weight difference may be undetectable or the final weight may sometimes be higher than its initial value. For reducing the error due to fluid absorption, polyethylene specimens are soaked into the lubricant until they reach saturation that takes days or sometimes weeks. For reference purposes, a loaded but not articulating control specimen is used for determining fluid sorption of a specimen [90, 93].

Gravimetric wear can be calculated as follows:

$$W_n = W_{\mathrm{an}} + S_n \tag{10}$$

where W_n is the net mass loss after n cycles of loading, W_{an} is the average uncorrected mass loss, and S_n is the average increase in mass of the control specimen over the same period.

By using the least-squares linear fit relationship between W_n and the number of loading cycles n, the average wear rate a_G can be calculated as follows:

$$W_n = a_G.n + b_n \tag{11}$$

where W_n is the net loss in mass after n cycles and b is a constant.

Although this method is effective to determine experimental wear amount in simulated conditions, it is not applicable to evaluate the clinical wear of retrieved prosthesis for which there is no prewear data available. Gravimetric method does not provide information about wear mechanisms, surface property changes, and plastic deformation of the component. Moreover, material transfer from the metal component or bone cement that is attached into the UHMWPE can cause a significant error while determining the weight loss both for in vivo and in vitro applications. These are the main limitations of this method [90, 94].

Volumetric methods such as the coordinate measuring machines (CMMs) and micro computed tomography (micro-CT) have recently been alternative methods to gravimetric method [42]. The use of CMM is standardized by ISO 14242-2:2000. This standard requires a CMM "with maximum axial-position error of measurement D":

$$D = 4 + 4L \times 10^{-6} \tag{12}$$

where D is in μm and L is the numerical value of the dimension in meters.

In addition, "Mesh spacing" must not be greater than 1 mm and "relocation of the test specimen" must not affect the measured volume more than 0.05%. For determining the wear amount of the prosthesis first of all initial, unworn geometry of the sample is measured and after wear test the measurement is repeated. Therefore, by comparing the initial and final geometries of the sample, the amount of volumetric wear is calculated [32]. These evaluation steps are applicable for simulator studies but in clinical application there is no initial geometrical data about retrieved prosthesis. At that point, measurements are taken from unworn regions of the retrieved prosthesis and by using collected data unworn geometry of the sample is predicted. A coordinate measuring machine, with 2 μm minimum accuracy, should be used for three-dimensional wear analysis [90, 95, 96]. An example of the CMM measurement of UHMWPE acetabular insert can be seen in **Figure 14**.

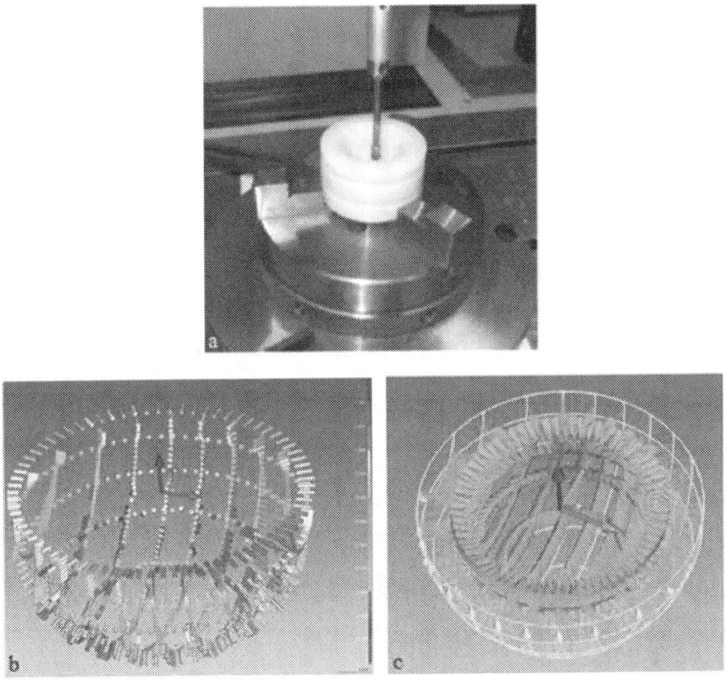

Figure 14. (a) CMM measurement of a UHMWPE acetabular insert, (b) defining measurement paths, and (c) inner surface of the prosthesis saturated with measurement points.

It is possible to analyze wear behavior of both hard-on-hard and hard-on-polymer retrieval prosthesis by using CMM. For evaluating the actual wear amount of polyethylene prosthesis, it is important to determine creep deformation. The CMM method provides advantages about defining wear volume, wear scars distribution, and creep deformation, but it is time consuming and shows uncertainty about the evolution of wear measurement [90, 97, 98]. Different surface characterization techniques such as mechanical, optical and laser profilometry, white light interferometry, and digital microscopy (see **Figure 15**) can be used for analyzing the wear behavior of the implant surfaces. These techniques may be applied both macroscopically and microscopically for characterizing damage modes such as burnishing, abrasion, scratching, pitting, plastic deformation, fracture, fatigue damage, and embedded debris [93, 97].

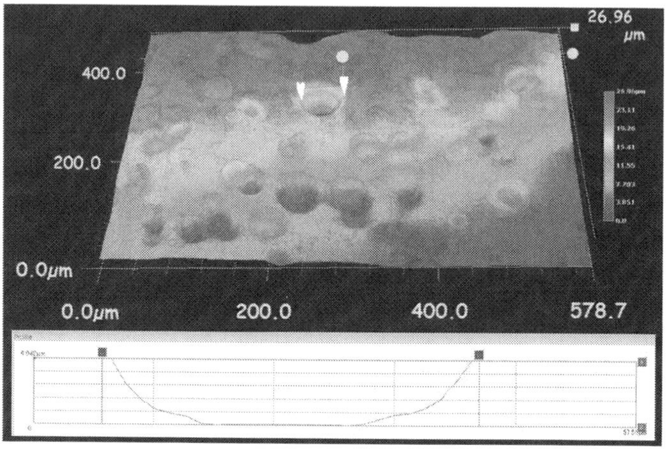

Figure 15. Determination of local wear damage on polyethylene sample by digital microscopy.

Surface profilometry is one of the most preferable techniques especially for the determination of wear in primitive wear tests such as pin-on-disk, pin-on-plate, and ball-on-disk (see **Figure 16**).

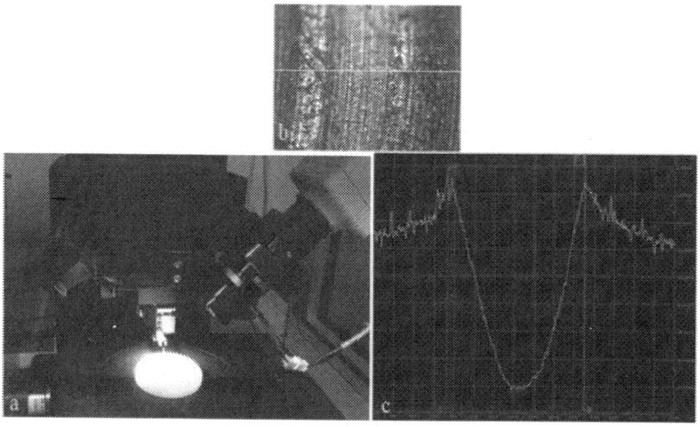

Figure 16. (a) Measurement of wear by mechanical surface profilometer, (b) wear track, and (c) wear track profile [78].

By using the cross-sectional area of wear track and its length the wear volume is calculated. Then, the wear factor (k) of the sample can be determined by using the following equation:

$$k = V / NS \tag{13}$$

where k is the wear factor (mm^3/(N m)), V is the wear volume (mm^3), N is the applied load (N), and S is the friction distance (m) [61, 99, 100].

For the determination of the wear amount of very low wearing metal-on-metal pairs, lubricant samples obtained during a wear test can be analyzed by inductively coupled plasma mass spectrometry (ICP-MS) to measure the metal ion concentration [93].

Author details

Binnur Sagbas

Address all correspondence to: bsagbas@gmail.com

Materials Science and Manufacturing Technologies Division, Department of Mechanical Engineering, Yildiz Technical University, Istanbul, Turkey

References

[1] Mang, T., Bobzin K., Bartels T., editors. Industrial Tribology: Tribosystems, Friction, Wear and Surface Engineering, Lubrication. Weinheim: WILEY-VCH Verlag GmbH & Co. KGaA; 2010. 672 p. ISBN: 978-3-527-32057-8

[2] Zhoua Z.R., Jin J.M. Biotribology: recent progresses and future perspectives. Biosurface and Biotribology. 2015;1(1):3–24. DOI: 10.1016/j.bsbt.2015.03.001

[3] Mattei L., DiPuccio F., Piccigallo B., Ciulli E. Lubrication and wear modelling of artificial hip joints: a review. Tribology International. 2011;44(5):532–549. DOI: 10.1016/j.triboint.2010.06.010

[4] Sagbas B., Durakbasa M.N., Sagbas M., Koyun A. Measurement and theoretical determination of frictional temperature rise between sliding surfaces of artificial hip joints. Measurement. 2014;51:411–419. DOI: 10.1016/j.measurement.2013.12.020

[5] Mathiaa T.G., Pawlus P., Wieczorowskic M. Recent trends in surface metrology. Wear. 2011;271(3–4): 494–508. DOI: 10.1016/j.wear.2010.06.001

[6] Xiong D., Ge S. Friction and wear properties of UHMWPE/Al2O3 ceramic under different lubricating conditions. Wear. 2011;250(1–12):242–245. DOI: 10.1016/S0043-1648(01)00647-0

[7] Ge S., Wang S., Gitis N., Vinogradov M., Xiao J. Wear behavior and wear debris distribution of UHMWPE against Si3N4 ball in bi-directional sliding. Wear. 2008;264(7–8):571–578. DOI: 10.1016/j.wear.2007.05.001

[8] Bhushan B., editor. Modern Tribology Handbook. 1st ed. Florida: CRC Press; 2000. 1760 p. ISBN: 9780849384035

[9] Derek J.W., editor. Modern Hip Resurfacing. Birmingham: Springer; 2009. 428 p. DOI: 10.1007/978-1-84800-088-9

[10] Abdelbary A., editors. Wear of Polymers and Composites. United Kingdom: Elsevier; 2014. 223 p. DOI: 10.1016/B978-1-78242-177-1.50010-2

[11] Jin Z.M., Stoneb M., Ingham E., Fishera J. v-Biotribology. Current Orthopaedics. 2006;20(1):32–40. DOI: 10.1016/j.cuor.2005.09.005

[12] Bhushan B., editors. Introduction to Tribology. 2nd ed. UK: John Wiley & Sons, Ltd; 2013. 711 p. DOI: 10.1002/9781118403259.fmatter

[13] Stachowiak G.W., Batchelor A.W., editors. Engineering Tribology. 3rd ed. USA: Elsevier; 2006. 801 p. DOI: 10.1016/B978-075067836-0/50000-6

[14] Davis J.R., editors. Surface Engineering for Corrosion and Wear Resistance. USA: ASM International; 2001. 279 p.

[15] Dowsan D., Shpenkov G.P., editors. Friction Surface Phenomena-Tribology Series 29. Netherlands: Elsevier; 1995. 359 p. ISBN: 0 444 81684 4

[16] Di Puccio F., Mattei L. Biotribology of artificial hip joints. World Journal of Orthopedics. 2015;6(1):77–94. DOI: 10.5312/wjo.v6.i1.77

[17] Ludema K.C. Friciton, Lubrication, Wear. Florida: CRC Press; 1996. 263 p. ISBN: 0-8493-2685-0

[18] Szeri A.Z. Fluid Film Lubrication. UK: Cambridge University Press; 2005. 427 p. ISBN: 0 521 48100 7

[19] Dumbleton J.H.. Tribology of Natural and Artificial Joints. Amsterdam: Elsevier Scientific Publishing Company; 1981. 472 p. DOI: ISM 0444-41898-9

[20] Hall, R.M., Unsworth, A. Friction in hip prostheses. Biomaterials. 1997;18(15):1017–1026. DOI: 10.1016/S0142-9612(97)00034-3

[21] Zhoua Z.R., Jin Z.M. Biotribology: recent progresses and future perspectives. Biosurface and Biotribology. 2015;1(1):3–24. DOI: 10.1016/j.bsbt.2015.03.001

[22] Blunt L., Bills P., Jiang X., Hardaker C., Chakrabarty G. The role of tribology and metrology in the latest development of bio-materials. Wear. 2009;266(3–4):424–431. DOI: 10.1016/j.wear.2008.04.015

[23] Trunfio-Sfarghiua A.M., Berthier Y., Meurissea M.H., Rieu J.P. Multiscale analysis of the tribological role of the molecular assemblies of synovial fluid. Case of a healthy joint and implants. Tribology International. 2007;40(10–12):1500–1515. DOI: 10.1016/j.triboint.2007.02.008

[24] Ghosh S., Choudhury D., Roy T., Moradi A., Masjuki H.H., Pingguan-Murphy B. Tribological performance of the biological components of synovial fluid in artificial joint implants. Science and Technology of Advanced Materials. 2015;16(4):1–12. DOI: 10.1088/1468-6996/16/4/045002

[25] Berrien, La S.J. Biotribology: Studies of the Effects of Biochemical Environments on the Wear and Damage of Articular Cartilage [thesis]. Blacksburg, Virginia: Virginia Polytechnic Institute and State University; 1999. 152 p. Available from: http://vtechworks.lib.vt.edu/bitstream/handle/10919/28289/ljbetd.pdf?sequence=1&isAllowed=y

[26] Ghosh S., Choudhury D., Das N.S., Pingguan-Murphy B. Tribological role of synovial fluid compositions on artificial joints—a systematic review of the last 10 years. Lubrication Science. 2014;26(6):387–410. DOI: 10.1002/ls.1266

[27] Tanimoto K., Kamiya T., Tanne Y., Kunimatsu R., Mitsuyoshi T., Tanaka E. et al.. Superficial zone protein affects boundary lubrication on the surface of mandibular condylar cartilage. Cell and Tissue Research. 2011;344(2):333–340. DOI: 10.1007/s00441-011-1156-z

[28] Trunfio-Sfarghiu A.M., Berthier Y., Meurisse M.H., Rieu J.P. Role of nanomechanical properties in the tribological performance of phospholipid biomimetic surfaces. Langmuir. 2008;24(16): 8765–8771. DOI: 10.1021/la8005234

[29] Wang A., Essner A., Schmidig G. The effects of lubricant composition on in vitro wear testing of polymeric acetabular components. Journal of Biomedical Materials Research B: Applied Biomaterials. 2004;68(1):45–52. DOI: 10.1002/jbm.b.10077

[30] Zywiel M.G., Sayeed S.A., Johnson A.J., Schmalzried T.P., Mont M.A. State of the art in hard-on-hard bearings: how did we get here and what have we achieved? Expert Review Medical Devices. 2011;8(2):187–207. DOI: 10.1586/erd.10.75

[31] Sagbas B., Durakbasa M.N. Effect of surface patterning on frictional heating of vitamin E blended UHMWPE. Wear. 2013;303(1–2):313–320. DOI: 10.1016/j.wear.2013.03.023

[32] Bills P., Blunt L., Jiang X. Development of a technique for accurately determining clinical wear in explanted total hip replacements. Wear. 2007;263(7–12):1133–1137. DOI: 10.1016/j.wear.2007.01.054

[33] Bhatt H., Goswami T. Implant wear mechanisms--basic approach. Biomedical Materials. 2008;3(4):1–9. DOI: 10.1088/1748-6041/3/4/042001

[34] Jacobs C.A., Christensen C.P., Greenwald A.S., McKellop H. Clinical performance of highly cross-linked polyethylenes in total hip arthroplasty. The Journal of Bone Joint Surgery American. 2007;89(12):2779–2786. DOI: 10.2106/JBJS.G.00043

[35] Bergstr J.S., Bischof J.E. An advanced thermomechanical constitutive model for UHMWPE. International Journal of Structural Changes in Solids—Mechanics and Applications. 2010;2(1):31–39.

[36] Vaidya C., Alvarez E., Vinciguerra J., Bruce D.A., DesJardins J.D. Reduction of total knee replacement wear with vitamin E blended highly cross-linked ultra-high molecular weight polyethylene. Proceedings of the Institution of Mechanical Engineers H. 2011;225(1):1–7. DOI: 10.1243/09544119JEIM774

[37] Tipper J.L., Galvin A.L., Williams S., Mc Ewen H.M., Stone M.H., Ingham E., et al.. Isolation and characterization of UHMWPE wear particles down to ten nanometers in size from in vitro hip and knee joint simulators. Journal of Biomedical Materials Research A. 2006;78(3):473–480. DOI: 10.1002/jbm.a.30824

[38] Bradford L., Baker D.A., Graham J., Chawan A., Ries M.D., Pruitt L.A. Wear and surface cracking in early retrieved highly cross-linked polyethylene acetabular liners. The Journal of Bone Joint Surgery American. 2004;86(6):1271–1282.

[39] Bracco P., Oral E. Vitamin E-stabilized UHMWPE for total joint implants: a review. Clinical Orthopaedic and Related Research. 2011;469(8):2286–2293. DOI: 10.1007/s11999-010-1717-6

[40] Oral E., Rowell S.L., Muratoglu O.K. The effect of alpha-tocopherol on the oxidation and free radical decay in irradiated UHMWPE. Biomaterials. 2006;27(32):5580–5587. DOI: 10.1016/j.biomaterials.2006.07.017

[41] Oral E., Wannomae K.K., Hawkins N., Harris W.H., Muratoglu O.K. α-Tocopherol-doped irradiated UHMWPE for high fatigue resistance and low wear. Biomaterials. 2004;25(24):5515–5522. DOI: 10.1016/j.biomaterials.2003.12.048

[42] Walter W.L., Yeung E., Esposito C. A review of squeaking hips. Journal of the American Academy of Orthopaedic Surgeons. 2010;18(6):319–326.

[43] Lusty P.J., Tai C.C., Sew-Hoy R.P., Walter W.L., Walter W.K, Zicat B.A. Third-generation alumina-on-alumina ceramic bearings in cementless total hip arthroplasty. The Journal of Bone Joint Surgery American. 2007;89(12):2676–2683. DOI: 10.2106/JBJS.F.01466

[44] Hua Z., Yan X. Liu D., Jin Z.M., Wang X., Liu L. Analysis of the friction-induced squeaking of ceramic-on ceramic. Tribology Letter. 2016;61(26):1–7. DOI: 10.1007/s11249-016-0644-4

[45] Silverman E.J., Ashley B., Sheth N.P. Metal-on-metal total hip arthroplasty: is there still a role in 2016?. Current Reviews in Musculoskeletal Medicine. 2016;9(1):93–96. DOI: 10.1007/s12178-016-9323-1

[46] Wimmer M.A., Fischer A., Buscher R., Pourzal R., Sprecher C., Hauert R., et al. Wear mechanisms in metal-on-metal bearings: the importance of tribochemical reaction layers. Journal of Orthopaedic Research. 2010;28(4):436–443. DOI: 10.1002/jor.21020

[47] Liao Y., Pourzal R., Wimmer M.A., Jacobs J.J, Fischer A., Marks L.D. Graphitic tribological layers in metal-on-metal hip replacements. Science. 2011;334(6063):1687–1690. DOI: 10.1126/science.1213902

[48] Cadossi M., Mazzotti A., Baldini N., Giannini S,. Savarino L. New couplings, old problems: is there a role for ceramic-on-metal hip arthroplasty? Journal of Biomedical Materials Research B Applied Biomaterials. 2016;104(1):204–209. DOI: 10.1002/jbm.b.33383

[49] Wierzcholski K.C. Friction forces for human hip joint lubrication at a naturally permeable cartilage. International Journal of Applied Mechanics and Engineering. 2006;11(3):515–527 .

[50] Brockett C., Williams S., Jin Z. M., Isaac G. and Fisher J. Friction of total hip replacements with different bearings and loading conditions. Journal of Biomedical Materials Research Part B: Applied Biomaterials. 2007;81B(2):508-515. DOI: 10.1002/jbm.b.30691

[51] Koizumi Y., Yan C., Li Y., Yamanaka K., Chiba A., Tanaka S.I., Hagiwara Y. Uneven damage on head and liner contact surfaces of a retrieved Co-Cr-based metal-on-metal hip joint bearing: an important reason for the high failure rate. Materials Science and Engineering: C.2016;62;532–543.. DOI: 10.1016/j.msec.2016.01.006

[52] Lu Z., McKellop H. Frictional heating of bearing materials tested in a hip joint wear simulator. Proceedings of the Institution of Mechanical Engineers, Part H: Journal of Engineering in Medicine. 1997;211(1):101–108. DOI: 10.1243/0954411971534728

[53] Yan Y., Neville A., Dowson D., Williams S., Fisher J. Effect of metallic nanoparticles on the biotribocorrosion behaviour of Metal-on-Metal hip prostheses. Wear. 2009;267(5–8):683–688. DOI: 10.1016/j.wear.2008.12.110

[54] Mazzucco D., Scott R., Spector M. Composition of joint fluid in patients undergoing total knee replacement and revision arthroplasty: correlation with flow properties. Biomaterials. 2004;25(18):4433–4445. DOI: 10.1016/j.biomaterials.2003.11.023

[55] Jalali-Vahid D., Jagatia M., Jin Z.M., Dowson D. Prediction of lubricating film thickness in UHMWPE hip joint replacements. Journal of Biomechanics. 2001;34(2):261–266. DOI: 10.1016/S0021-9290(00)00181-0

[56] Stewart T.D . Tribology of artificial joints. Orthopaedics and Trauma. 2010;24(6):435–440. DOI: 10.1016/j.mporth.2010.08.002

[57] Jin Z.M., Dowson D., Fisher J. Analysis of fluid film lubrication in artificial hip joint replacements with surfaces of high elastic modulus. Proceedings of the Institution of Mechanical Engineers, Part H: Journal of Engineering in Medicine. 1997;211:247–256. DOI: 10.1243/0954411971534359

[58] Liu H., Xiea D., Qianc L., Denga X., Lenga Y.X., Huang N. The mechanical properties of the ultrahigh molecular weight polyethylene (UHMWPE) modified by oxygen plasma. Surface and Coatings Technology. 2011;205(8–9):2697–2701. DOI: 10.1016/j.surfcoat.2010.08.120

[59] Young S.K., Lotito M.A, Keller T.S . Friction reduction in total joint arthroplasty. Wear. 1998;222(1):29–37. DOI: 10.1016/S0043-1648(98)00275-0

[60] Ito H., Kaneda K., Yuhta T., Nishimura I., Yasuda K., Matsuno T. Reduction of polyethylene wear by concave dimples on the frictional surface in artificial hip joints. The Journal of Arthroplasty. 2000;15(3):332–338. DOI: 10.1016/S0883-5403(00)90670-3

[61] Sagbas B., Durakbasa M.N. Surface texturing of vitamin E blended UHMWPE for reduction of wear. Acta Physica Polonica A. 2014;125(2):481–483. DOI: 10.12693/APhysPolA.125.481

[62] Sawano H., Warisawa S., Ishihara S. Study on long life of artificial joints by investigating optimal sliding surface geometry for improvement in wear resistance. Precision Engineering. 2009;33(4):492–498. DOI: 10.1016/j.precisioneng.2009.01.005

[63] Roy T., Choudhury D., Ghosh S., Mamat A.B., Pingguan-Murphy B. Improved friction and wear performance of micro dimpled ceramic-on-ceramic interface for hip joint arthroplasty. Ceramics International. 2015;41(1):681–690. DOI: 10.1016/j.ceramint.2014.08.123

[64] Qiu Y., Khonsari M.M. Experimental investigation of tribological performance of laser textured stainless steel rings. Tribology International. 2011;44(5):635–644. DOI: 10.1016/j.triboint.2011.01.003

[65] Roy T., Choudhury D., Mamat A.B., Pingguan-Murphy B. Fabrication and characterization of micro-dimple array on Al2O3 surfaces by using a micro-tooling. Ceramics International. 2014;40(1):2381–2388. DOI: 10.1016/j.ceramint.2013.08.009

[66] Ghosh S., Choudhury D., Roya T., Mamat A.B., Masjukid H.H., Pingguan-Murphy B. Tribological investigation of diamond-like carbon coated micro-dimpled surface under bovine serum and osteoarthritis oriented synovial fluid. Science and Technology of Advanced Materials. 2015;16(3):1–11. DOI: 10.1088/1468-6996/16/3/035002

[67] Bitounis D., Pourchez J., Forest V., Boudard D., Cottier M., Klein J.P. Detection and analysis of nanoparticles in patients: a critical review of the status quo of clinical nanotoxicology. Biomaterials. 2016;76:302–312. DOI: 10.1016/j.biomaterials.2015.10.061

[68] Kanaga Karuppiah K.S., Bruck A.L., Sundararajan S., Wang J., Lin Z., Xu Z.H., Li X. Friction and wear behavior of ultra-high molecular weight polyethylene as a function of polymer crystallinity. Acta Biomaterialia. 2008;4(5):1401–1410. DOI: 10.1016/j.actbio.2008.02.022

[69] Liu F., Leslie I., Williams S., Fisher J., Jin Z.M. Development of computational wear simulation of metal-on-metal hip resurfacing replacements. Journal of Biomechanics. 2008;41(3):686–694. DOI: 10.1016/j.jbiomech.2007.09.020

[70] Kurtz S.M, editor. UHMWPE Biomaterials Handbook. 2nd ed. China: Elsevier; 2009. 543 p. DOI: 10.1016/B978-0-12-374721-1.00037-7

[71] Fisher J., Mc Ewen H.M.J., Barnett P.I., Bell C., Stone M.H. Influences of sterilising techniques on polyethylene wear. The Knee. 2004;11(3):173–176. DOI: 10.1016/j.knee.2003.10.002

[72] Bragdon C., Jasty M., Muratoglu O., O'Connor D., Harris W. Third-body wear of highly cross-linked polyethylene in a hip simulator. The Journal of Arthroplasty. 2003;18(5): 553–561. DOI: 10.1016/S0883-5403(03)00146-3

[73] Wang A., Essner A. Three-body wear of UHMWPE acetabular cups by PMMA particles against CoCr, alumina and zirconia heads in a hip joint simulator. Wear. 2001;250(1–12):212–216. DOI: 10.1016/S0043-1648(01)00643-3

[74] Hirakawa K., Jacobs J.J., Urban R., Saito T. Mechanisms of failure of total hip replacements: lessons learned from retrieval studies. Clinical Orthopaedics and Related Research . 2004;420:10–17.

[75] Kim Y.H., Ritchie A., Hardaker C. Surface roughness of ceramic femoral heads after in vivo transfer of metal: correlation to polyethylene wear. Journal of Bone and Joint Surgery. 2005;87(3):577–582. DOI: 10.2106/JBJS.D.01790

[76] Mackay D., Gower A., Mawhinney B., Gregg P.J., McCaskie, A.W. Metallic instrument debris: a source of third-body wear particles? Journal of Arthroplasty . 2000;15(6):816–818. DOI: 10.1054/arth.2000.8096

[77] Willie B.M., Shea J.E., Bloebaum R.D., Hofmann A.A. Elemental and morphological identification of third-body particulate and calcium stearate inclusions in polyethylene components. Journal of Biomedical Materials Research. 2000;53(2): 137–142.

[78] Sagbas B. Measurement, analysis and metrological evaluation of physical magnitudes and geometrical features that show alteration with regard to friction in hip prosthesis [thesis]. Istanbul Turkey: Yildiz Technical University; 2013. 228 p.

[79] Stanczyk M., Telega J.J. Modelling of heat transfer in biomechanics: review, part II. Orthopaedics. Acta of Bioengineering and Biomechanics. 2002;4(2):1–31.

[80] Hu C.C., Liau J.J., Lung C.Y., Huang C.H., Cheng C.K. A two-dimensional finite element model for frictional heating analysis of total hip prosthesis. Materials Science and Engineering: C. 2001;17(1–2):11–18. DOI: 10.1016/S0928-4931(01)00328-9

[81] Liao Y.S., McKellop H., Lu Z., Campbell P., Benya P. The effect of frictional heating and forced cooling on the serum lubricant and wear of UHMW polyethylene cups against cobalt-chromium and zirconia balls. Biomaterials. 2003;24(18):3047–3059. DOI: 10.1016/S0142-9612(03)00148-0

[82] Bergmann G., Graichen F., Rohlmann A., Verdonschot N., van Lenthe G.H. Frictional heating of total hip implants. Part 1: measurements in patients. Journal of Biomechanics. 2001;34(4):421–428. DOI: 10.1016/S0021-9290(00)00188-3

[83] Bergmann G., Graichen F., Rohlmann A., Verdonschot N., van Lenthe G.H. Frictional heating of total hip implants. Part 2: finite element study. Journal of Biomechanics. 2001;34(4):429–435. DOI: 10.1016/S0021-9290(00)00234-7

[84] Uddin M.S., Majewski P. Frictional heating in hip implants—a review. Procedia Engineering. 2013;56:725–730. DOI: 10.1016/j.proeng.2013.03.185

[85] Rocchi M., Affatato S., Falasca G., Viceconti M. Thermomechanical analysis of ultra-high molecular weight polyethylene-metal hip prostheses. Proceedings of the Institution of Mechanical Engineers Part H: Journal of Engineering in Medicine. 2007;221:561–568. DOI: 10.1243/09544119JEIM137

[86] Sagbas B., Durakbasa M.N. Evaluating frictional temperature rise in sliding surface of artificial hip joint materials with different loading conditions. Acta Physica Polonica A. 2013;123:453–455. DOI: 10.12693/APhysPolA.123.453

[87] Pawlus P., Galda L., Dzierwa A., Koszela W. Abrasive wear resistance of textured steel rings. Wear. 2009;267(11):1873–1882. DOI: 10.1016/j.wear.2009.03.003

[88] Koszela W., Pawlus P., Galda L. The effect of oil pockets size and distribution on wear in lubricated sliding. Wear. 2007;263(7–12):1585–1592. DOI: 10.1016/j.wear.2007.01.108

[89] Langlois J., Zaoui A., Scemama C., Martell J., Bragdon C., Hamadouche M. Validation of a computer-assisted method for measurement of radiographic wear in total hip arthroplasty using all polyethylene cemented acetabular components. Journal of Orthopaedic Research. 2015;33(3):417–420. DOI: 10.1002/jor.22777

[90] Sagbas B., Durakbasa M.N. Measurement of wear in orthopedic prosthesis. Acta Physica Polonica A. 2012;121(1):131–134.

[91] McCalden R.W., Naudie D.D., Yuan X. Radiographic methods for the assessment of polyethylene wear after total hip arthroplasty. The Journal of Bone Joint Surgery. 2005;87(10):2323–2334. DOI: 10.2106/JBJS.E.00223

[92] Rahman L., Cobb J., Muirhead-Allwood S. Radiographic methods of wear analysis in total hip arthroplasty. Journal of the American Academy of Orthopaedic Surgeons. 2012;20(12):735–743. DOI: 10.5435/JAAOS-20-12-735

[93] Harper M.L., Sca M.A., Dooris A., Paré P.E. The fundamentals of biotribology and its application to spine arthroplasty. SAS Journal. 2009;3(4):125–132. DOI: 10.1016/j.esas.2009.11.004

[94] Carmignatoa S., Spinelli M., Affatato S., Savio E. Uncertainty evaluation of volumetric wear assessment from coordinate measurements of ceramic hip joint prostheses. Wear. 2011;270(9–10):584–590. DOI: 10.1016/j.wear.2011.01.012

[95] Muratoglu O.K., Rubash H.E., Bragdon C.R., Burroughs B.R., Huang A., Harris W.H. Simulated normal gait wear testing of a highly cross-linked polyethylene tibial insert. The Journal of Arthroplasty. 2007;22(3):435–444. DOI: 10.1016/j.arth.2006.07.014

[96] Blunt L.A., Bills P.J., Jiang X.Q., Chakrabarty G. Improvement in the assessment of wear of total knee replacements using coordinate-measuring machine techniques. Proceed-

ings of the Institution of Mechanical Engineers, Part H: Journal of Engineering in Medicine. 2008;222(3):309–318. DOI: 10.1243/09544119JEIM289

[97] Tuke M., Taylor A., Roques A., Maul C. 3D linear and volumetric wear measurement on artificial hip joints—validation of a new methodology. Precision Engineering. 2010;34(4):777–783. DOI: 10.1016/j.precisioneng.2010.06.001

[98] Estok D.M., Bragdon C.R., Plank G.R., Huang A., Muratoglu O.K, Harris W.H. The measurement of creep in ultrahigh molecular weight polyethylene: a comparison of conventional versus highly cross-linked polyethylene. Journal of Arthroplasty. 2005;20(2):239–243.

[99] Ma L., Rainforth W.M., Sun D., Wharton, J.A., Wood R.J.K . A '3-body' abrasion wear study of bioceramics for total hip joint replacements. Wear. 2009;267(11):2122–2131. DOI: 10.1016/j.wear.2009.08.040

[100] Pylios T., Shepherd D.E. Wear of medical grade silicone rubber against titanium and ultrahigh molecular weight polyethylene. Journal of Biomedical Materials Research Part B: Applied Biomaterials. 2008;84(2):520–523. DOI: 10.1002/jbm.b.30899

Chapter 7

Friction and Wear in Automotive Journal Bearings Operating in Today's Severe Conditions

David E. Sander, Hannes Allmaier and Hans-Herwig Priebsch

Additional information is available at the end of the chapter

http://dx.doi.org/Chapter DOI: 10.5772/64247

Abstract

A current trend in the transport sector seeks to increase the vehicle efficiency and to cut fuel consumption which leads to new technologies and advancements in modern and future combustion engines. Some of these technical progresses lead to highly stressed engine parts and new challenges arise, particularly for journal bearings. The increasing thermal and mechanical load caused by downsized and turbocharged engines, friction reduction by employing low-viscosity lubricants and other emission reduction measures—for utilizing stop-start systems—put additional stress on the crankshaft journal bearings. This contribution focuses on highly stressed journal bearings which operate in the boundary, mixed and hydrodynamic lubrication regime. Therefore, measurements on a journal bearing test-rig are performed which allow an extensive verification of the numerical investigation. For the numerical analysis of friction and wear, a mixed elasto-hydrodynamic simulation approach is developed, which considers the elastic deformation of the contacting components, the complex rheological behaviour of the lubricant and metal-metal contact if the lubricant is unable to separate the contacting surfaces. Both the rheological data and the surface roughness parameters are obtained from measurements. The current challenges are studied in four application-oriented examples and the influencing parameters on a reliable friction and wear prediction are explored.

Keywords: simulation, hydrodynamic bearing, plain bearing, mixed friction, elasto-hydrodynamic lubrication, non-Newtonian lubricant, conformal contact, stop-start

1. Introduction and outline

A plain journal bearing is a seemingly simple machine element where a hardened shaft typically rotates in a softer bearing shell. The two contacting surfaces are generally separated by a lubricant. In most technical applications, such as turbines or combustion engines, journal bearings are designed for a long life time. Hence, journal bearings mainly operate in pure hydrodynamic lubrication regime. However, especially in the automotive industry, a trend has evolved to cut fuel consumption and emissions derived from emission restrictions by legislation and customer satisfaction. A cost-effective potential to improve the efficiency is attributed to the reduction of friction in lubricated contacts. The friction coefficient finds its minimum in the transition between pure hydrodynamic lubrication and mixed lubrication regime, where already some metal-metal contact between the shaft and bearing shell occurs. Therefore, it is beneficial to operate journal bearings in this transition, but unfortunately wear occurs as metal-metal contact establishes and durability problems may follow. To design more efficient journal bearings, a detailed simulation approach is required which is suited to describe the complex behaviour of mixed elasto-hydrodynamic lubrication and is validated for a wide range of operating conditions.

In addition, beyond the consequences of friction reduction many new modes of engine operation stress the bearings. Therefore, this chapter begins with a general overview of the current challenges for journal bearings, especially in combustion engine application.

Subsequently, the requirements for the simulation model are elaborated from these challenges at the beginning of the next section. In the same section, the basic equations for the mixed elasto-hydrodynamic journal bearing simulation are presented. Additionally, the simplified layout of the journal bearing test-rig and the simulation model is described which is used for the study on journal bearing friction and wear.

The subsequent result section consists of different application examples and provides an overview of the current research by the authors. The first example analyses friction of dynamically loaded journal bearings under extreme loading conditions which lead to fluid film pressures above 2000 bar and to high shear rates. The complex properties of the lubricant are central to these results. The second example concentrates on the running-in wear of journal bearings until a steady state of operation is achieved. The third example focuses on severe metal-metal contact and on friction in mixed lubrication regime. Therefore, a static load is applied to the journal bearing and Stribeck curves are calculated. Finally, a transient start-stop simulation is performed in the final example.

This chapter concludes with a comprehensive summary of the studied results and presents a brief outlook on future research topics.

2. Current challenges for journal bearings in combustion engines

In today's automotive engines, journal bearings operate under severe conditions. Here, these conditions will be discussed in more detail as they are the prime subject for the presented study and also for future research. A more comprehensive review can be found in [1].

2.1. Turbocharged engines with high power density

Modern downsized combustion engines with turbocharger(s) achieve a high power output from small volume displacements and allow an efficiency improvement with lower emissions. At the same time, the high power density increases the thermal and mechanical load on the engine components. Additionally, the light-weight design of the engine further improves the vehicle performance. Hence, the engine and its component dimensions become smaller too. This implies even higher stresses on the components and the lubricated contacts. For instance, the big-end bearing of the connection rod has to resist specific loads above 100 MPa and a further increase is expected for future engines [2]. As a consequence, the minimum lubrication gap in the journal bearing decreases below 1 μm and metal-metal contact may occur at these operating conditions. Furthermore, the elastic deformations of the bearing shell and its surrounding parts have a major influence on bearing behaviour.

2.2. Stop-start system

One of the most widely used mechanisms to improve fuel economy, especially for urban driving, is the application of stop-start systems [3, 4]. By turning off the engine instead of operating it at idle (e.g. when waiting at crossings) significant fuel savings can be obtained [4, 5]. More recently, this procedure was extended to switching off the engine also during driving at very low loads which is commonly called engine coasting.

Stop-start systems have quickly become widely available, but despite their apparent simplicity the repeated stopping and starting of the engine present a big challenge for bearing manufacturers. When the engine is started the bearing has to overcome the boundary and mixed lubrication regime before a hydrodynamic film has formed, which completely separates the two contacting surfaces [6]. Thus, with the increasing number of stop-start cycles bearing wear will escalate.

2.3. Cylinder de-activation

Another fuel economy improvement is achieved through a cylinder de-activation technology (CDT) as a result of reduced air pumping losses. Further benefits are achieved due to an increased exhaust temperature under partial loading which yields an improved after-treatment efficiency for diesel engines. This will further help the three-way catalyst technology [7]. However, CDT affects the dynamic behaviour of the shaft and topics such as shaft bending and journal misalignment that are relevant for bearing design and bearing layout. The ability of the bearing shell to adapt its surface geometry and further to fit the deformed or misaligned shaft can become essential. Also, the friction power losses need to be investigated to show the overall efficiency benefits. A better understanding for the necessary redesign of the power train with CDT is the target in recent bearing simulation [8–10].

2.4. Ultralow-viscosity lubricants

The usage of low-viscosity lubricants in the automotive sector is certainly not new. However, the increasingly strict legislation forces the automotive industry to further increase engine efficiency. Here, the usage of a lower viscosity lubricant is considered to be a very economic measure to reach this goal [11]. However, the current trend goes to lubricants with a drastically reduced viscosity and new standards had to be defined accordingly [12]. With the new SAE standards 16, 12, 8 and even lower classes being targeted, pure hydrodynamic lubrication conditions will increasingly decline and be replaced by mixed lubrication near the minimum friction coefficient [13].

2.5. Complex rheological properties of modern lubricants

It is well known that modern multi-grade lubricants show a non-Newtonian behaviour. At high shear rates, which typically occur in automotive journal bearings, the viscosity significantly drops below the viscosity at low shear rate. To assess this shear thinning behaviour, a viscosity at high temperature and high shear rate (HTHS-viscosity) is standardized. It is shown that the HTHS-viscosity directly affects the mechanical efficiency of the engine in fleet tests [14, 15], which demonstrates the necessity to realistically consider the non-Newtonian behaviour of the lubricant in the simulation. There exists a further lubricant property counteracting the non-Newtonian behaviour in journal bearings, which increases the lubricant viscosity locally under load. This lubricant property is called the piezo-viscous effect and it is present in both mono and multi-grade lubricants [16].

Currently, the required lubricant data are not stated in the lubricant datasheets as such measurements are not part of common standards. Therefore, such rheological data are only sparse. The interdependency of the piezo-viscous effect and the non-Newtonian behaviour makes it difficult not only to measure them, but also to include these properties accurately in simulation. Currently, commonly used approaches use, for example, the Barus equation and the Cross equation to consider these effects in a simulation independently from each other and independent of the lubricant temperature (e.g., [17]). While these are useful approximations of the reality and considerably increase the quality of the results, the true lubricant behaviour is still considerably more complex [18–20].

2.6. New classes of surface textures and coatings

To increase the durability and to reduce the friction loss in journal bearings, new techniques have grown in popularity in the process of manufacturing journal bearings. While still two- and three-layer journal bearings dominate the automotive industry, new development trends bring refined surface structuring techniques, such as polymer coatings and diamond-like carbon (DLC) coatings.

Surface texturing, which is the intentional modification of the journal bearing surface with regular patterns (dimples, microgrooves), has the potential to increase the load-carrying capacity [21]. Such surface texturing methods themselves are not new and are, for example, widely used in other parts of the internal combustion engine, namely, in honing of the cylinder liners. For journal bearings it is experimentally shown that surface textures can affect their tribological properties [22, 23] as well as the damping properties and the compatibility of the journal bearing with contaminants [24, 25]. The theoretical understanding of the underlying effects of surface textures is currently limited [26]. So far it is commonly accepted that cavitation plays a crucial part for this effect [27–29] as well as that surface textures have an influence on local micro-hydrodynamics [26].

Polymer coatings are increasingly available on the market and can offer better durability in start-stop conditions [30, 31]. As these coatings are non-metallic materials, their behaviour in terms of wear and plastic deformation is distinctly different to metallic materials. These coatings require new approaches to describe their behaviour realistically in the simulation [32, 33]. DLC coatings are already widely used in engines today, like for piston pins or cam tappets, which are highly loaded contacts and dominantly operate in mixed lubrication. Crankshaft journal bearings dominantly operate in hydrodynamic lubrication regime and an ability to adapt the bearing surface is often required which makes DLC coatings apparently inappropriate. Nonetheless, it was recently shown that DLC have a potential for friction reduction for the journal bearings of the crank train [34].

3. Journal bearing simulation

Before the basic equations for the elasto-hydrodynamic journal bearing simulation are summarized, the requirements for the simulation approach to fulfil the previously highlighted challenges are explained. Then, the detailed lubricant properties under high pressure and high shear rate are presented. Finally, the layout of the journal bearing test-rig and the simulation model is described which is used for the study on journal bearing friction and wear.

3.1. Requirements for journal bearing simulation

The requirements for a reliable analysis of journal bearing friction and wear can be derived from the challenges described in Section 2.

- High load in a turbocharged and downsized engine deforms the engine parts elastically. Therefore, the elastic deformation needs to be considered in the simulation. For instance, an elastic deformation of the shaft can lead to a misalignment between journal and bearing and this affects the behaviour of the journal bearing. Furthermore, the high pressure in the lubricated contacts deforms the contacting surface which obviously affects the film thickness. Especially, the local elastic deformation of the significantly softer bearing shell—compared to the hardened shaft—needs to be considered.
- Due to the increasing specific pressure in journal bearings the rheological properties of the lubricant at high pressure play an important part. The viscosity of lubricants increases strongly at high hydrodynamic pressures and this piezo-viscous effect must be taken into account. In addition, modern automotive lubricants behave like highly non-Newtonian as a consequence of their complex composition. The effective viscosity varies within the lubrication gap due to locally different hydrodynamic pressures and shear rates. Both effects can have a major influence on the local viscosity in the journal bearing.

- Mixed lubrication and further the prediction of metal-metal contact become more crucial as the minimum film thickness decreases. To reliably predict friction and wear, an accurate description of metal-to-metal contact is required. Consequently, a contact model must be incorporated into the simulation model.
- A realistic description of the surface structure, which means both material and topography, of the journal bearings becomes increasingly important. Not only because journal bearings operate in mixed lubrication more frequently, but also because their surface properties become increasingly refined. Therefore, parameters for the simulation model must be derived from real measured surfaces.
- Any misalignment between journal and bearing can lead to severe metal-metal contact especially at the bearing edge. During the running in of the bearing, the bearing shells are able to adapt their geometry for a better fit between journal and bearing. This geometrical adaption can be larger than the minimum lubrication gap. Therefore, the adapted surface geometry needs to be considered in the simulation as metal-metal contact is largely overestimated otherwise [35, 36].
- The highly nonlinear behaviour of dynamically loaded journal bearings requires a transient calculation in time domain [33]. Especially to understand journal bearing behaviour during the starting and stopping of the shaft, a transient calculation is demanded.
- Thermal effects also change the lubricant viscosity and need to be considered in bearing simulation. In this study, the temperature influence is considered by deriving equivalent temperatures for the isothermal simulation. Of relevance but beyond the scope of this study is the consideration of the local temperature by thermal equations. This will be the focus of future research.
- Finally, the simulation approach needs to be validated by measurement results because approximated functions are used to describe the lubricant or statistically derived contact models are implemented.

3.2. Fundamentals of mixed elasto-hydrodynamic lubrication theory

The fundamental differential equation which describes the lubricant film in a journal bearing is the Reynolds equation. The Reynolds equation together with a suitable cavitation boundary condition and a coupling with the surrounding deformable bodies form the basis of the elasto-hydrodynamic lubrication (EHD) approach. The basic Reynolds equation considers smooth surfaces, but when the distance between the facing surfaces decreases their surface roughness will affect the oil flow. To take this micro-hydrodynamic effect into account, Patir and Cheng [37, 38] introduced flow factors to the Reynolds equation:

$$-\frac{\partial}{\partial x}\left(\phi_x \theta \frac{h^3}{12\eta}\frac{\partial p}{\partial x}\right)-\frac{\partial}{\partial y}\left(\phi_y \theta \frac{h^3}{12\eta}\frac{\partial p}{\partial y}\right)+\frac{\partial}{\partial x}\left(\theta h \frac{u_1+u_2}{2}\right)+\frac{\partial}{\partial x}\left(\theta \phi_s \frac{u_1+u_2}{2}\sigma s\right)+\frac{\partial}{\partial t}(\theta h)=0, \quad (1)$$

where x and y denote the circumferential and the axial direction. p is the hydrodynamic pressure and h is the oil film thickness which depends on x and y. Further, u_1 and u_2 denote the sliding speeds of the facing surfaces. The influence of surface roughness is considered by the pressure flow factors φ_x, φ_y and the shear flow factor φ_s. The oil viscosity η which is a function of x and y is considered to depend on the temperature, pressure and shear rate in this work. Finally, θ represents the fill ratio which is introduced to consider mass conserving cavitation in the model. The cavitation model is based on the Jakobsson-Floberg-Olsson (JFO) approach [39, 40].

If the fluid film cannot completely separate the two contacting surfaces anymore, single asperities interact and metal-metal contact occurs. The contact model according to Greenwood and Tripp [41] is used to estimate the metal-metal contact pressure by the following equation:

$$p_a = KE^* F_{5/2}(H_s), \tag{2}$$

where the composite elastic modulus E^* is a combined material parameter of the contacting surfaces, the elastic factor K depends on surface roughness, asperity radius and asperity density. $F_{5/2}$ is a form factor [35] which is a function of the dimensionless clearance parameter H_S.

The parameters for the contact model used in this study are derived from scans of the bearing surface and the shaft surface. The detailed surface parameters are described in [42, 43].

Friction losses in highly loaded journal bearings consist of hydrodynamic losses only or of both hydrodynamic losses and losses due to metal-metal contact. In this study, the friction losses are represented as friction torque which can be calculated by integrating the hydrodynamic shear stress τ_h and the asperity shear stress τ_a over the bearing surface:

$$M_{Friction} = r \iint_A (\tau_h + \tau_a) dx\, dy. \tag{3}$$

The hydrodynamic shear stress can be calculated by

$$\tau_h = \eta \cdot \frac{u_1 - u_2}{h} (\phi_f \pm \phi_{fs}) \pm \left(\phi_{fp} \frac{h}{2} \cdot \frac{\partial p}{\partial x} \right), \tag{4}$$

where + and – refer to the shell surface and the journal surface, respectively. φ_f, φ_{fs} and φ_{fp} are shear stress factors according to Patir and Cheng [37, 38]. According to the hydrodynamic shear stress the friction torque is proportional to the lubricant viscosity and it is essential to consider detailed lubricant properties to reliably assess friction:

$$\tau_a = \mu_{Bound} \cdot p_a, \tag{5}$$

The asperity shear stress is calculated by multiplying the asperity contact pressure with a boundary friction coefficient μ_{Bound}. In this study, the boundary friction coefficient is assumed to be constant. This simple approximation is able to yield reliable results for a wide range of operation as shown in the result section.

3.3. Lubricant properties at high pressure and high shear rate

The lubricant used in this study is a fully formulated low-viscous 0W20 hydrocarbon engine oil. It is a standard multi-grade lubricant which is available for the automotive market. The main properties of the lubricant are summarized in **Table 1**.

Density at 40°C	832.5 kg/m^3
Dynamic viscosity at 40°C	37.5 mPa s
Dynamic viscosity at 100°C	6.8 mPa s
HTHS-viscosity at 150°C and shear rate 10^6 1/s	2.7 mPa s

Table 1. Basic properties of the tested 0W20 lubricant.

The main lubricant properties show that the viscosity is strongly dependent on temperature. Various mathematical functions exist to consider the temperature dependency in simulation. In this study, the effect of temperature on viscosity is specified by the Vogel equation [44] which is very accurate for hydrocarbon lubricants,

$$\eta(T) = A \cdot e^{B/(T+C)}. \tag{6}$$

where T is the oil temperature and A, B and C are constants for a given lubricant. In dynamically loaded journal bearings, high pressures above 2000 bar can occur and the increase of viscosity can be very significant. Consequently, the viscosity increase with high pressure needs to be considered. Here, the simple Barus equation [45] is utilized which only requires the identification of a pressure viscosity coefficient α:

$$\eta(T,p) = \eta(T) \cdot e^{\alpha p}. \tag{7}$$

Due to the dynamic load and high operational speed, shear rates above 2 × 10^7 1/s can occur in journal bearings. At these conditions, a significant drop in viscosity can be identified. To describe the shear thinning effect, the Cross equation [46] is used:

$$\eta(T,p,\dot{\gamma}) = \eta(T,p) \cdot \left(r + \frac{1-r}{1+(K\dot{\gamma})^m} \right). \tag{8}$$

where K, r and m are coefficients for a given lubricant. All coefficients for Vogel, Barus and Cross equation are derived from measured rheological data of the investigated lubricant. The detailed derivation can be found in [47]. The values of the introduced coefficients are listed in **Table 2**.

A	0.0516 mPa s
B	1127.6°C
C	130.7°C
α	0.00095 1/bar
r	0.53
m	0.79
K	7.9×10^{-8} s

Table 2. Parameters for the rheological model of the lubricant.

The viscosity characteristics obtained from the derived parameters are also shown in **Figures 1** and **2**.

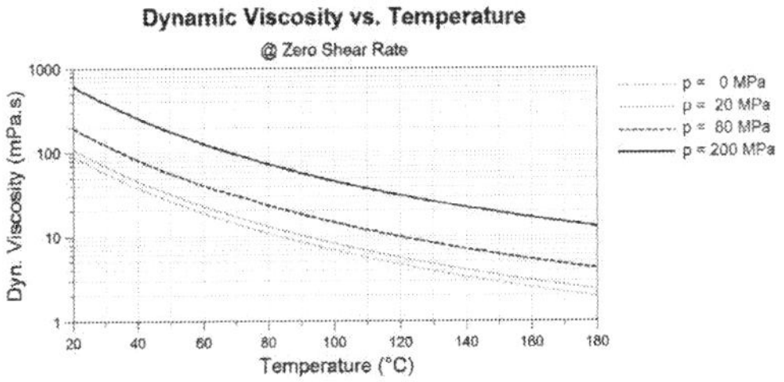

Figure 1. Viscosity-temperature dependency for the 0W20 lubricant at different pressures [48]; please note the logarithmic scale.

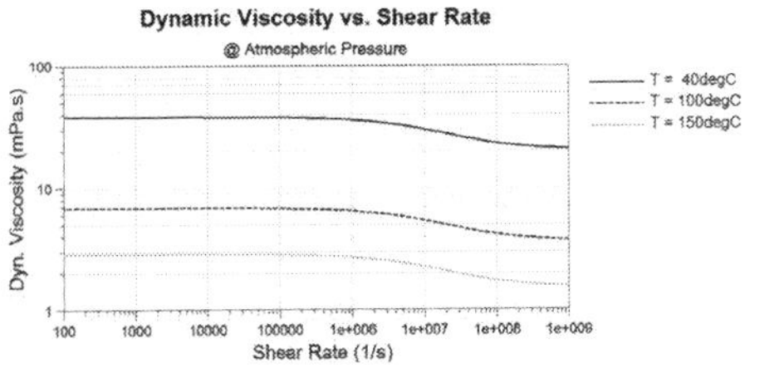

Figure 2. Viscosity-shear rate dependency for the 0W20 lubricant at different temperatures [48]; please note the logarithmic scale.

3.4. Journal bearing tests and the corresponding simulation model

Experimental results are used extensively in this study to validate the presented simulation approach. The experimental part is not the central topic of this work. Therefore, only a brief description of the journal bearing test-rig is provided. More details about the test-rig and the measurement equipment can be found in [43, 47, 49].

All tests are conducted on the journal bearing test-rig at KS Gleitlager[1] as shown in **Figure 3**. The test-rig enables a detailed analysis of journal bearing behaviour under static and dynamic loads for constant or transient shaft speed. The test-rig consists of a straight shaft (test shaft) which rests on two support brackets. Each support bracket contains a plain journal bearing and is fixed on a base. The test connecting rod with the test bearing is placed in between the two support brackets. An external load is applied onto the test connecting rod which is generated by an electromechanical high-frequency pulsator. The test shaft is driven by an elastically clutched electric motor.

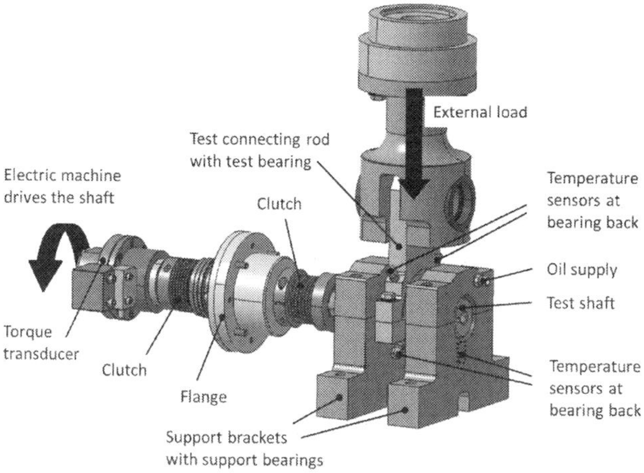

Figure 3. Configuration and components of the journal bearing test-rig [43].

The dimensions and oil supply design of the support bearings correspond to automotive main bearings with a 180° oil supply groove. The support bearings have a diameter of 54 mm and a width of 25 mm. The test journal bearing corresponds to a big-end bearing having an oil supply hole in the load-free (lower) shell. It has a diameter of 47.8 mm and a width of 17.2 mm.

The total friction torque caused by all three journal bearings is measured by a torque transducer which is located between the motor and the clutch.

[1] KS Gleitlager GmbH, Am Bahnhof 14, 68789 St. Leon-Rot, Germany.

For the simulation, the test-rig is modelled within the flexible multi-body solver AVL Excite Power Unit.[2] The multi-body simulation model consists of flexible bodies and joints which connects the bodies. The two support brackets with the support bearings and the test connecting rod including the test bearing are represented as mathematically condensed finite element (FE) structures [50]. Therefore, a pre-processing step needs to be performed with an FE solver to create the condensed bodies. The test shaft is also represented as elastic body but instead of using a condensed structure a simplified beam and disc body is generated.

The lubricated contacts between the test shaft and support bearings as well as between the test shaft and test bearing are modelled as elasto-hydrodynamic joints. The fluid film is calculated according to the averaged Reynolds equation discussed in the previous section (Eq. (1)). The numerical approach requires a discretization of the bearing surface. Therefore, 25 hydrodynamic (HD) nodes are defined in axial direction which are equally distributed in each bearing (support bearings and test bearing). The test bearing has 200 nodes in circumferential direction and the support bearings have 176 nodes. Every second HD node is directly coupled to the condensed FE model.

Because of the high nonlinearity of the model it is solved in the time domain using numerical time integration [33].

3.5. The consideration of temperature in the isothermal simulation method

The numerical approach assumes an isothermal bearing which means that the bearing temperature and the temperature of the lubricant are assumed to be constant in the lubrication gap. Certainly, the temperature changes under different operating conditions, for example, with varying shaft speed and bearing load. To consider the load- and speed-dependent temperature, an equivalent temperature is defined which is derived from measured temperatures at the back of the bearing shell (see sketch of the test-rig in **Figure 3**). Hence, an isothermal bearing simulation is performed while the temperature influence due to different operating conditions is still considered.

The background for this approach was introduced by previous research results from Allmaier et al. [51]. The authors investigated the thermal processes of journal bearings under high dynamic loads and compared the results with experimental data. It was shown that it is possible to predict the measured temperatures at the bearing shell with a complex thermo-elasto-hydrodynamic (TEHD) simulation approach. From these results, a simple equivalent bearing temperature relation was derived for the isothermal EHD simulation that is capable to predict the friction losses very accurately for a large range of different lubricants, journal speeds and loads as was shown in direct comparison to experimental data (see, in particular, [35] and [52], which is available online at www.intech.com[3]). This approach was initially suggested for main bearings with a distinct oil supply groove and was furthermore adapted to bearings with an oil supply hole in [47].

---

[2] AVL List GmbH, Advanced Simulation Technology, Hans-List-Platz 1, 8020 Graz Austria, www.avl.com.

[3] http://dx.doi.org/10.5772/51568.

4. Simulation results

In the result section, the behaviour of journal bearings under different operating conditions is discussed. Primarily, the bearings are excited with a dynamic load and they operate in the elasto-hydrodynamic lubrication regime. The elastic deformation and also the viscosity under high pressure and shear rate must be considered for a reliable prediction of friction and wear. Furthermore, severe metal-metal contact is generated by a static load and low shaft speed. This condition allows for the verification of the simulation approach in mixed lubrication regime. The presented results give an overview of current research by the authors. The majority of the results are taken from recent publications and, therefore, we refer to the original publications for further reading [42, 43, 47, 48, 53].

4.1. Impact of high pressure and shear thinning on journal bearing friction

The first application example discusses friction in dynamically loaded journal bearings. The bearings operate in the elasto-hydrodynamic regime with only a minor metal-metal contact. Due to the heavy load, maximum hydrodynamic pressures above 200 MPa arise in the bearings. At shaft speeds of up to 7000 rpm, high shear rates occur, which induce a noticeable drop of viscosity as will be shown.

To reliably predict friction in this condition a detailed description of the lubricant behaviour under high pressure and high shear rate becomes absolutely necessary. The influence of neglecting the piezo-viscous effect and the non-Newtonian behaviour on the accuracy of friction prediction in journal bearings is demonstrated.

4.1.1. Journal bearing conditions

Two dynamic load cases are investigated in this first study, one with a maximum external load of 40 kN and a second one with 80 kN. The dynamic characteristic of the load during a full load cycle is shown in **Figure 4**.

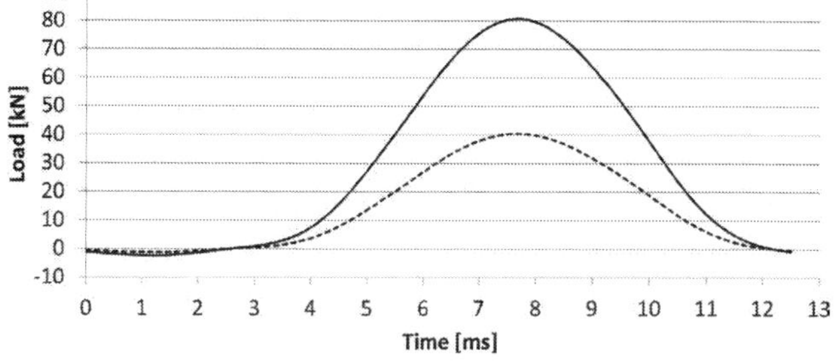

Figure 4. Dynamic load applied to the test bearing [47].

http://dx.doi.org/Chapter DOI: 10.5772/64247

The maximum specific loads in the test bearing which is the maximum load related to the projected bearing area, correspond to 50 MPa (40 kN load) and 100 MPa (80 kN load). The operational shaft speed starts at 1000 rpm and runs up to 7000 rpm with 1000 rpm increments. Each step is operated for 20 minutes. The run-up is followed by a run-down back to 1000 rpm. These conditions cover a wide range of operating conditions which are also typical for combustion engines.

4.1.2. Verification of the simulation results and the importance of the lubricant model

The figures in this section show a comparison of the simulation results and the measurement results. The simulation results are obtained with three different lubricant models. The results are arranged in such a way that the importance of lubricant properties in the elasto-hydrodynamic journal bearing simulation is highlighted. Therefore, we start with the simplest lubricant model $\eta(T)$ which is only dependent on temperature. Afterwards the piezo-viscous effect is added to the lubricant model $\eta(T,p)$ and, finally, the complete lubricant model also considers the shear thinning effect $\eta(T,p,\dot{\gamma})$.

Figure 5 shows the measured and calculated mean friction torque over shaft speed for the two different loads. The dashed lines represent the measured data. The error bars show the uncertainty which contains the error of the torque transducer and the standard error of the mean. As expected, the friction torque rises with increasing shaft speed and also with higher load. The influence of temperature can be clearly seen by comparing the 100 MPa results during the run-up and the run-down. At the run-down the bearing temperature is consistently 3°C higher compared to the run-up. Due to the lower viscosity at higher temperature, this small temperature difference reduces the friction torque by about 0.2 Nm.

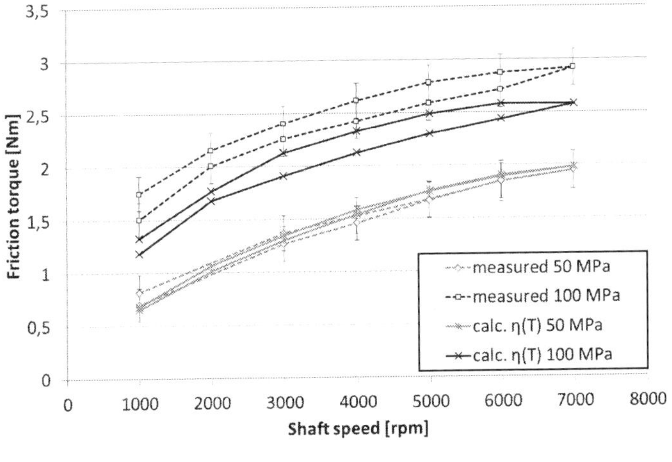

Figure 5. Comparison of the average friction torque measured on the test-rig with the simulated mean friction torque using the basic oil model $\eta(T)$ at 50 and 100 MPa specific loads [47].

The calculated average friction torque using the simple temperature-dependent viscosity model $\eta(T)$ is shown by solid curve in **Figure 5**. While the calculated friction torques at 50 MPa agree closely to the measured curve, the calculated friction torques at 100 MPa underestimate the measured torques over the entire speed range. However, the difference due to higher temperature at the run-down has been correctly identified. This indicates that the temperature consideration for the isothermal bearing calculation performs as intended.

The simulation results allow for a greater insight into the film formation and the occurring hydrodynamic pressure. At a specific load of 100 MPa a peak oil film pressure of around 200 MPa in the test bearing is identified independently from the shaft speed. As can be seen in **Figure 1**, the pressure-induced viscosity increase is significant for such high pressures. Consequently, the lubricant viscosity increases nearly sevenfold compared to the viscosity at ambient pressure. Therefore, calculations using the pressure-dependent oil model $\eta(T,p)$ are performed and the results are shown as solid curves in **Figure 6**.

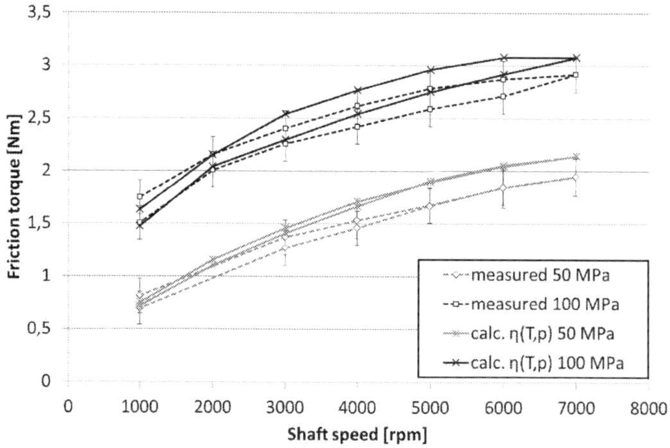

Figure 6. Comparison of the average friction torque measured on the test-rig with the simulated mean friction torque using the pressure-dependent oil model $\eta(T,p)$ at 50 and 100 MPa specific loads [47].

The friction torque increases for all operating conditions when considering the pressure-dependent oil model. The impact of the piezo-viscous effect depends on the applied load and is as expected, more relevant for the 100 MPa load than for the 50 MPa load. At low shaft speeds, the calculated friction torque agrees closely with the measured torque results for both the load cases. However, for shaft speeds above 3000 rpm the calculated torque overestimates increasingly the measured torque.

The relative sliding speed between shaft and bearing increases with higher shaft speed and, hence, higher shear rates occur. The shear rate increases nearly linearly with the shaft speed and reaches a maximum of 2.2×10^7 1\s at 7000 rpm [47]. At high shear rate like this, the viscosity drop due to the shear thinning effect is significant. For the studied lubricant a reduction of 25% is obtained (compare **Figure 2**). Finally, the results calculated with the complete oil model $\eta(T,p,\gamma;)$ are shown in **Figure 7**.

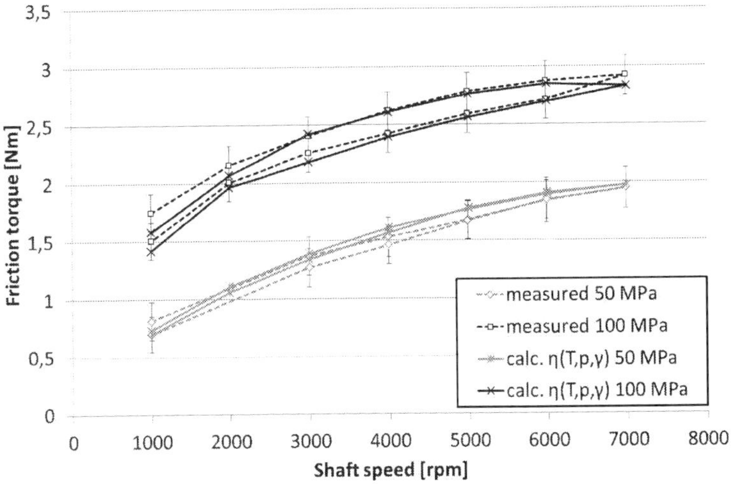

Figure 7. Comparison of the average friction torque measured on the test-rig with the simulated mean friction torque using the complete oil model $\eta(T,p,\gamma;)$ at 50 and 100 MPa specific loads [47].

As the shear rates increase with the shaft speed, the influence of the shear thinning effect is very small at low shaft speeds and becomes increasingly important for higher shaft speeds. At 1000 rpm the maximum shear rate is smaller by a factor of 10 compared to 7000 rpm. As a consequence, the influence of shear thinning on the friction torque is almost negligible for low shaft speed. The effect of shear thinning reduces the calculated friction torque particularly at high shaft speed. Above 5000 rpm, a torque difference of about 0.15 Nm is determined between the two lubricant models $\eta(T,p)$ and $\eta;(T,p,\gamma;)$ independently of the load case.

To summarize the first example, the calculated friction torques considering temperature-, pressure- and shear-rate-dependent viscosity match the measured friction torques within the measurement uncertainty for both the run-down and the run-up. A deviation may be identified at low shaft speeds during the run-up. It is important to note that new journal bearings are used for each test run. And therefore the journal bearing experiences a run-in process in the beginning of its operation. An adaption of the bearing surface in particular at the bearing edges takes place which causes higher friction losses during this time. The simulation considers already run-in bearings by including this adapted surface geometry. Consequently, differences between the measured and simulated results arise. This run-in process happens during the first and second step (1000 and 2000 rpm), when the measured friction torque is accordingly higher compared to the calculated friction torque. The wear process due to the run-in is discussed in more detail in the following example.

4.2. Edge loading and running-in wear in dynamically loaded journal bearings

After the friction tests under dynamic loading (see Section 4.1), the bearing shells were analysed and a worn region was identified at the bearing edges. In Section 4.1, these worn regions are considered in the simulation right from the start instead of a perfect cylindrical bearing surface. In this section, the worn surface geometry is calculated in a stepwise process by iteratively removing material from the initial perfect cylindrical bearing shell.

4.2.1. Running-in caused by edge loading and iterative wear calculation

When two contacting elements operate together for the first time, adaptions of their surfaces occur. These adjustments concern the geometrical conformity on both the macro- and micro-scale as well as changes of the mechanical and material properties [54]. Such an adjustment process takes place in the early stage of operation and is commonly called running-in. In hydrodynamic journal bearings a geometrical unconformity is caused by any misalignment between shaft and bearing shell. This unconformity originates from production tolerances, assembly deformation or thermal deviation. In this study, the applied heavy load bends the shaft elastically which leads to a misalignment and further, to metal-metal contact at the bearing edges. The principal manner is sketched in **Figure 8** on the left. In hydrodynamic journal bearings the softer shell generally adapts its surface to fit the harder journal during the running-in. When the running-in process has finished, metal-metal contact in the lubrication gap vanishes completely.

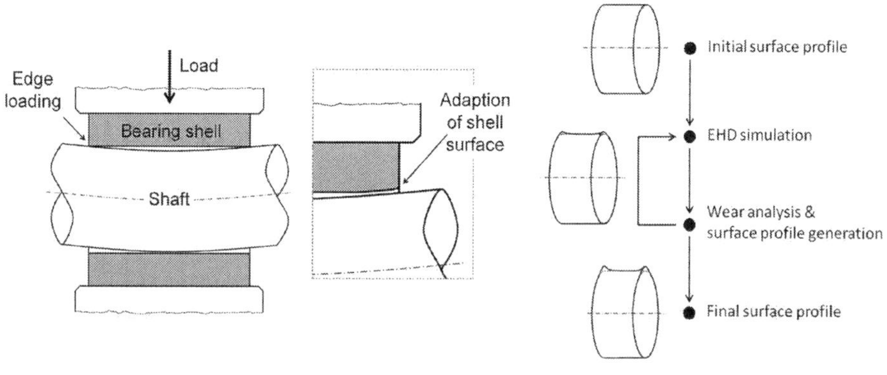

Figure 8. Edge loading caused by elastic bending of the shaft; the flowchart of the iterative surface profile generation on the right [42].

Figure 8 (right) shows the iterative scheme of the wear calculation. By starting from the initial perfectly cylindrical bearing shape of a new bearing, the wear equation from Archard [55] is used to calculate the wear depth for all HD nodes of the discretized journal bearing surface. Archard's wear equation can be written as

$$h_w = \frac{C}{H} \cdot W \cdot L = \frac{C}{H} \cdot t_{Step} \cdot \overline{W_L}, \tag{9}$$

where h_w denotes the wear depth, W the normal load, L the sliding distance, C the wear coefficient and H describes the hardness of the contact surface. The product of normal load and sliding distance can be substituted by a step time t_{Step} multiplied with a wear load W_L. The wear load is the averaged product of asperity contact pressure and relative sliding speed over a load cycle and is different for each (discretized) spot of the bearing surface.

In this wear process a maximum wear depth per iterative step is defined which is generated at the HD node with a maximum wear load. The wear depth of all other HD nodes can then be calculated and a new surface profile is generated. Additionally, the time required to achieve the worn geometry and the overall wear volume can be calculated for every step. With this adapted bearing surface another full elasto-hydrodynamic simulation step is started and the wear calculation is performed again. This process is repeated until the mixed lubrication, or more precisely the maximum asperity contact pressure, becomes insignificantly small. Wear on the journal surface is neglected because of the significantly higher hardness compared to the bearing shell surface.

Central to the wear analysis is the contact model which evaluates the metal-metal contact pressure according to the film thickness in the bearing. To reliably describe the contact pressure, the roughness parameters of the contact model need to be obtained from the real surfaces. Hence, the shaft and bearing shell surfaces are scanned and analysed. For a detailed derivation of the roughness parameters see [42].

4.2.2. Results of the iterative wear simulation

The running-in procedure is calculated for a shaft speed of 4000 rpm and a maximum specific load of 100 MPa (see previous example). Starting point of the wear process is the definition of the initial (cylindrical) journal bearing shape followed by an EHD simulation of a full load cycle. Metal-metal contact is observed when the applied load reaches its maximum. The test bearing in particular shows metal-metal contact at both bearing edges. The symmetrical appearance of the metal-metal contact is caused by the symmetric bearing layout and the according deflection curve of the shaft. **Figure 9** shows the distribution of the calculated metal-metal contact pressure on the left.

Subsequently, the wear load during a full load cycle is calculated and the worn surface geometry is generated. The generated profile is then used as bearing geometry in the subsequent simulation step. As a result, a reduction of the maximum metal-metal contact pressure is obtained. When the procedure is repeated 40 times, as show, it can be observed that the maximum metal-metal contact pressure gradually decreases. The evolution of the maximum contact pressure during the iterative process is shown in **Figure 9** (right). The stepwise pressure reduction is significant in the beginning of the wear process but becomes smaller from step to step. The step time to achieve the defined maximum wear depth per iterative step is also shown in **Figure 10**. The step time monotonically increases from 2.5 to 1050 s.

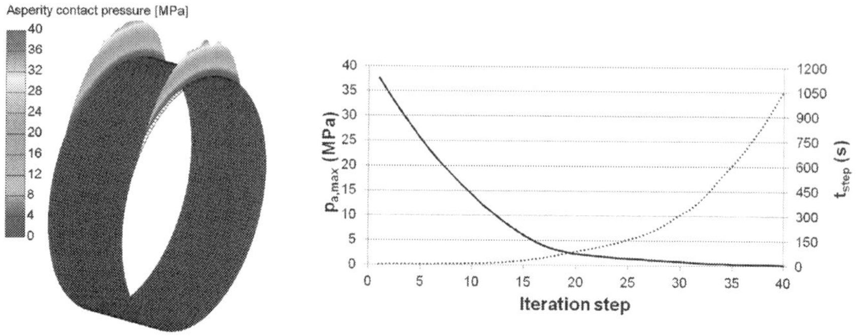

Figure 9. Calculated asperity contact pressure for the first step at the point of maximum load on the left; development of maximum asperity contact pressure and step time over the number of iterative steps [42].

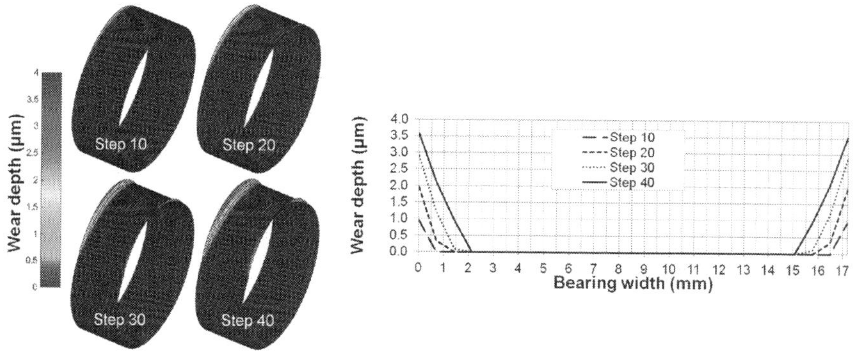

Figure 10. Surface profile on the left and a section view of the surface profile during the iterative running-in process at the position of maximum wear depth on the right [42].

Figure 10 shows the development of the surface profile for the test bearing. Wear concentrates on the edges of the upper bearing shell and shows a symmetrical appearance. With a higher number of iterative steps the worn area expands in axial and circumferential direction. And also the wear depth at the edges increases gradually and reaches a maximum of 3.6 µm after 40 steps.

Finally, the maximum metal-metal contact pressure and the wear volume are plotted over the time in **Figure 11**. Within the first 200 s of operation the maximum contact pressure drops sharply below 5 MPa. From there on the maximum asperity contact pressure steadily decreases to a minimum in the end. The wear volume strongly increases in the beginning and flattens out over time. A steady state of wear has not been identified during the investigated time period.

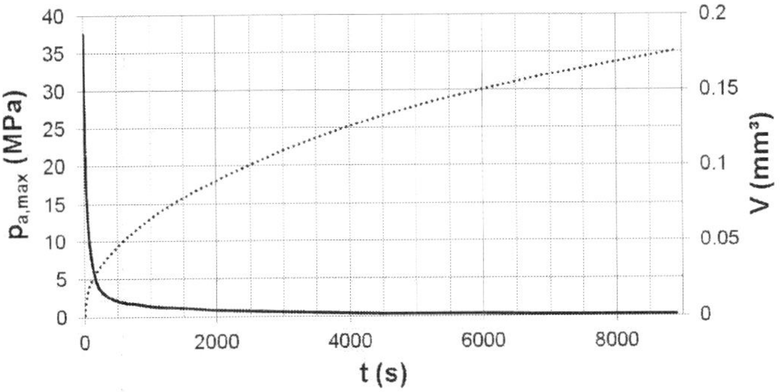

Figure 11. Maximum asperity contact pressure and the accumulated wear volume over time [42].

The quick reduction of asperity contact pressure in the beginning of the bearing operation verifies the statement from the previous example in Section 4.1: at low shaft speeds during the run-in a deviation between the measured and calculated friction torque was identified. This is caused by additional friction losses due to metal-metal contact of the run-in process. Metal-metal contact did not occur in the simulation because an already worn surface geometry was assumed.

In this section, the simulation of the wear process yielded a calculated final surface geometry that is a good match to the worn surface of the test bearing obtained after the dynamic tests. The comparison of wear depth as well as the extent of the worn area shows that the presented method is able to predict the worn surface geometry in highly stressed journal bearings (detailed comparison see [42]).

This section has also shown that the running-in process concentrates on rather small areas of the journal bearings and, therefore, metal-metal contact occurs only locally. The small amount of metal-metal contact only marginally affects the friction losses in the bearing. Hence, the investigated dynamical load is not well suited to analyse friction in the mixed lubrication regime. Consequently, the next section studies the mixed lubrication regime up to severe metal-metal contact.

4.3. Simulation of journal bearing friction in severe mixed lubrication

In the mixed lubrication regime, metal-metal contact occurs and becomes dominant which leads to a drastic rise in friction. Texture and roughness of the bearing surface has a major influence on friction in the mixed lubrication regime. In the simulation, the bearing surface is represented by the contact model which is responsible to reliably predict metal-metal contact pressure as well as contact area.

This section experimentally verifies the relatively simple simulation approach using Greenwood and Tripp contact model in combination with a constant friction coefficient. The input data for the contact model are derived from surface scans of the bearing. Furthermore, the influence of surface roughness on friction is discussed.

4.3.1. *Journal bearing conditions and experimental details*

To ensure the occurrence of metal-metal contact on the bearing test-rig and in contrast to the previous sections, a static load is applied to the test bearing. The tests start at high shaft speed and the shaft speed is slowly reduced to enable the journal bearing to enter the mixed lubrication regime. The results are represented by Stribeck curves that allow an identification of the different lubrication regimes, as shown in **Figure 12**. The Stribeck curve shows the friction torque over the shaft speed. The transition between the hydrodynamic lubrication regime and the mixed lubrication regime can be identified by a strong increase of the friction torque.

Two load cases are investigated in this section, the first with 8 kN static load and the second with a 4 kN load, which correspond to 10 MPa and 5 MPa specific load in the test bearing, respectively. In the beginning of each test, new bearing shells are placed into the test connecting rod and the support brackets. One test run consists of a constant speed-up to 6000 rpm followed by a constant speed-down until the shaft stops to rotate. One test run lasts for 12 min. Only the speed-down is recorded and evaluated.

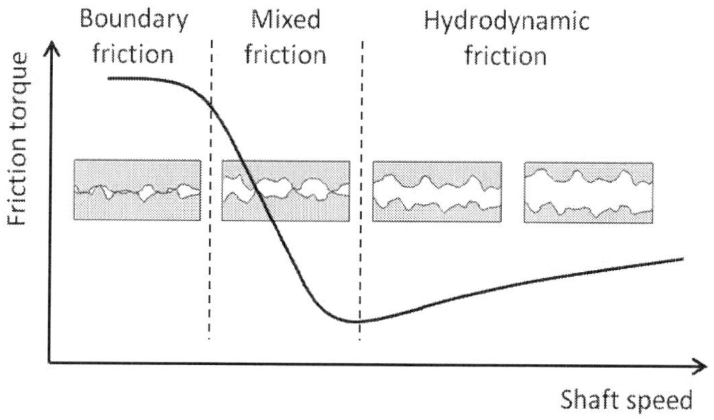

Figure 12. Sketch of a Stribeck curve with different lubrication regimes [43].

Surface	New	Worn
R_a (μm)	0.27	0.22
R_q (μm)	0.34	0.26
σ (μm)	0.28	0.23
δ (μm)	0.39	0.19

Table 3. Surface roughness and simulation input parameters for the contact model.

The bearing surface is scanned by a white light interferometer and the input parameters for the contact model are derived. For evaluating the influence of surface roughness, the bearing shell is measured before the test run and a worn bearing shell is scanned too. The main properties are summarized in **Table 3**. The arithmetic average R_a of the bearing shell surface decreases from 0.27 μm for the new bearing to 0.22 μm for the worn bearing. A similar decrease can be seen for the root mean square R_q, which is 0.34 μm for the new and 0.26 μm for the worn bearing. The asperity summit roughness σ and the mean summit height δ are relevant for the contact model. Additional details about the contact model can be found in [43]. In contrast to the previous example, a cylindrical geometry of the journal bearings is considered in the simulation. The difference between the new and the worn surface only concerns the surface roughness parameters.

Beside the parameters for the contact model, a suitable boundary friction coefficient that considers the presence of friction modifying additives [56] is required to predict friction. Here, the boundary friction coefficient is chosen to be constant μ_{Bound} = 0.02. The limits of this simplification are discussed in the results.

4.3.2. *Verification of the simulation results and the influence of surface roughness*

Figure 13 shows a comparison of the simulation results and the measurement results. On the left the results for the 8 kN load and on the right for 4 kN are presented. The dashed line shows the measurement results and the grey area indicates the uncertainty band of torque and speed measurement.

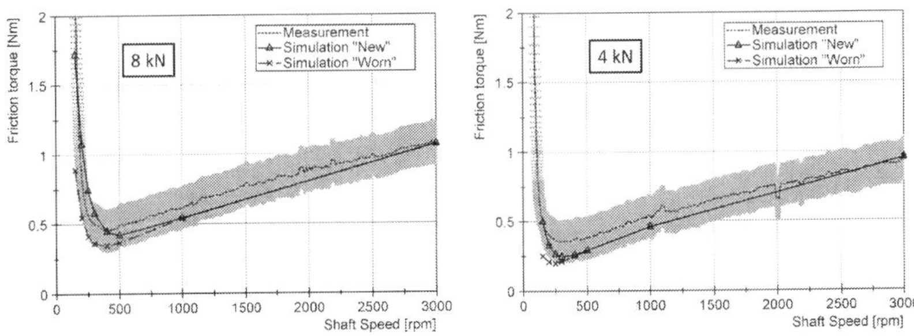

Figure 13. Comparison between the calculated and measured friction torque over shaft speed at a load of 8 and 4 kN [43].

At maximum shaft speed the bearing operates in pure hydrodynamic lubrication regime and a friction torque of about 1.1 Nm arise for the 8 kN load. By reducing the shaft speed the friction torque decreases to a minimum at around 400 rpm. Below 400 rpm the bearing operates in mixed lubrication regime and the friction torque increases abruptly. Similar behaviour can be seen at a load of 4 kN. In the hydrodynamic regime the friction torque is 0.15 Nm lower compared to the 8 kN load and also the transition between hydrodynamic and mixed lubrication regime is shifted to a lower shaft speed.

The two other curves represent the calculated friction torque with the new bearing surface roughness (solid curve) and the worn bearing surface roughness (dashed curve). In the hydrodynamic lubrication regime both surfaces predict identical friction torques. A difference can be seen at low speed, when the new bearing surface predicts an earlier increase of friction torque compared to the worn surface. However, at 8 kN load the results calculated with both surfaces lie within the measurement uncertainty but regarding the transition speed the calculation with the worn surface appears to be more suitable. At 4 kN the results with the new surface match the transition speed.

It can be expected that a modification of the surface roughness occurs at the beginning of the operation. As mentioned earlier, a speed-up is performed before the measurements take place. During the speed-up metal-metal contact already occurs and the surface may smoothen already. For a higher load more metal-metal contact develops and a quicker surface modification can be expected. The quicker adaption at 8 kN load eventually explains the good match of measurement and simulation with the worn surface roughness.

The comparison of measurement and simulation shows a good agreement in hydrodynamic lubrication regime and mixed lubrication regime. Hence, the presented roughness data observed from surface scans in combination with the simple contact model are suitable for the friction prediction in journal bearings operation under severe conditions. Also, the chosen constant friction coefficient μ_{Bound} is suitable for both loads to predict friction with a great accuracy. Only at a very low shaft speed when the bearing operates in boundary friction regime, the friction torque is underestimated. Here, a higher boundary friction coefficient would be more applicable.

4.4. Friction analysis of journal bearings during starting and stopping

A transient start-stop simulation is performed in the final example. The journal rests in the bearing in the beginning of the simulation. When the journal starts to rotate the breakaway torque has to be overcome. Afterwards, the bearing operates in mixed lubrication regime until the lift-off speed is reached.

4.4.1. Experimental details and journal bearing resting conditions

A static load of 2.5 kN is applied to the test connecting rod. The resting test shaft is constantly accelerated to a maximum speed of about 260 rpm. The duration of the speed-up is 3 s. Subsequently, the shaft is constantly decelerated over 3 s until the shaft stops. This start-stop cycle was repeated 6000 times. In between each cycle was a recovery phase of 3 s.

The simulation model from previous examples is modified in a way that the shaft follows a predefined rotation to analyse the start-stop behaviour. The parameters for the contact model and the boundary friction coefficient remain unchanged. However, before the shaft starts to rotate a static condition needs to be established. Therefore, an initialization phase is included where the static load is gradually applied to the bearing and the shaft moves slowly towards the bearing shell until it reaches its resting position. **Figure 14** (left) shows the bearing force and the minimum film height in the lubrication gap.

During the first 0.1 s, the static load is constantly increased until 2.5 kN is reached. At the beginning of the initialisation phase, the shaft is located at the centre of the bearing. As the load increases the shaft subsides within the bearing and the minimum radial distance between shaft and bearing (minimum film height) decrease. After 0.3 s, the minimum distance is reached. At resting position, the journal is mainly supported by metal-metal contact which can be seen in **Figure 14** (right). While the test bearing shows an equally distributed contact pressure all along the bearing axis, the support bearing show an asymmetric distribution. This asymmetric distribution is caused by the elastic deformation of the shaft due to the static load (deflection curve). The actual starting of the shaft happens after 0.5 s. The shaft is then accelerated and decelerated again until the shaft completely stops to rotate.

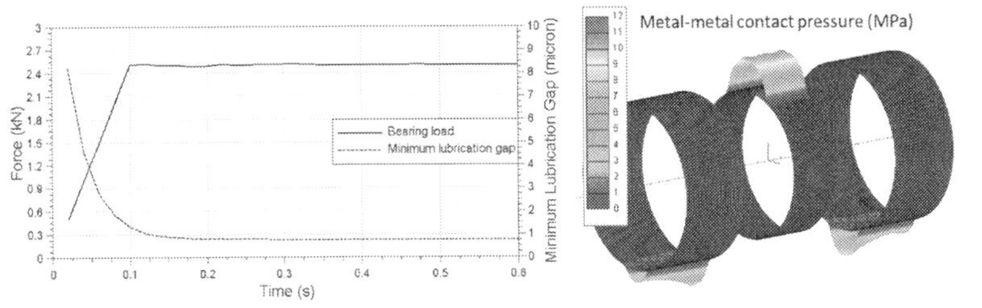

Figure 14. Load of the test bearing and minimum film height during the initial phase of the start-stop cycle on the left; metal-metal contact pressure distribution after 0.5 s in the test bearing and the two support bearings [53].

4.4.2. Verification of the simulation results and the influence of surface roughness

The measurement and simulation results are summarized in **Figure 15**. The dashed black curve shows the rotational speed of the shaft; it reaches a top speed of 260 rpm. The solid black curve represents the measured torque. The results were taken from the 3000th cycle and, therefore, a running-in of the bearing has already occurred. Shortly before the shaft begins to rotate the torque abruptly increases to a maximum. This peak, or breakaway torque, indicates the transition between static and sliding friction. Once in motion the bearing enters the mixed lubrication regime and the friction torque decreases quickly. After 1.5 s, the friction torque becomes a minimum as the bearings enter the hydrodynamic lubrication regime. At stopping the measured torque does not show any increase until 6.2 s when the recording of the data ended. The measured signal shows strong fluctuations which are caused by the dynamic behaviour of the test-rig. It can be expected that the impulse at the breakaway cause this strong fluctuations. Bouyer and Fillon [57] observed a similar behaviour and referred to the slip-stick phenomenon.

The solid grey curve shows the calculated torque considering a new bearing surface. When the shaft starts to rotate, it raises identically with the measured friction torque. Also the magnitude of the breakaway torque is equal to the measured one. After the breakaway both curves drop parallel as they reach mixed lubrication. The hydrodynamic regime is reached after 2 s. In the end of the start-stop cycle the simulation predicts an increase of friction torque before the shaft completely stops. The fluctuations are not present in the simulation because the rotational speed of the shaft is specified.

Similar to the previous result section, the influence of surface roughness is analysed by simulation. The friction torque calculated with a worn surface roughness is shown as dashed grey line in **Figure 15**. The breakaway torque remains unchanged compared to the results with the new bearing surface because at rest the shaft is completely supported by metal-metal contact and, hence, the breakaway torque is only defined by boundary friction coefficient, μ_{Bound}. In the mixed lubrication regime, the friction torque reduces quickly and pure hydrodynamic lubrication is reached after 1.5 s. At top speed both surface roughnesses yield a similar torque in the simulation. A major difference can be seen in the behaviour during the slowdown and the stopping of the shaft. The calculation with the worn bearing surface roughness shows a clear reduction of maximum friction torque which also fits better with the measured friction torque.

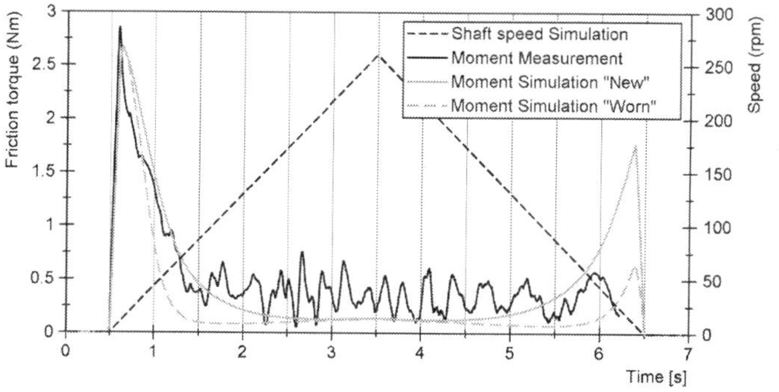

Figure 15. Measured and calculated friction torque during a start-stop cycle; results from simulation are obtained for new and worn bearing surface roughness [53].

Overall, the calculated friction torque follows the starting and stopping behaviour of the shaft in the bearing very closely. Especially the breakaway torque and the transitions between the boundary and mixed lubrication regime as well as between the mixed and hydrodynamic lubrication regime are identified. The model considering the new bearing surface roughness overestimates the friction torque during stopping of the shaft.

5. Conclusion and outlook

This chapter analyses the behaviour of automotive journal bearings under severe loading conditions. The severe loading conditions force the bearing to operate in the boundary, mixed and elasto-hydrodynamic lubrication regime. Therefore, an extensive simulation approach is employed which considers elastic deformation of the components, the rheological properties of the lubricant under high pressure and at high shear rate, and a contact model to calculate metal-metal contact. Both the lubricant properties and the surface roughness parameters for the contact model are experimentally identified. The simulation approach is comprehensively validated on the basis of measurements on a journal bearing test-rig.

The first example shows the importance of considering the piezo-viscous and the non-Newtonian effect of the lubricant to reliably predict friction in highly loaded journal bearings. If both the effects are neglected, a discrepancy of up to 15% can occur in comparison to the measured results for the studied operating conditions. The second example discusses the running-in behaviour of misaligned journal bearings which is caused by the elastic bending of the shaft. Therefore, an iterative simulation approach is presented which removes material from the bearing shell step by step. Friction in mixed lubrication is analysed in the third example where the Stribeck curves are calculated and compared to the measured results. The influence of surface roughness on mixed friction is investigated. Finally, a transient start-stop cycle is calculated in the fourth example. Before the shaft starts to rotate, a breakaway torque has to be surmounted. The bearing shortly operates in mixed lubrication before the contacting surfaces are completely separated. Especially at the stopping of the shaft an influence of surface roughness can be identified.

Future challenges for journal bearing simulation include the precise wear prediction and the determination of wear coefficients for different bearing materials. In this context, the consideration of polymeric sliding layers in the simulation presents a new challenge for the future. While mixed lubrication can be described numerically very accurately as shown in this chapter, its escalation up to bearing failure is still beyond the scope of the presented work although relative estimations are enabled in a comparative sense based on the results. However, the numerical prediction of bearing failure requires a deeper understanding of the thermal processes in the journal bearing, which will be the subject of future work.

Acknowledgements

This work was supported by the "COMET-Competence Centers for Excellent Technologies Programme" of the Austrian Federal Ministry for Transport, Innovation and Technology (bmvit); the Austrian Federal Ministry of Science, Research and Economy (bmwfw); the Austrian Research Promotion Agency (FFG); and the Province of Styria and the Styrian Business Promotion Agency (SFG). Furthermore, we acknowledge the partial financial support provided by the Austrian Science Fund (FWF) (P27806-N30).

Author details

David E. Sander*, Hannes Allmaier and Hans-Herwig Priebsch

*Address all correspondence to: david.sander@v2c2.at

Virtual Vehicle Research Center, Graz, Austria

References

[1] Allmaier H, Offner G. Current challenges and frontiers for the EHD simulation of journal bearings: a review. SAE Technical Paper. 2016;2016-01-1856..

[2] Damm K, Pucher K, Skiadas A, Witt M. Sputter bearings for highly charged diesel engines. MTZ Worldwide. 2015;76(5):26–30.

[3] Bishop J, Nedungadi A, Ostrowski G, Surampudi B, et al. An engine start/stop system for improved fuel economy. SAE Technical Paper. 2007; 2007-01-1777. doi: 10.4271/2007-01-1777

[4] Silva C, Ross M, Farias T. Analysis and simulation of "low-cost" strategies to reduce fuel consumption and emissions in conventional gasoline light-duty vehicles. Energy Conversion and Management. 2009;50(2):215–222.

[5] Fonseca N, Casanova J, Valdés M. Influence of the stop/start system on CO2 emissions of a diesel vehicle in urban traffic. Transportation Research Part D: Transport and Environment. 2011;16(2):194–200.

[6] Mokhtar MOA, Howarth RB, Davies PB. Wear characteristics of plain hydrodynamic journal bearings during repeated starting and stopping. Tribology Transactions. 1977;20(3):191–194. doi:10.1080/05698197708982833

[7] Wilcutts M, Switkes J, Shost M, Tripathi A. Design and benefits of dynamic skip fire strategies for cylinder deactivated engines. SAE Technical Paper. 2013; 2013-01-0359. doi:10.4271/2013-01-0359

[8] Mohammadpour M, Rahmani R, Rahnejat H. The effect of cylinder de-activation on thermo-friction characteristics of the connecting rod bearing in the New European Drive Cycle (NEDC). SAE Technical Paper. 2014; 2014-01-2089.

[9] Mohammadpour M, Rahmani R, Rahnejat H. Effect of cylinder deactivation on the tribo-dynamics and acoustic emission of overlay big end bearings. Proceedings of the Institution of Mechanical Engineers, Part K: Journal of Multi-body Dynamics. 2014;228(2):138–151.

[10] Shahmohamadi H, Rahmani R, Rahnejat H, Garner CP, Dowson D. Big end bearing losses with thermal cavitation flow under cylinder deactivation. Tribology Letters. 2015;57(2):1–17. doi:10.1007/s11249-014-0444-7

[11] Holmberg K, Andersson P, Nylund NO, Mäkelä K, Erdemir A. Global energy consumption due to friction in trucks and buses. Tribology International. 2014;78:94–114.

[12] Covitch M, Brown M, May C, Selby T. Extending SAE J300 to viscosity grades below SAE 20. SAE Technical Paper. 2010; 2010-01-2286.

[13] Ligier JL, Noel B. Friction reduction and reliability for engines bearings. Lubricants. 2015;3:569–596.

[14] Macián V, Tormos B, Bermúdez V, Ramírez L. Assessment of the effect of low viscosity oils usage on a light duty diesel engine fuel consumption in stationary and transient conditions. Tribology International. 2014;79:132–139.

[15] Macián V, Tormos B, Ruiz S, Miró G. Low viscosity engine oils: study of wear effects and oil key parameters in a heavy duty engine fleet test. Tribology International. 2016;94:240–248.

[16] Allmaier H, Priestner C, Reich FM, Priebsch HH, Forstner C, Novotny-Farkas F. Predicting friction reliably and accurately in journal bearings—the importance of extensive oil-models. Tribology International. 2012;48:93–101.

[17] Knauder C, Allmaier H, Sander DE, Salhofer S, Reich FM, Sams T. Analysis of the journal bearing friction losses in a heavy-duty diesel engine. Lubricants. 2015;3:142–154.

[18] Bair S. Measurements of real non-Newtonian response for liquid lubricants under moderate pressures. Proceedings of the Institution of Mechanical Engineers, Part J: Journal of Engineering Tribology. 2001;215(3):223–233.

[19] Bair S. The shear rheology of thin compressed liquid films. Proceedings of the Institution of Mechanical Engineers, Part J: Journal of Engineering Tribology. 2002;216(1):1–17.

[20] Bair S. High pressure rheology for quantitative elastohydrodynamics. Elsevier Science, Amsterdam; 2007.

[21] Costa HL, Hutchings IM. Some innovative surface texturing techniques for tribological purposes. Proceedings of the Institution of Mechanical Engineers, Part J: Journal of Engineering Tribology. 2015;229:429–448. doi: 10.1177/1350650114539936

[22] Lu X, Khonsari MM. An experimental investigation of dimple effect on the Stribeck curve of journal bearings. Tribology Letters. 2007;27:169–176.

[23] Henry Y, Bouyer J, Fillon M. An experimental analysis of the hydrodynamic contribution of textured thrust bearings during steady-state operation: a comparison with the untextured parallel surface configuration. Proceedings of the Institution of Mechanical Engineers, Part J: Journal of Engineering Tribology. 2015;229:362–375. doi: 10.1177/1350650114537484

[24] Dadouche A, Conlon M, Dmochowski W, Koszela W, Galda L, Pawlus P. Experimental evaluation of steady-state and dynamic performance of hydrodynamic journal bearings: plain vs. textured surface. In: 10th EDF/Prime workshop; Poitiers. 2011.

[25] Dadouche A, Conlon MJ. Operational performance of textured journal bearings lubricated with a contaminated fluid. Tribology International. 2016;93:377–389.

[26] Gropper D, Wang L, Harvey TJ. Hydrodynamic lubrication of textured surfaces: a review of modeling techniques and key findings. Tribology International. 2016;94:509–529.

[27] Braun MJ, Hannon WM. Cavitation formation and modelling for fluid film bearings: a review. Proceedings of the Institution of Mechanical Engineers, Part J: Journal of Engineering Tribology. 2010;224:839–863. doi:10.1243/13506501JET772

[28] Qiu Y, Khonsari MM. Experimental investigation of tribological performance of laser textured stainless steel rings. Tribology International. 2011;44:635–644.

[29] Zhang J, Meng Y. Direct observation of cavitation phenomenon and hydrodynamic lubrication analysis of textured surfaces. Tribology Letters. 2012;46:147–158.

[30] George J, Brock R. Polymeric engine bearings for hybrid and start stop applications. SAE Technical Paper. 2012; doi:10.4271/2012-01-1966

[31] Ferreira M, Silva A, Praça M, Costa S. Polymeric coated lead free bronze bearings for high durability in medium duty diesel engines. SAE Technical Paper. 2014; doi: 10.4271/2014-36-0405

[32] Gudin D, Mian O, Sanders S. Experimental measurement and modelling of plain bearing wear in start–stop applications. Proceedings of the Institution of Mechanical Engineers, Part J: Journal of Engineering Tribology. 2013;227(5):433–446. doi: 10.1177/1350650112471287

[33] Offner G, Knaus O. A generic friction model for radial slider bearing simulation considering elastic and plastic deformation. Lubricants. 2015;3:522–538.

[34] Bobzin K, Brögelmann T. Minimizing frictional losses in crankshaft bearings of automobile powertrain by diamond-like carbon coatings under elastohydrodynamic lubrication. Surface & Coatings Technology. 2015;290:100–109.

[35] Allmaier H, Priestner C, Six C, Priebsch HH, Forstner C, Novotny-Farkas F. Predicting friction reliably and accurately in journal bearings—a systematic validation of simulation results with experimental measurements. Tribology International. 2011;44:1151–1160.

[36] Priestner C, Allmaier H, Priebsch HH, Forstner C. Refined simulation of friction power loss in crank shaft slider bearings considering wear in the mixed lubrication regime. Tribology International. 2012;46(1):200–207. 10.1016/j.triboint.2011.03.012

[37] Patir N, Cheng H. An average flow model for determining effects of three-dimensional roughness on partial hydrodynamic lubrication. ASME Journal of Lubrication Technology. 1978;100:12–17.

[38] Patir N, Cheng H. Application of average flow model to lubrication between rough sliding surfaces. ASME Journal of Lubrication Technology. 1979;101(2):220–229.

[39] Jakobsson B, Floberg L. The finite journal bearing, considering vaporization. Gumperts Förlag. 1957.

[40] Krasser J. Thermoelastohydrodynamische Analyse dynamisch belasteter Radialgleitlager [dissertation]. Graz University of Technology; 1996.

[41] Greenwood J, Tripp J. The contact of two nominally flat rough surfaces. Proceedings of the Institution of Mechanical Engineers. 1970;185(1):625–633.

[42] Sander DE, Allmaier H, Priebsch HH, Reich FM, Witt M, Skiadas A, Knaus O. Edge loading and running-in wear in dynamically loaded journal bearings. Tribology International. 2015;92:395–403.

[43] Sander DE, Allmaier H, Priebsch HH, Witt M, Skiadas A. Simulation of journal bearing friction in severe mixed lubrication – validation and effect of surface smoothing due to running-in. Tribology International. 2016;96:173–183.

[44] Vogel H. The law of the relation between the viscosity of liquids and the temperature. Physikalische Zeitschrift. 1921;22:645-646.

[45] Barus C. Isothermals, isopiestics and isometrics relative to viscosity. American Journal of Science. 1893;266:87-96.

[46] Cross MM. Rheology of non-Newtonian fluids: a new flow equation for pseudoplastic systems. Journal of Colloid Science. 1965;20(5):417–437.

[47] Sander DE, Allmaier H, Priebsch HH, Reich FM, Witt M, Füllenbach T, Skiadas A, Brouwer L, Schwarze H. Impact of high pressure and shear thinning on journal bearing friction. Tribology International. 2015;81:29–37.

[48] Allmaier H, Sander DE, Priebsch HH, Witt M, Füllenbach T, Skiadas A. Non-Newtonian and running-in wear effects in journal bearings operating under mixed lubrication. Proceedings of the Institution of Mechanical Engineers, Part J: Journal of Engineering Tribology. 2015;230(2):135–142.

[49] Damm K, Skiadas A, Witt M, Schwarze H. Gleitlagererprobung Anhand der Forderungen des Automobilmarkts. ATZextra. 2010;15(1):54–63.

[50] Offner G. Friction power loss simulation of internal combustion engines considering mixed lubricated radial slider, axial slider and piston to liner contacts. Tribology Transactions. 2013;56(3):503–515.

[51] Allmaier H, Priestner C, Reich FM, Priebsch HH , Novotny-Farkas F. Predicting friction reliably and accurately in journal bearings-extending the simulation model to TEHD. Tribology International. 2013;58:20–28.

[52] Allmaier H, Priestner C, Sander DE, Reich F. Friction in automotive engines. In: H Pihtili, editor. Tribology in engineering. Intech; 2013. p. 149–184. http://dx.doi.org/10.5772/51568

[53] Sander DE. A validated elasto-hydrodynamic simulation for journal bearings operating in severe conditions [unpublished doctoral dissertation]. Graz, Austria: Graz University of Technology; 2016.

[54] Blau PJ. On the nature of running-in. Tribology International. 2006;38(11):1007–1012.

[55] Archard J. Contact and rubbing of flat surfaces. Journal of Applied Physics. 1953;24(8): 981–988.

[56] Morina A, Neville A. Understanding the composition and low friction tribofilm formation/removal in boundary lubrication. Tribology International. 2007;40:1696–1704.

[57] Bouyer J, Fillon M. Experimental measurement of the friction torque on hydrodynamic plain journal bearings during start-up. Tribology International. 2011;44:772–781.

Section 3

Lubrication Technology

Chapter 8

The Tribology and Chemistry of Phosphorus-Containing Lubricant Additives

David W. Johnson

Additional information is available at the end of the chapter

http://dx.doi.org/10.5772/62208

Abstract

Phosphate esters, thiophosphate esters and metal thiophosphates have been used as lubricant additives for over 50 years. Recently, phosphorus-containing ionic liquids have emerged as a new class of lubricant additives. While the use of phosphorus compounds has been extensive, a detailed knowledge of how they work has been a much more recent development. In this chapter, the use of phosphate esters, thiophosphate esters, metal thiophosphates and phosphorus-containing ionic liquids as antiwear or extreme pressure additives is discussed. The primary emphasis will be on how they form a protective film, which is both durable and reduces friction. The first part of the chapter discusses the use of alkyl phosphates, triaryl phosphates and metal-containing thiophosphate esters with primarily iron- and steel-based bearing materials. The second part of the chapter examines phosphorus-containing ionic liquids and the challenges posed by new bearing materials with different surface chemistries.

Keywords: phosphate ester, thiophosphates, ZDDP, zinc dialkyldithiophosphate, anti-wear additive, turbine engines, refrigeration

1. Introduction

Lubricant additives are a critical part of all lubricant systems because they impart properties to the formulated lubricant that the basestock does not possess. Additives allow longer lubricant lifetimes by eliminating certain modes of basestock decomposition. For example, when a lubricant is used, it is exposed to temperatures much higher than the bulk oil temperature for short periods of time. This occurs in the presence of oxygen, which can cause oxidation of the lubricant. Additives can reduce the decomposition of the basestock by scavenging free radicals formed in the initial stages of the reaction. Aircraft lubricants typically contain the following groups of additives:

- Boundary lubrication additives (antiwear, lubricity and extreme pressure)
- Antioxidants
- Anticorrosion additives
- Antifoaming additives
- Metal deactivators
- Viscosity index improvers

Choices of additives are frequently limited based on the thermal stability of the additive, needs for long lifetimes and tendencies of the additives to form deposits [1].

Organophosphorus compounds are found in a range of formal oxidation states from -3 to +5. Of these oxidation states, compounds in the -3, +3 and +5 oxidation states have particularly important industrial applications [2]. The structures of some of these compounds are shown in **Figure 1**.

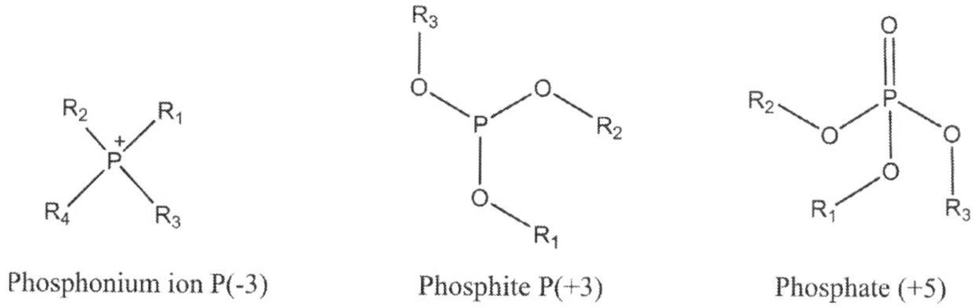

Figure 1. Phosphorus compounds in various oxidation states.

The thermal and oxidative stability of phosphorus compounds, particularly phosphorus in the +5 oxidation state, has led to a number of important industrial applications [3]. The structure of phosphates, for example (**Figure 1**), shows a fully oxidized phosphorus at the center of the molecule. The possibility of exchanging the oxygen atom for either a sulfur or a nitrogen atom greatly increases the variety of compounds as does the ability to include various organic groups attached to the oxygen, nitrogen or sulfur atoms. While organophosphites and organophosphonium ions have important applications, by far the phosphates have found the greatest use.

Initial industrial applications of organic phosphates include fire retardants, especially for the film industry [4] and environmentally degradable pesticides due to their toxicity to insects, but not animals and their relatively high reactivity in the environment [5]. Phosphate esters have been used as surfactants and as brightening agents in detergents [6]. Phosphate esters with a variety of structures and reactivities are commercially available and have found important industrial application. Commonly available compounds include aromatic phosphate esters, alkyl phosphate esters, thiophosphates and a number of metal-containing phosphate esters, including zinc dialkyldithiophosphates (ZDDP) [7].

The development of lubricant additives began in response to the need for higher performance lubricants. Many of the initial hydrocarbon lubricating oils contained sulfur and phosphorus-containing species that in hindsight were naturally occurring antioxidants and antiwear agents. These naturally occurring additives however varied substantially as the crude oil sources changed. When it was initially suggested that synthetic additives should be added, they were regarded as an expensive component of unknown use. As time progressed, phosphorus-containing additives became a required part of formulated lubricants in order to meet more stringent specifications [8]. The purpose of this chapter is to review the use of the various types of phosphate esters as lubricant additives and to illustrate what is known about how they function as lubricant additives.

2. Zinc dialkyldithiophosphates and other metal dithiophosphates

Lubricant additives containing zinc, phosphorus and sulfur originate with a series of patents filed in the 1940s by Lubrizol [9], American Cyanamid [10] and Union Oil [11] claiming a composition that functioned as an antioxidant and a corrosion inhibitor. These additives, now known as ZDDPs, were originally prepared by the reaction of phosphorus pentasulfide (P_4S_{10}) with one or more alcohols to give the appropriate dialkyldithiophosphoric acid. The acid was then neutralized by the addition of zinc oxide to give the product as shown in **Figure 2** [12].

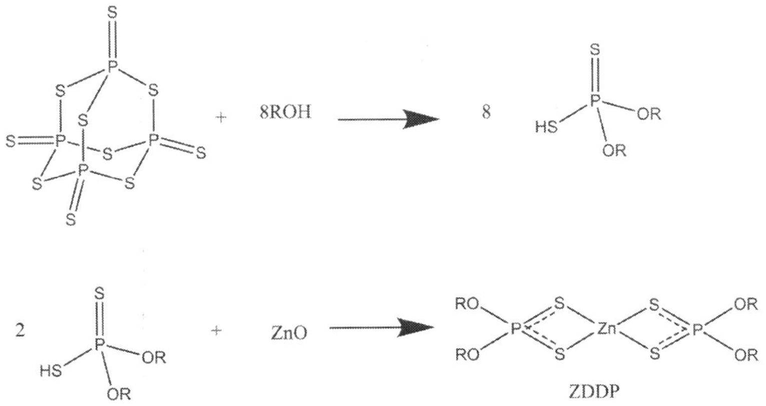

Figure 2. Method for the preparation of ZDDP.

ZDDPs have been prepared with a wide variety of alcohols in order to tune physical properties, including oil solubility and viscosity and chemical reactivity of the resultant additive. In addition, the chemical properties can be adjusted by altering the ratio of zinc oxide to the dialkyldithiophosphoric acid resulting in acidic, neutral and basic ZDDPs [13].

ZDDP was initially applied in industrial oils as an antioxidant. ZDDP appears to react rapidly with peroxy radicals in the oils [14]. In 1955, it was shown that ZDDP could also act as an antiwear additive [15], which led to its primary use in the automotive industry [16]. The use of ZDDP in aerospace and refrigeration industries did not develop because of their need for an ashless additive which made ZDDP unsuitable [17].

2.1. The structure of ZDDP in solution

The interaction of ZDDP with metal surfaces is closely related to its structure in solution. In solution, ZDDP can act as an antioxidant [18], increasing its utility, but the primary use is as an antiwear additive. In solution, acidic and neutral ZDDP exists as an equilibrium between a monomeric and a dimeric form, with the dimer being dominant in toluene or chloroform solution [19]. Quantum mechanical models have suggested a mechanism for the interconversion with two different transition states. The first transition state contains a four-membered ring with two bridging sulfur atoms and the second contains a six-membered ring which leads to the final product that contains an eight-membered ring [20]. Basic ZDDP, in contrast to acidic and neutral ZDDP, incorporates excess zinc oxide into a structure that contains three ZDDP molecules and a zinc oxide [21]. Polymeric forms are observed in the solid state [22], and larger molecules are observed in solution by dynamic light scattering [23]. The structures of these forms are included in **Figure 3**.

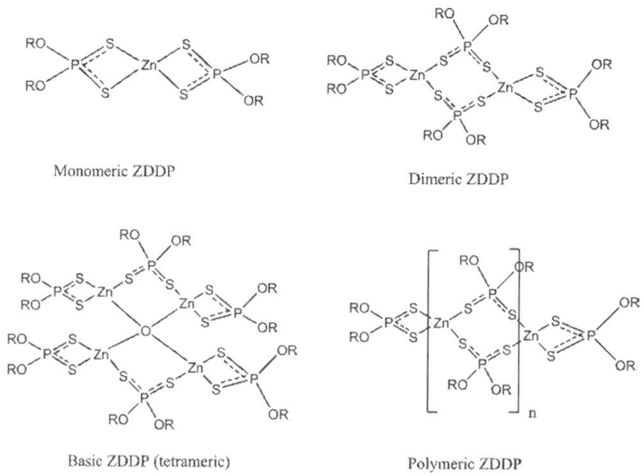

Figure 3. Structures of some of the observed forms of ZDDP.

While there are several structures observed in low-temperature solution, it appears that at high temperatures, basic ZDDP converts to the same structure as neutral ZDDP and gives identical wear test performance [24]. It is also thought that oxidation of the thiophosphate to phosphate is a part of the degradation mechanism [25].

2.2. The reactions of ZDDP at metal surfaces

All antiwear additives operate by forming a protective film on the bearing surface. That film acts to protect the surface during periods of elastohydrodynamic and boundary lubrication when asperity-asperity contact is a major cause of wear. The same film also protects the surface during start-up when there is no lubricant flowing through the system [26]. The composition of the surface film, how it forms and how it can be modified, has been the object of a great deal of research.

An initial model for the species found on metal surfaces begins with the study of the solid state structures of ZDDP. Several structures have been observed, including polymeric [27] and binuclear [28] structures depending on the alkyl groups present in the ZDDP. Solid state nuclear magnetic resonance (NMR) studies have indicated that structures not seen crystallographically exist and that these structures differ from the solution structure [29]. X-ray absorption near-edge spectroscopy (XANES) and 31-P NMR indicated the formation of a surface film and even when there was no phosphorus remaining in solution, the film continued to form. This is an indication that the decomposition of a species in solution precedes the formation of the film [30].

Metal coupons, when they are immersed in a solution of ZDDP dissolved in a lubricant or nonpolar solvent rapidly form a thermal film on the surface of the coupon. Analysis of the films by several techniques indicates that they are not simple zinc or iron compounds such as zinc phosphate, zinc phosphide and iron phosphate [31]. The thermally generated films have similar compositions to the coatings formed inside tribological junctions. Wear reduction is found to be the greatest when a thermal film is formed first, followed by friction between the surfaces (i.e., rubbing motions of a pin on disk) in a ZDDP containing oil to generate the tribofilm [32]. The outermost layer of the thermally generated film is zinc polyphosphate, which is converted to zinc pyrophosphate or zinc orthophosphate deeper in the film [33]. The tribofilms differ from the thermal films because a significant amount of the zinc is replaced by iron [34].

A number of different techniques have been used to study the tribofilms formed by ZDDP. Studies have demonstrated that phosphate must be present in the oil for a wear-resistant surface and once the phosphate is depleted, sulfur causes corrosion inside the wear scar [35]. ZDDP has also been shown to stop the reduction of surface metal oxides the lubricant, forming an easily sheared layer that prevents direct contact between the two surfaces [36]. A commonly proposed mechanism for the reduction of wear involves the formation of pads that cover most of the surface of the bearing. The formation of a linkage isomer [37] that ultimately binds to the metal surface [38] is the initial step, followed by a series of reactions that eliminates the majority of the alkyl substituents and most of the sulfur. These reactions leave a deposit of zinc polyphosphate at the surface. Iron diffusion into this layer gives a mixed zinc polyphosphate/iron polyphosphate layer with increased iron closer to the surface [39]. The pads appear to be bound to the surface through a thin layer of mixed zinc and iron sulfide [40]. The film thickness eventually stabilizes at a thickness of 40–150 nm [41]. The film is shown schematically in **Figure 4**. The film itself has the interesting property of responding to increased pressure through an increase in the modulus of elasticity, which is referred to as smart material behavior [42].

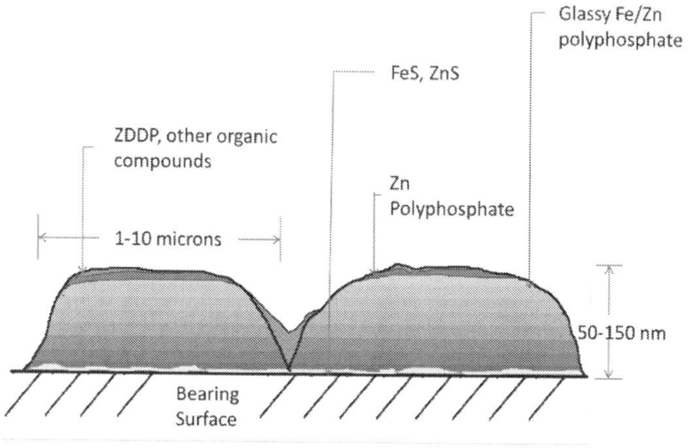

Figure 4. Layered structure of surface film formed from ZDDP.

3. Dithiophosphate esters and related materials

Current environmental standards have significantly reduced the allowed particulate emissions for diesel engine. Metal containing additives frequently clog filters and phosphorus additives can poison catalysts. Lubricants and lubricant additives have been identified as a possible contributor to metals, phosphorus and sulfur in the exhaust stream. In order to reduce particulate emissions, but still take advantage of the advantages of ZDDP like additives, a number of compounds based on thiophosphate and phosphorothionate compounds have been developed as ashless lubricant additives [43]. The structures of some of the ashless thiophosphate-based additives are shown in **Figure 5**.

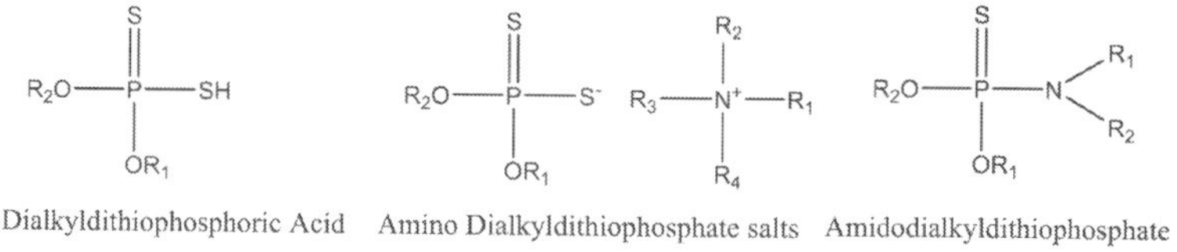

Figure 5. Structures of some ashless dithiophosphate additives.

A number of lubricant additives have been developed for automotive applications based on O,O'-dialkyldithiophosphoric acids. These compounds are simply ZDDP without the zinc. A major concern with these additives is the acidity of the S-H group. In order to reduce the total acid number of the additive, the compound can be neutralized with a quaternary ammonium hydroxide forming the ammonium salt [44]. The degree of substitution on the ammonium ion gives good solubility in the oil and leads to potential for property optimization. These amine phosphate additives, actually a type of ionic liquid discussed later, offer some of the highest load carrying capabilities, however suffer from reduced hydrolytic stability and higher resulting acidity during use compared to other additives [45].

Another common class of additives is the amidothiophosphates. They are prepared by the addition of amines to an appropriate thiophosphoryl chloride [46]. Other additives are prepared by the addition of sulfur across the double bond of acrylic acid or an acrylate ester [47]. An alternate approach is to use hydrogen peroxide to oxidize and then dimerize the alkyldithiophosphate [48].

Thermal films form from dialkyldithiophosphates at about 150°C. The films consist of a mixture of iron (II) polyphosphate and iron (II) sulfate [49]. Under tribological conditions, the ashless compounds form thicker films and give better overall performance than ZDDP under the same conditions. The ashless additives also gave shorter chain polyphosphates than ZDDP, but an average film thickness of 400 nm [50].

Triaryl phosphorothionates are another class of molecules that have received significant attention as antiwear additives. These molecules are structurally similar to the phosphate esters, except a doubly bound sulfur atom replaces the oxygen atom. Triphenyl phosphorothionate, the parent compound in the series, has received the most study. In solution, triphenyl phosphorothionate is oxidized thermally to form triphenyl phosphate which reacts rapidly with the bearing surface. There appear to be no solid products formed from this reaction [51]. In the presence of iron or steel, a metal catalyzed reaction converts triphenylphosphorothionate to triphenyl phosphate, and the decomposition of triphenylphosphate results in the production of a multilayered solid film on iron or iron oxide [52]. The wear debris analyzed by transmission electron microscopy was found to include a mixture of Fe_3O_4 particles and amorphous material. A mechanism for the reduction of wear included solid particles trapped within the tribofilm. Antiwear additives that produced a film with the smallest number of iron oxide particles resulted in the greatest reduction of wear [53].

4. Phosphate esters

The initial application of phosphate esters as lubricant additives was more than 60 years ago. While initially developed as antioxidants for aircraft engines, phosphate esters have found increasing use in both automobile engines [54] and refrigeration compressors [55]. Trialkyl phosphates were initially preferred because of their high reactivity [56], but the reactivity also limited their use to moderate temperatures and mild lubrication conditions [57]. Triaryl phosphates and particularly tricresyl phosphate have been known to reduce friction and wear under boundary lubrication conditions since the 1940s [58]. With many applications, especially aviation, moving to synthetic basestocks, the need for effective antiwear additives increased.

In the case of neopentyl polyol basestocks, the formation of a surface film was more important because its presence also reduced the reactivity of the basestock [59]. Although concerns about the volatility and toxicity of the triaryl phosphates have come to light, triaryl phosphates continue to be used. These concerns however have brought about the development of more highly substituted phosphate esters such as butylated triphenyl phosphate as a replacement for tricresyl phosphate. The larger alkyl groups decrease the volatility of the phosphate ester and also make the formation of the *ortho* isomer more difficult.

4.1. Mechanism for the modification of metal surfaces by phosphate esters

The knowledge of the interaction of phosphate esters with iron-based surfaces begins with the work on vapor phase lubrication by Klaus [60]. A wide range of techniques have been used to

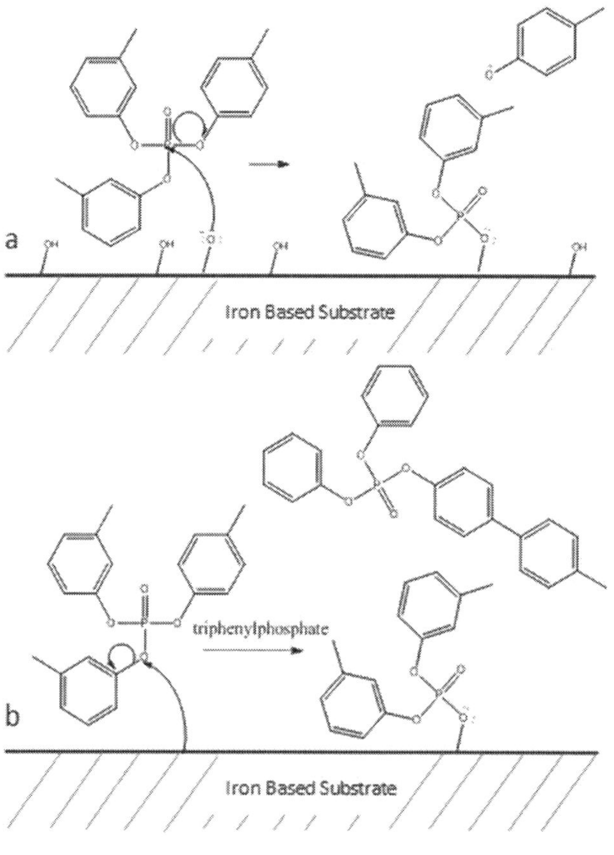

Figure 6. Reaction of phosphate esters with oxidized (a) and reduced (b) metal surfaces.

the reactions and the structure of the films formed on the surface, including Fourier transform infrared spectroscopy (FT-IR) [61], nuclear magnetic resonance(NMR) [62], and surface methods including x-ray photoelectron spectroscopy (XPS) [63], Auger electron spectroscopy (AES) [64] and x-ray absorption near edge spectroscopy(XANES) [65]. These techniques have demonstrated that tricresyl phosphate forms a multilayer film on steel surfaces which acts as a lubricious polymer. Later work by Forster [66] demonstrated that the diffusion of iron through the phosphate film was the rate-determining step in film formation and caused a lubricious coating to be maintained under wear. Iron is estimated to diffuse at a rate of 1–3×10^{-16} cm^2/sec through the phosphate glass [67].

Phosphate esters are known to react more rapidly with iron oxides than with metallic iron on steel surfaces. In the presence of an oxidized metal surface, a surface oxide or hydroxide initially adds to the phosphate, breaking a P-O bond and resulting in a surface bound dialkyl phosphate and releasing the substituted phenol. Continued reaction releases the remaining substituted phenols leaving an iron polyphosphate. If there is limited oxide or hydroxide at the surface, the C-O bond of the phosphate can be cleaved releasing the aryl group which then adds to an aryl group bound to another phosphate resulting in the same bound diaryl phosphate and a higher molecular weight soluble phosphate ester [68] as is shown in **Figure 6**.

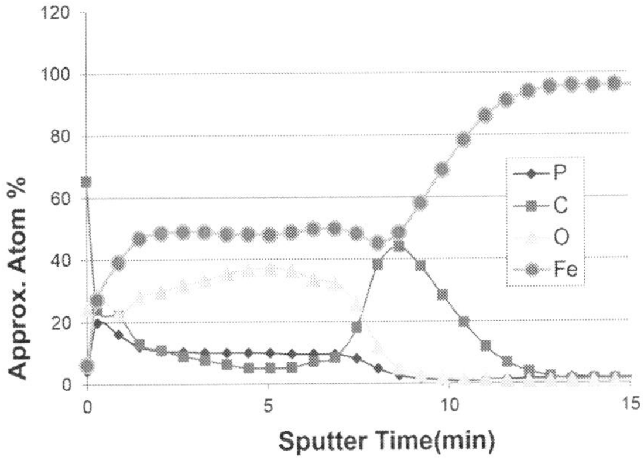

Figure 7. Auger depth profile of a film formed by the deposition of BTPP onto an iron foil at 425°C under nitrogen (sputter rate 1.5 nm/min).

An Auger depth profile showing the composition of the film formed is shown in **Figure 7**. Immediately adjacent to the iron surface, a carbon layer that has been characterized as either amorphous carbon or low-order graphite is observed under both thermal and tribological conditions. A layer of iron polyphosphate then extends to very near the surface of the coating. At the surface of the bearing, a layer containing bound alkyl groups is found as is seen in **Figure 8**. The bound alkyl groups help the surface retain some liquid lubricant upon standing resulting in additional reduction in wear on engine startup before adequate lubricant flow is achieved.

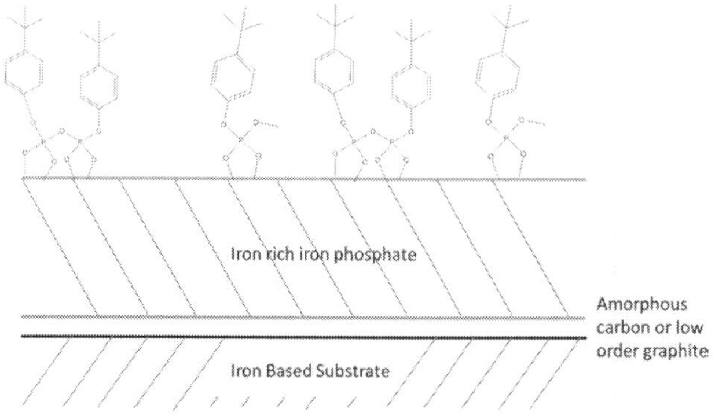

Figure 8. The layered structure of the film formed based on x-ray photoelectron spectroscopy and Auger depth profiling.

4.2. Safety of phosphate esters

One of the major perceived advantages of the phosphate esters in addition to their effectiveness is their relative safety. They are nonflammable [69], break down in the environment to phenols and phosphate and they exhibit only low-order toxicity with LD_{50}s in the range of 10–20 g/kg [70]. There are still concerns about delayed neuropathy that has been associated with certain isomers of the triaryl phosphates, as well as products that might be derived from the decomposition of the triaryl phosphates at high temperatures. There is evidence that a paralytic condition known as organophosphate-induced delayed neuropathy (OIDN) can develop [71].

A recent concern is that triaryl phosphates can enter the cabin as a part of the bleed air, especially given the air force shift from liquid oxygen carried on the plane to oxygen generated on board the aircraft [72]. Tricresyl phosphate has been reported in cabin air on aircraft at levels below toxicity thresholds [73]. Methods have been developed to detect and confirm human exposure to tri (o-cresyl) phosphate [74]. In a limited study of healthy pilots, none had evidence of tri (o-cresyl) phosphate exposure [75]. The contradictory reports indicate there are significant gaps in our knowledge of aerotoxic syndrome. These observations have accelerated the search for safer phosphate ester additives.

The general mechanism for the toxicity of organophosphates is through the inhibition of acetylcholinesterase [76]. This is the same mechanism that is associated with chemical warfare agents and organophosphorus insecticides. Considering the high LD_{50} associated with phosphate ester lubricant additives, it is unlikely that they operate by this mechanism. Organophosphate induced delayed neuropathy (OIDN) is an alternate type of toxicity that results in cramping, loss of coordination and may result in paralysis within 1–2 weeks after exposure. Victims recover only very slowly and the recovery is seldom complete [77].

Of the compounds used as lubricant additives, tricresyl phosphate containing the *ortho* (*o*-TCP) isomer is of particular concern for the development of OIDN. Exposures to *o*-TCP have occurred in the cases of adulterated Jamaica ginger [78] and cooking oil adulterated with turbine engine oil [79]. Aerotoxic syndrome and OIDN have recently focused on *o*-TCP present as one of the isomers present in the additive even though less than 2% of the TCP is permitted to be the *ortho* isomer [80].

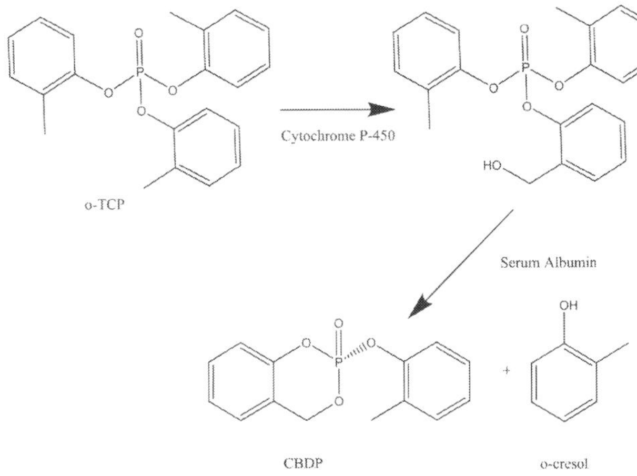

Figure 9. Conversion of *o*-TCP to CBDP, and irreversible inhibitor of human butyrylcholinesterase and human acetylcholinesterase.

The mechanism of OIDN by *o*-TCP is thought to begin with the benzylic oxidation of the *ortho* methyl group to give a benzyl alcohol by cytochrome P-450. The alcohol then cyclizes in a reaction catalyzed by serum albumin with the release of cresol as is shown in **Figure 9**. The cyclization product 2-(*o*-cresyl)-4H-1,3,2-dibenzodioxaphosphoranone (CBDP) is an irreversible inhibitor of human butyrylcholinesterase and human acetylcholinesterase [81].

In addition to the concern for the toxicity of the *ortho* isomers of TCP, there is concern about the products formulated lubricant components react with each other under extreme conditions. In particular, trimethylolpropane esters (a component of many neopentyl polyol ester-based basestocks) react with *o*-TCP to form neurotoxic compounds in fires [82, 83]. Currently available TCP contains only *meta* and *para* isomers, however there is evidence that isomerization to form *ortho* isomers is possible under engine conditions. Another approach is to remove the possibility of reaction at benzylic protons by the use of t-butyl groups as substituents on the aromatic rings.

5. Phosphorus-containing ionic liquids

In many lubricant applications, there is a need for increased durability, increased service intervals, reductions in friction and reductions in emissions. Phosphorus-containing ionic liquids have demonstrated potential in solving some of these problems [84]. Ionic liquids have very low vapor pressures and tend to be soluble in a wide range of nonpolar liquids. They can be prepared with differing properties, depending on the nature of the anion and the cation, modification of which can involve a number of central atoms. Since the topic of this work is phosphorus containing species, only species containing phosphorus in either the anion or the cation will be considered.

In order to have an ionic compound that is liquid at room temperature, large irregularly shaped ions are required. Frequently, these compounds incorporate long organic substituent groups which also impart solubility in organic liquids. Many of these ionic liquids are miscible with hydrocarbon-based lubricants. Compounds can contain phosphorus in the cation, anion or both ions. Phosphorus-containing anions have been shown to be the most effective [85]. These compounds also have a tendency to interact strongly with metal surfaces to form a thin protective phosphorus-containing film under boundary lubrication conditions [86]. It has also been shown that phosphorus-based ionic liquids provide a decrease in friction and wear in used diesel engine lubricants where the additives had been depleted but an increase in wear was observed when the additives had not been depleted. This would indicate interference between phosphorus-containing ionic liquids and existing antiwear additives [87].

Ionic liquids can contain phosphorus either as part of the cation (a phosphonium ion) or the anion (a phosphate ion). Studies of alkyl imidazolium phosphate ionic liquids formed a phosphate containing layer on steel contacts that was thought to also incorporate the imidazolium ion which reduced friction [88]. Some initial investigations implied that only phosphate containing ionic liquids would form a lubricious film, and studies of phosphonium-based ionic liquids with nonphosphate anions have been shown to also form a phosphorus-containing film and ionic liquids with a nonphosphate anion performed better under high loads [89]. It appears that all phosphorus-containing ionic liquids operate by similar mechanisms where a tribo-boundary film is formed as a top layer and beneath that layer is a plastic deformation zone [90].

6. Advanced bearing materials

The desire for more energy efficient and more powerful turbine engines for military and aerospace applications has led a push for lighter bearing materials that can withstand higher temperatures and are more durable. Greater durability and temperature stability can be achieved by the development of either harder metal alloys or ceramic bearing materials. These more exotic materials however may not function properly with the current lubricant and additive packages.

6.1. Carburized metal alloys

One approach to achieve higher speeds and higher temperatures particularly for military aerospace applications has led to the development of a new series of metal alloys. These alloys have a bulk composition that is somewhat similar to existing bearing materials, however they

http://dx.doi.org/10.5772/62208

are either carburized or nitrided during the heat treating step. This added step forms a hard coating on the surface of the bearing that extends into the top 50–100 μM of the material. Carburization during heat treatment consists of heating the material in the presence of a reactive carbon species such as methane of carbon monoxide. The process converts the surface of the material from the normal oxides and hydroxides to a surface that contains mainly carbides. Similarly, the nitriding process converts the surface to nitrides by heating the material in an atmosphere of a reactive nitrogen species such as nitrous oxide.

The primary concern associated with carburized steels is due to the change in surface chemistry from oxides and hydroxides to carbides. Phosphorus-containing additives typically react with the oxide surface to displace one or more of the organic groups, forming a layer that contains an iron phosphate polymer that is durable and lubricous. In the absence of the oxides, it is unknown whether the phosphates will bind. An additional concern is the known reactivity of certain metal carbides with phosphate esters and lubricant esters [91]. Initial work has shown that phosphate esters react with the nonheat treated Pyrowear 675 rods under similar conditions as many other steel coupons [92]. The reactivity of the carburized steels with ester-based lubricants and phosphate esters are ongoing.

6.2. Ceramic bearings

Ceramic bearings have been investigated for a number of high-temperature applications in turbine engines. In general, unlubricated use has resulted in poor performance, but with an appropriate lubricant good performance was achieved [93]. Phosphate esters were found not to react with either silicon nitride or silicon carbide under conditions similar to those seen in steel bearings. At higher temperatures, a film is formed but it is ineffective in extending the life of the bearing [94]. In order to form an effective coating, a pretreatment which introduces iron onto the surface is required [95].

7. Conclusion

Phosphate esters and related compounds with sulfur and metals added have found wide application because of their ability to form thermal and tribological films on a wide range of materials. The mechanisms of film formation differ for different members of this series, but they all result in the formation of a tribofilm that is both lubricious and durable. The ability to make changes in the organic groups and the atoms forming the connections to the phosphorus atom allow compounds with a range of physical properties and reactivities to be prepared. A greater understanding of additive reactivity, as it pertains to the formation of durable films on metal surfaces, will allow the development of more effective phosphorus-containing additives. Although environmental and toxicity concerns are present, additives based on the chemistry of phosphorus-containing additives are likely to find application well into the future and the development of new materials has the promise to reduce those concerns.

Author details

David W. Johnson

Address all correspondence to: djohnson1@udayton.edu

Department of Chemistry, University of Dayton, Dayton, OH, USA

References

[1] Snyder CE, Gschwender LJ, Sharma SK. Long-term additive trends in aerospace applications. In: Rudnick LR, Ed., Lubricant additives: chemistry and applications, 2nd edn., CRC Press, 2009. p. 637–646.

[2] Corbridge DEC. Phosphorus: chemistry, biochemistry and technology, 6th edn., CRC Press, 2013. ISBN: 978-1439840887.

[3] Marino MP, Placek DC. Phosphate esters. In: Rudnick LR, Ed., Synthetic lubricants and high performance functional fluids. CRC Press, 2002, p. 103–140.

[4] Moy P. Aryl phosphate ester fire-retardant additive for low-smoke vinyl applications. Journal of Vinyl and Additive Technology 2004;10:187–192. DOI: 10.1002/vnl.20028.

[5] Barr DB, Bravo R, Weerasekera G, Caltabiano LM, Whitehead Jr. RD, Olsson AO, Caudill SP, Schober SE, Pirkle JL, Sampson RJ, Needham LL. Concentrations of dialkyl phosphate metabolites of organophosphorus pesticides in the U.S. population. Environmental Health Perspectives. 2004;112:186–200.

[6] Oldenhoveda de Guertechin L. Surfactants: classification. In: Broze G., Ed., Handbook of detergents, Surfactant Science Series, Volume 82, Marcel Dekker, Inc., 1999. p. 18–20.

[7] McDonald RA. Zinc dialkyldithiophosphates. In: Rudnick LR., Ed., Lubricant additives: chemistry and applications, Chemical industries, Volume 124, 2nd edn., CRC Press, p. 51–62.

[8] Spikes H. The history and mechanism of ZDDP. Tribology Letters 2004;17:469–489. DOI: 10.1023/B:TRIL.0000044495.26882.b5.

[9] Asseff PA. Lubricants suitable for internal-combustion engines. U.S. Patent 2261047. Assigned Lubrizol, Inc.

[10] Cook EW, Thomas WD Jr. Lubricating-oil addition agents. U.S. Patent 2344392. Assigned America Cyanamide.

[11] Fruehler HC. Antioxidant for lubricating oils. U.S. Patent 2364283. Assigned Union Oil.

[12] Fruehler HC. Antioxidant for lubricating oils. U.S. Patent 2364284. Assigned Union Oil.

[13] Varlot K, Kasrai M, Martin JM, Vacher B, Bancroft GM, Yamaguchi ES, Ryason PR. Antiwear film formation from neutral and basic ZDDP, influence of the reaction temperature and concentration. Tribology Letters. 2000;8:9–16. DOI: 10.1023/A: 1019162529554.

[14] Burn AJ. The mechanism of the antioxidant action of zinc dialkyl dithiophosphates. Tetrahedron. 1966;22:2153–2161. DOI: 10.1016/S0040-4020(01)82135-9.

[15] Bidwell JB, Williams RK. The new look in lubricating oils. SAE Transactions 1955;63:349–361. DOI: 10.4271/550258.

[16] Stanley CS, Larson, R. The performance of zinc dithiophosphates as lubricating oil additives. SAE Journal 1958;107C:107–120. DOI: 10.4271/580110.

[17] Cavestri RC. Potentially useful polyolester lubricant additives: an overview of antioxidants, antiwear and antiseize compounds. 1996:171–172. International Refrigeration and Air Conditioning Conference. Paper 319. http://docs.lib.purdue.edu/iracc/319.

[18] Barnes AM, Bartle KD, Thibon VRA. A review of zinc dialkyldithiophosphates (ZDDPS): characterization and role in the lubricating oil. Tribology International 2001;34:389–395. DOI: 10.1016/S0301-679X(01)00028-7.

[19] Harrison PG, Kikabhai T. Proton and phosphorus-31 nuclear magnetic resonance study of zinc(II) O,O'-dialkyl dithiophosphates in solution. Journal of the Chemical Society Dalton Transactions. 1987:807–814. DOI: 10.1039/DT9870000807.

[20] Harrison JJ, Chan CY, Onopchenko A, Pradhan AR, Petersen M. Neutral zinc(II) O,O-dialkyldithiophosphates – variable temperature 31P NMR and quantum chemical study of the ZDDP monomer-dimer equilibrium. Magnetic Resonance in Chemistry 2008;46:115–124. DOI: 10.1002/mrc.2130.

[21] Armstrong DR, Ferrari ES, Roberts KJ, Adams, D. An investigation into the molecular stability of zinc di-alkyl-di-thiophosphates(ZDDPs) in relation to their use as anti-wear and anti-corrosion additives in lubricating oils. Wear 1997;208:138–146. DOI: 10.1016/S0043-1648(96)07332-2.

[22] Haiduc I, Sowerby DB, Shao-Fang L. Stereochemical aspects of phosphor-1,1-dithiolato metal complexes (dithiophosphates, dithiophosphinates): coordination patterns, molecular structures and supramolecular associations—I. Polyhedron 1995;14:3389–3472. DOI: 10.1016/0277-5387(95)00108-5.

[23] Yamaguchi ES, Primer RL, Aragon SR, Labrador EQ. Dynamic light scattering studies of neutral diisobutyl zinc dithiophosphate. Tribology Transactions 1997;40:330–338. DOI: 10.1080/10402009708983662.

[24] Yamaguchi ES, Ryason PR, Labrador EQ, Hansen TP. Comparison of the relative performance of neutral and basic ZnDTP salts. Tribology Transactions 1996;39:220–224. DOI: 10.1080/10402009608983524.

[25] Peng P, Hong SZ, Lu WZ. The degradation of zinc dialkyldithiophosphate additives in fully formulated engine oil as studied by P-31 NMR spectroscopy. Lubrication Engineering 1994;50:230–235. DOI: 10.1016/j.phpro.2015.05.055.

[26] Vizintin J. Oil surface: additive reaction mechanism. In: Totten GE, Liang H, Eds., Surface modification and mechanisms. Marcel Decker, Inc., New York, N.Y., 2004, 243–298.

[27] Lawton SL, Kokotailo GT. The crystal and molecular structures of zinc and cadmium O,O-diisopropylphosphorodithiolates. Inorganic Chemistry 1969;8:2410–2421. DOI: 10.1021/ic50081a033.

[28] Ivanov AV, Antzutkin ON, Larsson AC, Kritijkos M, Forsling W. Polycrystalline and surface O,O'-dialkyldithiophosphate zinc(II) complexes: preparation, 31P CP/MAS NMR and single crystal x-ray diffraction studies. Inorganica Chimica Acta 2001;315:26–35. DOI: 10.1016/S0020-1693(01)00293-6.

[29] Larsson AC, Ivanov AV, Forsling W, Antzutkin ON, Abraham AE, deDios AC. Correlations between 31P chemical shift anisotropy and molecular structure in polycrystalline O,O'-dialkyldithiophosphate zinc(II) and nickel(II) complexes: 31PCP/MAS NMR and ab initio quantum mechanical calculation studies. Journal of the American Chemical Society. 2005;127:2218–2230. DOI: 10.1021/ja0306112.

[30] Suominen-Fuller ML, Kasrai M, Bancroft GM, Fyfe K, Tan KH. Solution decomposition of zinc dialkyl dithiophosphate and its effect on antiwear and thermal film formation by x-ray absorption spectroscopy. Tribology International 1998;31:627–644. DOI: 10.1016/S0301-679X(98)00084-X.

[31] Bird RJ, Galvin GD. The application of photoelectron spectroscopy to the study of E.P. films on lubricated surfaces. Wear. 1976;37:143–167. DOI: 10.1016/0043-1648(76)90188-5.

[32] Bancroft GM, Kasrai M, Fuller M, Yin Z, Fyfe K, Tan KH. Mechanisms of tribochemical film formation: stability of tribo- and thermally-generated ZDDP films. Tribology Letters 1997;3:47–51. DOI: 10.1023/A:1019179610589.

[33] Bec S, Tonck A, Georges JM, Coy RC, Bell JC, Roper GW. Relationship between mechanical properties and structure of zinc dithiophosphate antiwear films. Proceedings of the Royal Society (London) A. 1999;455:4181–4203. DOI: 10.1098/rspa.1999.049.

[34] Piras FM, Rossi A, Spencer ND. Combined in situ (ATR FT-IR) and ex situ (XPS) study of the ZnDTP-iron surface interaction. Tribology Letters 2003;15:181–191. DOI: 10.1023/A:1024800900716.

[35] Masuko M, Ohkido T, Suzuki A, Ueno T. Fundamental changes in friction and wear characteristics due to ZnDTP deterioration in simulating engine oil degradation during use. In: Transient processes in tribology. Proceedings of the 30th Leeds-Lyon Sympo-

sium on Tribology, Lyon, France, 2–8 September 2003, Elsevier, Philadelphia, PA, USA, 2004, p. 359–356.

[36] Rhodes KL, Stair PC. The surface chemistry of zinc dialkyldithiophosphate, an antiwear additive on oxidized iron and steel foils. Journal of Vacuum Science and Technology A 1988;6:971–974. DOI: 10.1116/1.575042.

[37] Fuller M, Yin Z, Kasrai M, Bancroft GM, Yamaguchi ES, Ryason PR, Willermet PA, Tan KH. Chemical characterization of tribochemical and thermal films generated from neutral and basic ZDDPs using X-ray absorption spectroscopy. Tribology International 1997;30:305–315. DOI: 10.1016/S0301-679X(96)00059-X.

[38] Willermet PA, Dailey DP, Carter III RO, Schmitz PJ, Zhu W. Mechanism of formation of antiwear films from zinc dialkyldithiophosphates. Tribology International 1995;28:177–187. DOI: 10.1016/0301-679X(95)98965-G.

[39] Martin JM. Antiwear mechanisms of zinc dithiophosphate: a chemical hardness approach. Tribology Letters 1999;6:1–8. DOI: 10.1023/A:1019191019134.

[40] Yin Z, Kasrai M, Fuller M, Bancroft GM, Fyfe K, Tan KH. Application of soft X-ray absorption spectroscopy in chemical characterization of antiwear films generated by ZDDP. Part I: The effects of physical parameters. Wear 1997;202:172–191. DOI: 10.1016/S0043-1648(96)07272-9.

[41] Fuller ML, Fernandez LR, Massoumi GR, Lennard WN, Kasrai M, Bancroft GM. The use of X-ray absorption spectroscopy for monitoring the thickness of antiwear films from ZDDP. Tribology Letters 2000;8:187–192. DOI: 10.1023/A:1019195404055.

[42] Shakhvorostov D, Muserm MH, Song Y, Norton PR. Smart materials behavior in phosphates: role of hydroxyl groups and relevance to antiwear films. Journal of Chemical Physics 2009;131:044704. DOI: 10.1063/1.3182854.

[43] Spikes H. Low and zero sulfated ash, phosphorus and sulfur anti-wear additives for engine oils. Lubrication Science 2008;20:103–136. DOI: 10.1002/ls.57.

[44] Borshchevskii SB, Shabanova EV, Markov AA, Rebrov YI. Antifriction and antiwear properties of dialkyldithioaminephosphates. Chemistry and Technology of Fuels and Oils. 1984;20:503–506; Translated from Khimiya i Tekhnologiya Topliv i Masel, No. 10, pp., October, 1984, 23–25. DOI: 10.1007/BF00725414.

[45] Fu X, Liu W, Xue Q. The application research on series of ashless P-containing EP and AW additives. Industrial Lubrication and Tribology 2005;57:80–83. DOI: 10.1108/00368790510583393.

[46] Zaskal'ko PP, Parfenova VA, Markov AA, Lesninova VA, Belov PS. Amidothiophosphates – effective antiwear and extreme pressure additives for lubricating oils. Chemistry and Technology of Fuels and Oils. 1976;12:58–60. Translated from Khimiya I Tekhnologiyi Topliv I Masel. 1976, 45–48.

[47] Boreshchevskii SB, Levitina IS, Shabanova EV, Kotova GG. Additives based on dithiophosphoric acids and unsaturated compounds for lubricating oils. Chemistry and Technology of Fuels and Oils. 1992;27:326–328. Translated from Khimiya i Tekhnologiya Topliv i Masel. 1992, 24–26.

[48] Chen X, Kim B, Eisenbaumer, RL, Aswath PB. Synthesis and antiwear behavior of ashless alkylthioperoxydiphosphates. Tribology: Materials, Surfaces and Interfaces 2012;6:121–133. DOI: 10.1179/1751584X12Y.0000000009.

[49] Najman MN, Kasrai M, Bancroft GM. Chemistry of antiwear films for ashless thiophosphate oil additives. Tribology Letters 2004;17:217–229. DOI: 10.1023/B:TRIL.0000032448.77085.f4.

[50] Kim BH, Mourhatch R, Aswath PB. Properties of tribofilms formed with ashless dithiophosphate and zinc dialkyldithiophosphate under extreme pressure conditions. Wear 2010;268:579–591. DOI: 10.1016/j.wear.2009.10.004.

[51] Mangolini F, Rossi A, Spencer ND. Reactivity of triphenylphosphorothionate in lubricant oil solution. Tribology Letters 2009;35:31–43. DOI: 10.1007/s11249-009-9429-3.

[52] Mangolini F, Rossi A, Spencer ND. Influence of metallic and oxidized iron/steel on the reactivity of triphenyl phosphorothionate in oil solution. Tribology International 2011;44:670–683. DOI: 10.1016/j.triboint.2010.02.009.

[53] Kim BH, Jiang JC, Aswath PB. Mechanism of wear at extreme load and boundary conditions with ashless anti-wear additives: analysis of wear surfaces and wear debris. Wear 2011;270:181–194. DOI: 10.1016/j.wear.2010.10.058.

[54] Haycock R, Caines AJ, Haycock RF, Hillier JE. Automotive lubricants reference book, 2nd edn, John Wiley and Sons, p. 333–335. ISBN: 9781860584718.

[55] Schnur NE. Blended polyol ester lubricants for refrigerant heat transfer fluids. US Patent Number US6551523 B1. Assignee: Cognis Corporation.

[56] Yamamoto Y, Hirano F. Scuffing resistance of phosphate esters. Wear 1978;50:343–348. DOI: 10.1016/0043-1648(78)90078-9.

[57] Yamamoto Y, Hirano F. Effect of different phosphate esters on frictional characteristics. Tribology International 1980;13:165–169. DOI: 10.1016/0301-679X(80)90034-1.

[58] Beek O, Givens JW, Williams EC. On the mechanism of boundary lubrication. II. Wear prevention by addition agents. Proceedings of the Royal Society A 1940;177:103–118. DOI: 10.1098/rspa.1940.0113.

[59] Cottington RL, Ravner H. Interactions in neopentyl polyol ester-tricresyl phosphate—Iron system at 500 F. ASLE Transactions 1969;12:280–286. DOI: 10.1080/05698196908972273.

[60] Graham EE, Klaus EE. Lubrication from the vapor phase at high temperatures. ASLE Transactions 1986;29:229–234. DOI: 10.1080/05698198608981682.

[61] Sasaki K, Inauoshi N, Tashiro K. Friction-induced dynamic chemical changes of tricresyl phosphate as lubricant additive observed under boundary lubrication with 2D fast imaging FTIR-ATR spectrometer. Wear 2010;268:911–916. DOI: 10.1016/j.wear.2009.12.017.

[62] Bansal V, Sastry MIS, Sarpal AS, Jain SK, Srivastava SP, Bhatnagar AK. Characterization of nitrogen and phosphorous components in a multifunctional lubricant additive by NMR and IR techniques. Lubrication Engineering 1997;53:17–26.

[63] Jiusheng J, Wenqi R, Tianhui R, Xingguo F, Weimi L. Tribological properties of phosphate esters as additives in rape seed oil. Journal of Synthetic Lubrication 2003;20:151–158. DOI: 10.1002/jsl.3000200206.

[64] Saba CS, Forster NH. Reactions of aromatic phosphate esters with metals and their oxides. Tribology Letters 2002;12:135–146. DOI: 10.1023/A:1014081523491.

[65] Nicholls M, Najman MN, Zhang Z, Kasrai M, Norton PR, Gilbert PUPA. The contribution of XANES spectroscopy to tribology. Canadian Journal of Chemistry 2007;85:816–830. DOI: 10.1139/v07-093.

[66] Forster NH. Rolling contact testing of vapor phase lubricants—part III: surface analysis. Tribology Transactions 1999;42:1–9. DOI: 10.1080/10402009908982183.

[67] Sheaffer SK. High temperature reaction of aryl phosphate esters on an iron film. Tribology Transactions 2003;46:332–338. DOI: 10.1080/10402000308982634.

[68] Johnson DW, Morrow SJ, Forster N, Saba CS. Vapor-phase lubrication: reaction of phosphate ester vapors with iron and steel. Chemistry of Materials 2002;14:3767–3775. DOI: 10.1021/cm010921o.

[69] Gschwender LJ, Kramer DC, Lok BK, Sharma SK, Snyder Jr. CE, Sztenderowicz ML. Liquid lubricants and lubrication. In: Bhushan B., Ed., Modern tribology handbook, CRC Press, 2001, p. 361–382.

[70] Johannsen FR, Wright PL, Gordon DE, Levinskas GJ, Radue RW, Graham PR. Evaluation of delayed neurotoxicity and dose–response relationships of phosphate esters in the adult hen. Toxicology and Applied Pharmacology 1977;41:291–304. DOI: 10.1016/0041-008X(77)90030-8.

[71] Schopfer LM, Furlong CE, Lockridge O. Development of diagnostics in the search for an explanation of aerotoxic syndrome. Analytical Biochemistry 2010;404:64–74. DOI: 10.1016/j.ab.2010.04.032.

[72] Michaelis S. Contaminated aircraft cabin air. Journal of Biological Physics and Chemistry 2011;11:132–145. DOI: 10.4024/41111/11-4-abs1.jbpc.11.04.

[73] Deboer J. Antello A, van der Ween I, Brandsma S, Lammertse N. Tricresyl phosphate and the aerotoxic syndrome of flight crew members. Chemosphere 2015;119:558–561. DOI: 10.1016/j.chemosphere.2015.05.015.

[74] Johnson D, Carter MC, Crow BS, Isenberg SL, Graham LA, Erol HA, Watson CM, Pantazides BG, van dern Schans MJ, Langenberg JP, Noort D, Blake T, Thomas JD, Johnson RC. Quantitation of ortho-cresyl phosphate adducts to butyrylcholinesterase in human serum by immunomagnetic-UHPLC-MS/MS. Journal of Mass Spectrometry. 2015;50:683–692. DOI: 10.1002/jms.3576.

[75] Tacal O, Schopfer LM. Healthy F-16 pilots show no evidence of exposure to tri-ortho-cresyl phosphate through on-board oxygen generating system. Chemico-Biological Interactions. 2014;215:69–74. DOI: 10.1016/j.cbi.2014.03.004.

[76] Brown MA, Brix KA. Review of health consequences from high, intermediate and low-level exposure to organophosphorus nerve agents. Journal of Applied Toxicology 1998;18:393–408. DOI: [N8]10.1002/(SICI)1099-1263(199811/12)18:6<393::AID-JAT528>3.0.CO;2-0.

[77] Senanayake N. Tri-cresyl phosphate neuropathy in Sri Lanka: a clinical and neurophysiological study with a three year follow up. Journal of Neurology, Neurosurgery and Psychiatry 1981;44:775–780.

[78] Kidd JG, Langworthy OR. Jake paralysis: paralysis following the ingestion of Jamaica ginger extract adulterated with tri-ortho-cresyl phosphate. Bulletin of the Johns Hopkins Hospital 1933;52:39–65.

[79] Smith HV, Spaulding JMK. Outbreak of paralysis in Morocco due to ortho-cresyl phosphate poisoning. Lancet 1959:2:1019–1021.

[80] MIL-PRF-7808L, performance specification; lubricating oil, aircraft turbine engine, synthetic base (02 May 1997) [Superseeding MIL-L-7808K]. 1997. Available from: http://everyspec.com/MIL-PRF/MIL-PRF-000100-09999/MIL-PRF-7808L_5699/ [accessed 2016-01-31].

[81] Carletti E, Schopfer LM, Colletier JP, Froment MT, Nachon F, Weik M, Lockridge O, Masson P. Reaction of cresyl salingenin phosphate, the organophosphorus agent implicated in aerotoxic syndrome with human cholinesterases: mechanistic studies employing kinetics, mass spectrometry and x-ray structure analysis. Chemical Research in Toxicology 2011;24:797–808. DOI: 10.1021/tx100447k.

[82] Rubey WA, Streibich RC, Bush J, Centers PW, Wright RL. Neurotoxin formations from pilot-scale incineration of synthetic ester turbine lubricants with triaryl phosphate additive. Archives of Toxicology 1996;70:508–509. DOI: 10.1007/s002040050306.

[83] Centers PW. Potential neurotoxin formation in thermally degraded synthetic ester turbine lubricants. Archives of Toxicology 1992;66:679–680. DOI: 10.1007/BF01981509.

[84] Somers AE, Howlett PC, MacFarlane DR, Forsyth M. A review of ionic liquid lubricants. Lubricants 2013;1:3–21. DOI: 10.3390/lubricants1010003.

[85] Zhou Y, Dyck J, Graham TW, Lou H, Leonard DN, Qu J. Ionic liquids composed of phosphonium cations and organophosphate, carboxylate, and sulfonate anions as lubricant antiwear additives. Langmuir. 2014;30:13301–13311. DOI: 10.102/la5032366.

[86] Barnhill WC, Qu J, Luo H, Meyer III HM, Ma C, Chi M, Papke BL. Phosphonium-organophosphate ionic liquids as lubricant additives: effects of cation structure on physiochemical and tribological characteristics. Applied Materials and Interfaces 2014;6:22585–22593. DOI: 10.1021/am506702u.

[87] Anand M, Hadfield M, Viesca JL, Thomas B, Hernandez-Battez A, Austen S. Ionic liquid as tribological performance improving additive for in-service and used fully formulated diesel engine lubricants. Wear. 2015;334–335:67–74. DOI: 10.1016/j.wear.2015.01.055.

[88] Zhang L, Feng D, Xu B. Tribological characteristics of alkylimidazolium diethyl phosphate ionic liquids as lubricants for steel-steel contact. Tribology Letters. 2009;34:95–101. DOI: 10.1007/s11249-009-9412z.

[89] Somers AE, Howlett PC, Sun J, MacFarlane DR, Forsyth M. Transition in wear performance for ionic liquid lubricant under increasing load. Tribology International 2010;40:27–284. DOI: 10.1007/s11249-010-9695-0.

[90] Qu J, Bansal DG, Yu B, Howe JY, Luo H, Dai S, Li H, Blau PJ, Bunting BG, Mordukhovich G, Smolenski DJ. Antiwear performance and mechanism of an oil-miscible ionic liquid as a lubricant additive. Applied Materials and Interfaces 2012;4:997–1002. DOI: 10.1021/am201646k.

[91] Johnson DW, Hils JE, Forster, N. Interaction of polyol esters and phosphate esters with metal carbides. Tribology Letters 2011;42:223–232. DOI: 10.1007/s11249-011-9766-x.

[92] Johnson DW, Bachus M, Hils JE. Interaction between lubricants containing phosphate ester additives and stainless steels. Lubricants 2013;1:48–60. DOI: 10.3390/lubricnats1020048.

[93] Zaretsky, EV. Ceramic bearings for use in gas turbine engines. Journal of Materials Engineering and Performance 1989;11:237–253. DOI: 10.1007/s11665-013-0726-5.

[94] Bertrand PA. Reactions of tricresyl phosphate with bearing materials. Tribology Letters 1997;3:367–377.

[95] Graham EE, Morales W, Patel N. Method of providing ceramic article with wear resistant coating. US Patent 5738913, 1998. Assignee: Cleveland State University.

Chapter 9

Modelling Cavitation in (Elasto)Hydrodynamic Lubrication

Andreas Almqvist and Peter Wall

Additional information is available at the end of the chapter

http://dx.doi.org/10.5772/63533

Abstract

In this chapter we will present a derivation of a mathematical model describing how cavitation influences the pressure distribution in a thin lubricant film between two moving surfaces. The main idea in the derivation is to first describe the influence of cavitation on the mass flow and thereafter using a conservation law for the mass. This leads to a nonlinear system with two complementary variables: one is the pressure distribution and the other is related to the density, i.e. a nonlinear complementarity problem (NLCP). The proposed approach is used to derive a mass conserving cavitation model considering that density, viscosity and film thickness of the lubricant depend on the pressure. To demonstrate the applicability and evaluate the proposed model and the suggested numerical implementation, a few model problems are analysed and presented.

Keywords: Cavitation, Reynolds equation, Thin film flow, Complementarity problem, Hydrodynamic lubrication

1. Introduction

A central problem in hydrodynamic lubrication is to model the pressure in a lubricant (fluid) between two surfaces which are in relative motion. In this chapter we consider the full film regime, i.e. when the surfaces are fully separated by the lubricant. The pressure and the velocity field can be modelled using the Navier-Stokes equations and the continuity equation. However, in real applications the distance between the surfaces, h, is extremely small in relation to the size of the surfaces. This fact can be used in order to derive a simplified two-dimensional equation for the hydrodynamic pressure, p. Indeed, if the velocity of the upper surface is v_u and the velocity of the lower surface is v_l, then under the assumptions outlined in e.g. [1], the following Reynolds equation is obtained

$$\frac{\partial}{\partial t}(\rho h) = \nabla \cdot \left(\frac{\rho h^3}{12\mu} \nabla p - \frac{1}{2} \rho h (\mathbf{v}_u + \mathbf{v}_l) \right), \tag{1}$$

where μ and ρ are the viscosity and density of the lubricant. A generalization of this thin film approximation which considers the effects of non-Newtonian fluids can e.g. be found in the *Encyclopaedia of Tribology*, see the entry [2] by Larsson.

A fluid cannot sustain large tensile stress and it is known that when the pressure becomes too low the continuous film will rupture and air bubbles will be formed. This phenomenon is known as cavitation and has a huge impact on the hydrodynamic performance. In areas where cavitation takes place the pressure is usually treated as constant and is assumed to have the same value as the saturated vapour pressure, i.e., the pressure at which cavitation starts. This is commonly referred to as the cavitation pressure, here denoted by p_c.

The common practice in the field is to build mathematical models that consider hydrodynamic cavitation based on the Reynolds equation (1). These models rests on the assumption that the pressure can be regarded as constant in areas where cavitation takes place and this constant level of pressure is typically also assumed to be the saturated vapour pressure, i.e., the pressure at which cavitation starts. The most important pioneering works are the papers presenting the Jakobsson-Floberg-Olsson (JFO) cavitation boundary conditions [3–5], Elrod's work [6] comprising these boundary conditions in one universal equation for the unknown pressure (or saturation) and its corresponding finite difference scheme and then Vijayaraghavan's generalization [7] of Elrod's results. These results have been frequently used to study the effects of cavitation in real applications and they have also been subject for many further generalizations.

In Elrod's approach the lubricant is treated as incompressible while Vijayaraghavan and Keith assumed that the relation between the density and the pressure is of the form

$$\rho = \rho_c \exp((p - p_c)/\beta), \tag{2}$$

where ρ_c is the density at cavitation pressure p_c, and β is the bulk modulus of the lubricant. One example of a generalization of the results in [6] and [7] is the paper [8] by Sahlin et al., who developed a cavitation algorithm for arbitrary pressure-density relationships. The case considering incompressible lubricants was further developed by Bayada et al. [9, 10]. Their results put the finite difference scheme in [6] into a framework which makes further mathematical analysis possible. Moreover, these works consider both cavitation and the effects of surface roughness, where the surface roughness is analyzed by homogenization. In [10, 11] these ideas were extended to also include elastohydrodynamic effects, i.e. elastic deformation of the surfaces caused by the hydrodynamic pressure. In particular this means that the film thickness, h, depends on the pressure.

A significant progress in mathematical modeling of cavitation in hydrodynamic lubrication was recently presented by Giacopini et al. [12]. They reformulated the model presented in [9, 10] for incompressible fluids. The model in [9, 10] includes two unknowns, namely the pressure and the saturation (or density) while the unknowns in the reformulation are the pressure and a new unknown variable related to the saturation. The major advantage with the reformulation is that the two unknowns are complementary, i.e. their product is zero, in the whole domain. This implies that the discretized system of equations becomes a linear complementarity problem (LCP) which can be readily solved by standard numerical methods as e.g. the Lemke's pivoting algorithm [13, 14]. The idea of using complementarity was further developed by Bertocci et al. [15]. They present a very comprehensive cavitation model, which discretized assumes the form of a nonlinear complementarity problem (NLCP).

In the work by Almqvist et al. [14], a new approach for modeling cavitation was presented. The model was derived by first considering how the mass flow is influenced by cavitation and thereafter using the law of conservation of mass. This immediately leads to a linear complementarity problem formulation. This is in contrast to the approach in [12, 15] where the pressure–density relationship is inserted directly into the Reynolds equation, thereafter it is necessary to argue for that certain terms can be cancelled before they arrive at a complementarity formulation. Moreover, in addition to [12] the approach in Almqvist et al. [14] also covers compressible fluids. In the particular case when the pressure–density relationship is as in (2), a change of variables was introduced which transforms the problem such that the discrete formulation is an LCP.

This chapter extends the mathematical modelling presented in [14] to also include more general pressure–density relationships and allow the viscosity and film thickness to depend on the pressure. In particular this means that the model can be used to efficiently study elastohydrodynamic lubrication where cavitation is present. To demonstrate the applicability and evaluate the proposed model and the associated numerical solution method, four model problems are analysed.

2. Mathematical model

In this section, a model considering cavitation in thin film flow between two surfaces in relative motion is developed. The flow is regarded as compressible, the viscosity can be shear rate and pressure dependent (non-Newtonian and piezo-viscous) and elastic deformation of the contacting surfaces can be considered.

Let us start by defining the (three dimensional) fluid domain Ω between the two lubricated surfaces h_l (lower) and h_u (upper):

$$\Omega = \{(x,z) | 0 \le x_i \le l_i, h_l \le z \le h_u\}. \tag{3}$$

The fluid domain Ω is schematically illustrated in **Figure 1** (left).

The hydrodynamic pressures that develop may be large enough to deform the surfaces. This implies that the film thickness $h = h_u - h_l$ depends on the pressure, i.e., $h = h(x,t,p(x,t))$. The most frequently used approach for describing the pressure dependence of the film thickness, is based on standard contact mechanics results for line and point loading of elastic half-spaces. These are detailed in Chapters 2 and 3 in the book by Johnson [16] and in [2]. In addition, when the surfaces move relative to each other Ω varies in time.

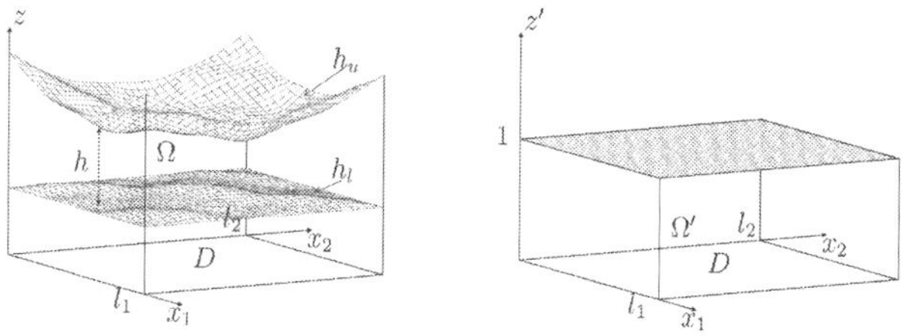

Figure 1. Schematic illustration of the original (left) and the transformed (right) domain.

The classical thin film approximation, presented in e.g. [1, p. 147], can be obtained by employing the scaling

$$X_1 = x_1 / l_1,\ X_2 = x_2 / l_2,\ Z = z / h_0,\ T = t / t_0,\ \bar{\mu} = \mu / \mu_0,\ \bar{\rho} = \rho / \rho_0,$$
$$\bar{u} = u / u_0,\ \bar{v} = v / v_0,\ \bar{w} = w / w_0,\ P = p \Big/ \frac{\mu_0 u_0 l_1}{h_0^2}. \tag{4}$$

More precisely, using the fact that $h_0 << l_i$ and neglecting terms of the order h_0/l_i and $(h_0/l_i)^2$ leads to that $p = p(x,t)$ and the following relationship between the pressure and the velocity field in the lubricant

$$\frac{\partial}{\partial z}\left(\mu(x,z,t)\frac{\partial \mathbf{u}}{\partial z}(x,z,t)\right) = \nabla p(x,t), \tag{5}$$

where $\mathbf{u} = (u,v)$ and $\nabla = (\partial / \partial x_1, \partial / \partial x_2)$. Remark: an alternative, one-dimensional, thin film approximation was introduced in [17]. Expressed in the terminology used here, the difference between the models is a consequence of Barus' pressure–viscosity relationship; $\mu = \mu_0 exp(\alpha p)$ and the scaling of the vertical velocity component;

$$\bar{w} = w / \left(u_0 \frac{h_0}{l_1} \right).$$

In the following, the viscosity is assumed to be on the form $\mu(x,z,t) = \mu_0 g(z,p(x,t))$, where the reference parameter μ_0 is typically taken as the viscosity at ambient pressure and g is a function which can be used to describe the variation of the viscosity across the film (z-dependence) and with the pressure. It should be noted that both μ, ρ and h may depend on the fluid pressure.

As pointed out above, Ω varies with time. In order to simplify the analysis, a transformation of the height coordinate z is introduced that leads to a fixed, time independent, domain Ω'. Indeed, let

$$z' = \frac{z - z_1}{h} \tag{6}$$

then the corresponding solution domain becomes

$$\Omega' = \{(x,z) | 0 \le x_i \le l_i, 0 \le z' \le 1\}, \tag{7}$$

irrespectively of the shape of the surfaces h_l and h_u. The domain Ω and its corresponding dimensionless form Ω' are depicted in **Figure 1**. In this case, the correspondence to Eq. (5) becomes

$$\frac{1}{h^2(x,t)} \frac{\partial}{\partial z'} \left(\mu(x,z',t) \frac{\partial \mathbf{u}}{\partial z'}(x,z',t) \right) = \nabla p(x,t). \tag{8}$$

Integrating Eq. (8) twice with respect to z' and by assuming no slip at the surfaces, i.e., $\mathbf{u}(x,1,t) = \mathbf{v}_u(x,t)$ at the upper surface and $\mathbf{u}(x,0,t) = \mathbf{v}_l(x,t)$ at the lower surface, yields the following closed form expression for the velocity field

$$\mathbf{u}(x,z',t) = \left(f_1(x,z',t) - \frac{f_1(x,1,t)}{f_0(x,1,t)} f_0(x,z',t) \right) h^2(x) \nabla p(x) \tag{9}$$

$$+ \frac{f_0(x,z',t)}{f_0(x,1,t)} \left(\mathbf{v}_u(x,t) - \mathbf{v}_l(x,t) \right) + \mathbf{v}_l(x,t),$$

where

$$f_i(x,z',t)=\int_0^{z'}\frac{s^i}{\mu(x,s,t)}ds \tag{10}$$

and $\mathbf{v}_u = (u_u(x,t),v_u(x,t))$ and $\mathbf{v}_l = (u_l(x,t),v_l(x,t))$.

The flow is regarded as compressible according to the "arbitrary" pressure–density relation

$$\rho(p)=\rho_c f(p). \tag{11}$$

The function $f(p)$ is strictly increasing for $p \geq p_c$ and it satisfies $f(p) = 1$ when $p \leq p_c$.

With the aim set to derive an Reynolds type of equation for the pressure in the lubricant, the analysis continues by first formulating an expression for the mass flow and thereafter requiring continuity of the mass flow. By using Eq. (9) for the velocity field and Eq. (11) the mass flow, **q** becomes

$$\mathbf{q}(x,t)=\int_0^1\rho(p(x,t))\mathbf{u}(x,z',t)h(x,t)dz' \tag{12}$$

$$=\frac{\rho h}{2}a\mathbf{v}-\frac{\rho h^3}{12\mu_0 b}\nabla p,$$

where $\mathbf{v} = \mathbf{v}_u + \mathbf{v}_l = (u_u + u_l, v_u + v_l)$,

$$a(x,t)=\begin{pmatrix} a_1 & 0 \\ 0 & a_2 \end{pmatrix}, \tag{13}$$

$$a_1(x,t) = 2(u_u+u_l)^{-1}\left(\frac{\int_0^1 f_0(x,z',t)\,dz'}{f_0(x,1,t)}(u_u-u_l)-u_l\right),$$

$$a_2(x,t) = 2(v_u+v_l)^{-1}\left(\frac{\int_0^1 f_0(x,z',t)\,dz'}{f_0(x,1,t)}(v_u-v_l)-v_l\right), \tag{14}$$

$$\frac{1}{b(x,t)} = \int_0^1 12\mu_0\left(f_1(x,z',t)-\frac{f_1(x,1,t)}{f_0(x,1,t)}f_0(x,z',t)\right)dz'.$$

http://dx.doi.org/10.5772/63533

With this expression for $\mathbf{q}$, the requirement for conservation of mass reads

$$\frac{\partial(\rho h)}{\partial t} = -\nabla \cdot \mathbf{q}. \tag{15}$$

In order to incorporate the effect of cavitation, we assume that the following holds. In the full film zones, the pressure is larger than the cavitation pressure and the density is expressed by Eq. (11). In the cavitation zones, the pressure equals the cavitation pressure and the density is interpreted as degree of saturation δ. The saturation δ is an unknown in the cavitation zones and satisfies that $0 \le \delta \le 1$ while $\delta = 1$ in the full film zones. Hence the pressure–density relationship in both the full film and the cavitated zones becomes

$$\rho(p) = \rho_c \begin{cases} f(p), & p > p_c \\ \delta, & p = p_c \end{cases}. \tag{16}$$

Note that both p and δ are unknowns.

For computational purposes it is beneficial to introduce a change of variables $\bar{p}$ and η that satisfy the complementarity condition $\bar{p}\eta = 0$ in the whole domain. This can be accomplished by defining $\bar{p}$ and η as

$$\bar{p} = p - p_c, \quad \bar{p} \ge 0 \tag{17}$$

and

$$\eta = 1 - \delta = \begin{cases} 0, & \bar{p} > 0 \\ 1 - \delta, & \bar{p} = 0 \end{cases}. \tag{18}$$

In this notation, Eq. (16) becomes

$$\rho(\bar{p}) = \rho_c \begin{cases} f(\bar{p} + p_c), & \bar{p} > 0 \\ 1 - \eta, & \bar{p} = 0 \end{cases}. \tag{19}$$

Now, because of complementarity, this piecewise definition of ρ can be rewritten as a single expression valid throughout the whole domain, i.e.,

$$\rho = \rho_c\left((1-\eta) + f(\bar{p} + p_c) - 1\right) = \rho_c\left(f(\bar{p} + p_c) - \eta\right) \tag{20}$$

which then leads to the following expression for the mass flow

$$\mathbf{q} = \frac{\rho_c\left(f\left(\bar{p}+p_c\right)-\eta\right)h}{2}a\mathbf{v} - \frac{\rho_c\left(f\left(\bar{p}+p_c\right)-\eta\right)h^3}{12\mu_0 b}\nabla p$$

$$= \rho_c\left(\frac{fh}{2}a\mathbf{v} - \frac{fh^3}{12\mu_0 b}\nabla\bar{p} - \frac{\eta h}{2}a\mathbf{v}\right), \quad \bar{p} \geq 0. \tag{21}$$

Preservation of mass flow is ensured by inserting Eq. (21) and Eq. (20) into the continuity equation Eq. (15), which leads to the following mass preserving cavitation model:

$$\frac{\partial\left(fh-\eta h\right)}{\partial t} = -\nabla\cdot\left(\frac{fh}{2}a\mathbf{v} - \frac{fh^3}{12\mu_0 b}\nabla\bar{p} - \frac{\eta h}{2}a\mathbf{v}\right)$$
$$\bar{p} \geq 0, \quad 0 \leq \eta \leq 1, \quad \bar{p}\eta = 0. \tag{22}$$

The system in Eq. (22) can be solved numerically by the LCP-based solution procedure described in Section 3. When the solution $(\bar{p},\eta)$ has been obtained, the fluid pressure p and the saturation δ can be found from (17) and (18).

3. Numerical solution procedure

In this section, we present a numerical solution procedure for the cavitation model Eq. (22) such that the standard theory for linear complementary problems (LCP) can be applied.

Let us start by introducing the notation

$$\alpha = \frac{fh^3}{12\mu_0 b} \quad \text{and} \quad \beta = \left(\beta_1, \beta_2\right) = \frac{h}{2}a\mathbf{v},$$

while keeping in mind that f, h, a and b depend on p. In this notation the system Eq. (22) reads

$$\frac{\partial\left(fh\right)}{\partial t} - \frac{\partial\left(\eta h\right)}{\partial t} = \nabla\cdot\left(f\beta - \alpha\nabla\bar{p}\right) - \nabla\cdot\left(\eta\beta\right)$$
$$\bar{p} \geq 0, \quad 0 \leq \eta \leq 1, \quad \bar{p}\eta = 0. \tag{23}$$

A spatial finite difference discretization of the problem Eq. (23) can be obtained by dividing the domain into a uniform rectangular grid with $N_1 \times N_2$ elements of size $\Delta x_1 \times \Delta x_2$. The following notation is adopted $x_1^i = i\,\Delta x_1$, $x_2^j = j\,\Delta x_2$, where $i = 0,\ldots,N_1$, $j = 0,\ldots,N_2$ and

$$u_{i,j} := u\left(x_1^i, x_2^j, t\right).$$

Since the finite difference approximations of the partial derivatives w.r.t. x_1 and x_2 are obtained in a similar manner, the details (given below) are only provided for the partial derivatives w.r.t. x_1. The problem at hand, is elliptic in the full film domain, where $\eta = 0$. In the caviatted regions, where $\bar{p} = 0$, we observe that the equation is hyperbolic in η. A central difference scheme is therefore used to approximate the derivatives of the first term in the right hand side of Eq. (23) while an upwind difference scheme is employed for the second term.

In order to present the finite differences we introduce the notation

$$a_{i\pm 1/2,j} = \frac{a_{i\pm 1,j} + a_{i,j}}{2}$$

and the approximation

$$\left.\frac{\partial u}{\partial x_1}\right|_{i+1/2,j} \approx \frac{u_{i+1,j} - u_{i,j}}{\Delta x_1} \text{ and } \left.\frac{\partial u}{\partial x_1}\right|_{i-1/2,j} \approx \frac{u_{i,j} - u_{i-1,j}}{\Delta x_1}.$$

In this notation, the first term in the right hand side of Eq. (23) becomes:

$$\frac{\partial}{\partial x_1}\left(f\beta_1 - \alpha\frac{\partial \bar{p}}{\partial x_1}\right)$$

$$\approx \frac{(f\beta_1)_{i+1,j} - (f\beta_1)_{i-1,j}}{2\Delta x_1} - \frac{\alpha_{i+1/2,j}\left.\frac{\partial \bar{p}}{\partial x_1}\right|_{i+1/2,j} - \alpha_{i-1/2,j}\left.\frac{\partial \bar{p}}{\partial x_1}\right|_{i-1/2,j}}{\Delta x_1}$$

$$\approx \frac{(f\beta_1)_{i+1,j} - (f\beta_1)_{i-1,j}}{2\Delta x_1}$$

$$-\frac{\left(\frac{\alpha_{i+1,j}+\alpha_{i,j}}{2}\right)\left(\frac{\bar{p}_{i+1,j}-\bar{p}_{i,j}}{\Delta x_1}\right)-\left(\frac{\alpha_{i-1,j}+\alpha_{i,j}}{2}\right)\left(\frac{\bar{p}_{i,j}-\bar{p}_{i-1,j}}{\Delta x_1}\right)}{\Delta x_1}$$

$$=\frac{(f\beta_1)_{i+1,j}-(f\beta_1)_{i-1,j}}{2\Delta x_1}$$

$$-\frac{1}{\Delta x_1^{\,2}}\left(\alpha_{i-1/2,j}\bar{p}_{i-1,j}-\left(\alpha_{i-1/2,j}+\alpha_{i+1/2,j}\right)\bar{p}_{i,j}+\alpha_{i+1/2,j}\bar{p}_{i+1,j}\right). \tag{24}$$

The upwind discretization of the second term reads

$$\frac{\partial}{\partial x_1}(\eta\beta_1)\approx\frac{\eta_{i,j}\beta_{1i,j}-\eta_{i-1,j}\beta_{1i-1,j}}{\Delta x_1}. \tag{25}$$

By using Eq. (24), Eq. (25) and the corresponding finite difference approximations of the partial derivatives w.r.t. x_2, the spatially discretized (which is still continuous in time) form of the system Eq. (23) can be written as

$$\left(\frac{\partial(fh)}{\partial t}\right)_{i,j}-\left(\frac{\partial(\eta h)}{\partial t}\right)_{i,j}=\frac{(f\beta_1)_{i+1,j}-(f\beta_1)_{i-1,j}}{2\Delta x_1}+\frac{(f\beta_2)_{i,j+1}-(f\beta_2)_{i,j-1}}{2\Delta x_2}$$

$$-\frac{1}{\Delta x_1^{\,2}}\left(\alpha_{i-1/2,j}\bar{p}_{i-1,j}-\left(\alpha_{i-1/2,j}+\alpha_{i+1/2,j}\right)\bar{p}_{i,j}+\alpha_{i+1/2,j}\bar{p}_{i+1,j}\right)$$

$$-\frac{1}{\Delta x_2^{\,2}}\left(\alpha_{i,j-1/2}\bar{p}_{i,j-1}-\left(\alpha_{i,j-1/2}+\alpha_{i,j+1/2}\right)\bar{p}_{i,j}+\alpha_{i,j+1/2}\bar{p}_{i,j+1}\right) \tag{26}$$

$$-\frac{(\eta\beta_1)_{i+1,j}-(\eta\beta_1)_{i-1,j}}{2\Delta x_1}-\frac{(\eta\beta_2)_{i,j+1}-(\eta\beta_2)_{i,j-1}}{2\Delta x_2},$$

together with the conditions

$$\bar{p}_{i,j}\geq 0,\quad 0\leq\eta_{i,j}\leq 1,\quad \bar{p}_{i,j}\eta_{i,j}=0.$$

Since h and f depend on p, a numerical solution procedure can be posed as follows:

1. Guess a pressure distribution.
2. Compute the coefficients f, h α and β, in order to linearize Eq. (26) and pose it on the form

$$\bar{p} = q + M\eta, \quad \bar{p} \geq 0, \quad 0 \leq \eta \leq 1, \quad \bar{p} \cdot \eta = 0,$$

where the vector q and the matrix M are constants and where the unknowns $\bar{p}$ and η are vectors with $(N_1 - 1) \cdot (N_2 - 1)$ elements.

3. Solve the now obtained linear complementarity problem corresponding to Eq. (26), with the Lemke algorithm [18].

To fully discretize the problem at hand several approaches can be applied. For instance, first order forward (explicit) or backward (implicit) Euler, the second order implicit Crank-Nicolson method.

4. Numerical examples

In this section, the numerical solution procedure, for the cavitation model in Eq. (22), developed in Section 3, is examined by considering four different one-dimensional slider bearing examples. In all four examples, only the lower surface is moving, i.e. $\mathbf{v} = (u_l, 0)$, the lubricant is assumed to be Newtonian, the length of the bearing is L and the load the bearing supports is w_0. In the first example, results obtained with the present approach are compared to results for the one-dimensional parabolic slider in [8]. Two different pressure–density relationships are examined. The second example extends the first one in the sense that besides cavitation the bearing surfaces are also assumed to be linear elastic. In particular, this means that the film thickness in addition depends on the pressure. The film thickness, in this case, consists of three parts; a parameter related to the minimum film thickness h_0, the bearing geometry g, and the elastic deformation of the bearing surfaces. The Boussinesq-Cerruti half space solution is used to model the elastic deformation [2]. Indeed, h is given by

$$h(x, p(x)) = h_0 + g(x) - \frac{4}{\pi E'} \int_{-\infty}^{\infty} \ln|x - s| \, p(s) \, ds, \tag{27}$$

where

$$\frac{2}{E'} = \frac{1 - \nu_l}{E_l} + \frac{1 - \nu_u}{E_u}, \tag{28}$$

where v_l and v_u are the Poisson's ratios for the lower and upper surfaces and E_l and E_u are the corresponding Young's modulus.

The third example considers a double parabolic slider, which in addition to the single parabolic slider, exhibits reformation and highlight that mass is conserved.

The fourth and last example considers a quadruple parabolic slider bearing. The reason for choosing this configuration is to test the hypothesis that an elastically deformable bearing, in general, does not generate more film than the corresponding rigid one.

In all examples, the initial (undeformed) bearing geometry consists of 1, 2 or 4 parabolic parts. These bearing geometries, g_n, can then be described via the L-periodic auxiliary function G

$$g_n(x) = G(x/n), \quad n = 1,2 \text{ or } 4, \tag{29}$$

where

$$G(x) = h_1\left(\frac{2}{L}\right)^2\left(x-\frac{L}{2}\right)^2, \quad 0 \le x \le L. \tag{30}$$

In **Table 1**, parameters common to all four examples are listed. These are the bearing parameters L and h_1, the speed of the lower surface u_l, the applied load w_0, the cavitation pressure p_c, the dynamic viscosity μ and the boundary conditions for the normalized density at $x = 0$ and $x = L$, denoted by ρ_0/ρ_c and ρ_L/ρ_c, respectively.

Example 1

In this example, a model problem with rigid surfaces and two different Newtonian lubricants is considered in order to compare with previous results presented in [8]. Indeed, a single parabolic slider bearing of length L with rigid surfaces and film thickness of the form

$$h(x) = h_0 + g_1(x), \tag{31}$$

is considered here. Two different Newtonian lubricants are studied, one which obeys the constant bulk modulus pressure–density relationship;

$$\rho = \rho_c \exp((p - p_c)/\beta) \tag{32}$$

while the other one obeys the Dowson-Higginson pressure–density relationship;

$$\rho = \rho_c \frac{C_1 + C_2(p - p_c)}{C_1 + (p - p_c)}. \tag{33}$$

The parameters for the two different lubricants are given in **Table 2**. As in [8], the case with constant bulk modulus, β, was solved for a fixed film thickness, h_β, where $h_0 = h_1 = 25.4$ μm. The pressure solution, p_β, obtained in this way corresponds to a load carrying capacity $w_0 \approx 117$ kN/m. In the case with the Dowson-Higginson compressibility, force-balance was included and the load carrying capacity obtained for the constant bulk modulus case was used as the applied load. This resulted in the film thickness h_{DH} with the minimum $h_0 = 25.8$ μm and the pressure p_{DH}. In **Figure 2** the film thickness and pressure distributions corresponding to the two different cases are depicted. It can be seen that the pressure distribution p_β is in good agreement with the results in [8].

L	h_1	u_l	w_0	p_c	μ	ρ_0/ρ_c	ρ_L/ρ_c
76.2 mm	25.4 μm	4.57 m/s	117 kN/m	100 kPa	0.039 Pas	1.0001	1

Table 1. Parameters common to all four examples.

β	C_1	C_2
0.069 GPa	2.22 GPa	1.66

Table 2. Parameters specific to Example 1.

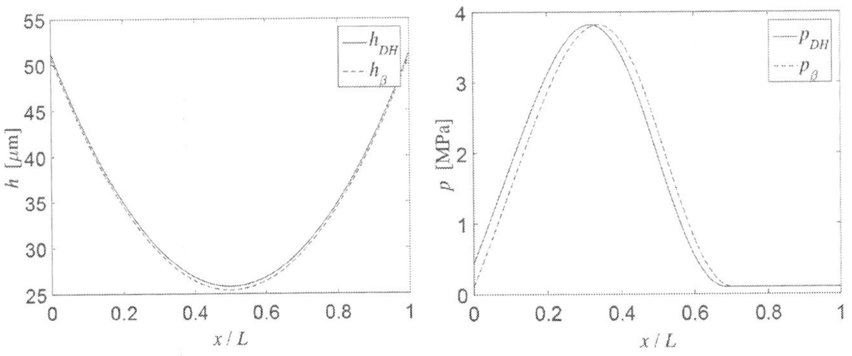

Figure 2. Film thickness (left) and pressure distributions (right) for two lubricants obeying constant bulk modulus β and Dowson-Higgison pressure-density relationships, respectively. The solutions correspond to the single parabolic slider defined in Example 1.

Example 2

This example extends the previous one, by including elastic deformation of the bearing surfaces. To illustrate the effect of surface deformation, the combined elastic modulus, E', was chosen to be 10% of the one for two steel surfaces with Poisson's ratio v and Young's modulus E, listed in **Table 3**. The lubricant was assumed to obey the constant bulk modulus, pressure –density relationship, with β = 3.34 as listed in **Table 3**. Both of the bearings carry the same load as in Example 1, i.e., w_0 = 117 kN/m.

β	v	E
3.34 GPa	0.3	210 GPa

Table 3. Parameters specific to Examples 2, 3 and 4.

In **Figure 3**, the film thickness h_e and pressure distribution p_e for the elastic bearing are depicted together with the solutions h_r and p_r for the rigid bearing. In particular, it can be seen that the elastic deformation leads to a thicker film, a smaller zone of cavitation and a lower maximum pressure. The minimum film thickness for the rigid bearing is 25.8 μm and for the bearing with elastic surfaces it is 30.4 μm.

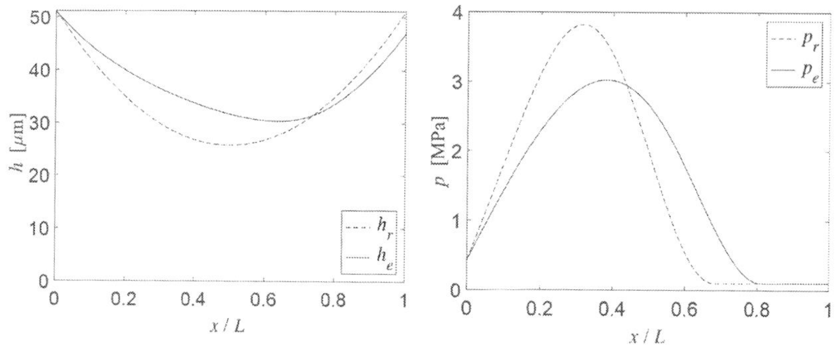

Figure 3. Film thickness (left) and pressure distributions (right) for the rigid and the elastic single parabolic slider bearings in Example 2.

Example 3

The only difference between the problem studied in this example and the one studied in Example 2, is that the bearing geometry now corresponds to a double parabolic slider, g_2, defined through Eq. (29). This example was chosen to illustrate that the proposed LCP-based numerical method renders a solution, which corresponds to a mass conserving flow.

The film thickness h_e and pressure distribution p_e for the elastic bearing are depicted together with the solutions h_r and p_r for the rigid bearing in **Figure 4**. In a mass conserving flow in a rigid double parabolic slider (described with g_2), the maximum compression occurs twice, once underneath each of the two parabolas. Since the compression is equal in these two points, the pressure, which is the maximum pressure, is also equal in these two points. It is clear from **Figure 4** that the pressure solution for the rigid bearing fulfils these conditions.

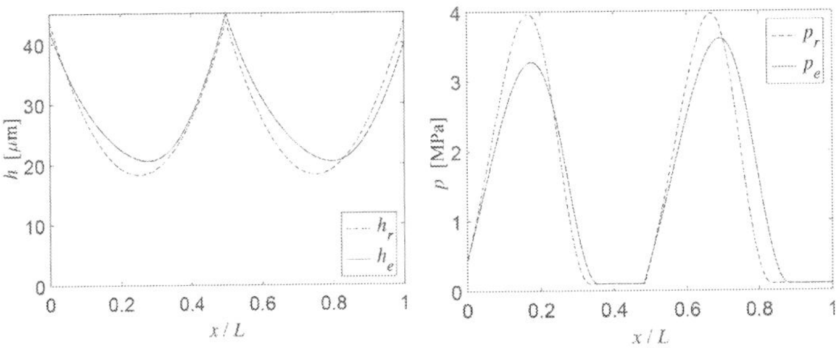

Figure 4. Film thickness (left) and pressure distributions (right) for the rigid and the elastic double parabolic slider bearings in Example 3.

Also, as in Example 1, the bearing with elastic surfaces generates a thicker lubricant film, a smaller zone of cavitation and a lower maximum pressure.

Example 4

The only difference between this example and Examples 2 and 3, is that the bearing geometry now corresponds to a quadruple parabolic slider, g_4. In both Examples 2 and 3, the elastic bearing generated thicker films, smaller zone of cavitation and lower maximum pressures than the corresponding rigid ones. This example was constructed to show that this is not true in general. Indeed, as shown in **Figure 5**, the film thickness h_e for the elastic bearing is thinner, some of the zones of cavitation become longer and the pressure underneath the first and the last parabola is higher than for the rigid bearing.

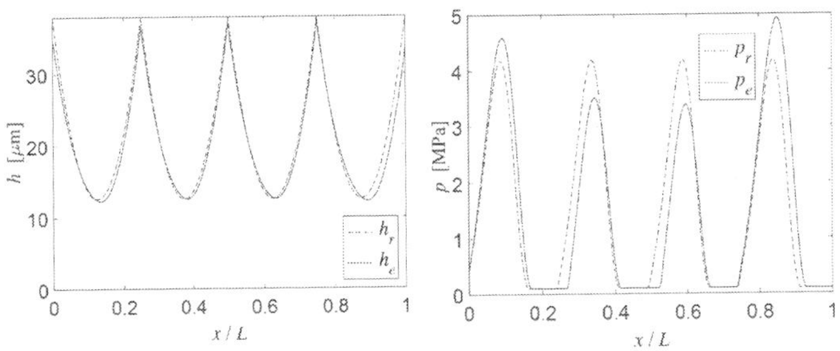

Figure 5. Film thickness (left) and pressure distributions (right) for the rigid and the elastic quadruple parabolic slider bearings in Example 4.

5. Concluding remarks

A new mathematical model for thin film lubrication has been derived. The model is quite general, e.g., it accounts for lubricant compressibility, cavitation, pressure dependent viscosity, non-Newtonian rheology and elastic deformation of the surfaces. The main novelty of the model is that cavitation of a compressible fluid is considered via a formulation of the mass flow, which ultimately results in a complementarity problem. Hence, standard methods developed for linear complementarity problems can be used in the numerical solution procedure. The model's applicability has been demonstrated in several numerical examples.

Acknowledgements

The authors want to acknowledge The Swedish Research Council (VR) for the financial support through the grants no. 2013-4978 and 2014-4894.

Author details

Andreas Almqvist[1*] and Peter Wall[2]

*Address all correspondence to: andreas.almqvist@ltu.se

1 Machine Elements, Luleå University of Technology, Luleå, Sweden

2 Mathematical Sciences, Luleå University of Technology, Luleå, Sweden

References

[1] B. J. Hamrock. Fundamentals of Fluid Film Lubrication. McGraw-Hill, Inc, New York, 1994.

[2] R. Larsson. EHL governing equations. In Q. J. Wang and Y.-W. Chung, editors, Encyclopedia of Tribology, pp. 827–832. Springer, USA, 2013.

[3] L. Floberg. Cavitation boundary conditions with regard to the number of streamers and tensile strength of the liquid. Cavitation and related phenomena in lubrication, ImechE, England, pp. 31–36, 1974.

[4] B. Jakobsson and L. Floberg. The finite journal bearing, considering vaporization. Gumperts Förlag, Sweden, 1957.

[5] K.-O. Olsson. Cavitation in dynamically loaded bearings. Scandinavian Univ. Books, Sweden, 1965.

[6] H. G. Elrod. A cavitation algorithm. Journal of Tribology, 103:350–354, 1981.

[7] D. Vijayaraghavan and T. G. Keith Jr. Development and evaluation of a cavitation algorithm. STLE Tribology Transactions, 32(2):225–233, 1989.

[8] F. Sahlin, A. Almqvist, R. Larsson, and S. Glavatskih. A cavitation algorithm for arbitrary lubricant compressibility. Tribology International, 40(8):1294–1300, 2007.

[9] G. Bayada, S. Martin, C. Vézquez, et al. Two-scale homogenization of a hydrodynamic elrod-adams model. Asymptotic Analysis, 44(1):75–110, 2005.

[10] G. Bayada, S. Martin, and C. Vézquez. An average flow model of the Reynolds roughness including a mass-flow preserving cavitation model. Journal of Tribology, 127(4):793–802, 2005.

[11] G. Bayada, S. Martin, and C. Vazquez. Micro-roughness effects in (elasto)hydrodynamic lubrication including a mass-flow preserving cavitation model. Tribology International, 39(12):1707–1718, 2006.

[12] M. Giacopini, M. T. Fowell, D. Dini, and A. Strozzi. A mass-conserving complementarity formulation to study lubricant films in the presence of cavitation. Journal of Tribology, 132(4):041702, 2010.

[13] R. W. Cottle, J. S. Pang, and R. E. Stone. The linear complementarity problem. Number 60. Society for Industrial Mathematics, 2009.

[14] A. Almqvist, J. Fabricius, R. Larsson, and Peter Wall. A new approach for studying cavitation in lubrication. Journal of Tribology, 136(1):011706-1–011706-6, 2014.

[15] L. Bertocchi, D. Dini, M. Giacopini, M. T. Fowell, and A. Baldini. Fluid film lubrication in the presence of cavitation: a mass-conserving two-dimensional formulation for compressible, piezoviscous and non-Newtonian fluidsbaya. Tribology International, 67:61–71, 2013.

[16] K. L. Johnson. Contact Mechanics. Cambridge University Press, Cambridge, UK, 1984.

[17] K. R. Rajagopal and A. Z. Szeri. On an inconsistency in the derivation of the equations of elastohydrodynamic lubrication. Proceedings of the Royal Society of London. Series A: Mathematical, Physical and Engineering Sciences, 459(2039):2771–2786, 2003.

[18] A. Almqvist, A. Spencer, and P. Wall. A pivoting algorithm solving linear complementarity problems. http://www.mathworks.com/matlabcentral/fileexchange/41485, 2013.

Chapter 10

Preparation of Nickel-Based Nanolubricants and Investigation of their Tribological Behavior

Yujuan Zhang, Shengmao Zhang, Pingyu Zhang,
Guangbin Yang and Zhijun Zhang

Additional information is available at the end of the chapter

http://dx.doi.org/10.5772/64592

Abstract

In situ surface-modification technique is adopted in present research to fabricate a series of Ni nanoparticles as well as Cu@Ni nanoparticles with different size and morphology. The correlation among the composition, structure, size, and morphology and tribological properties of as-synthesized additives were explored, and the friction-reducing, antiwear, and worn surface self-healing mechanisms of the additives were discussed. It was found that Ni nanoparticles with a smaller size show higher surface activity and can readily deposit on the sliding surface and form a stable and continuous protective layer thereon. Compared with sphere-like and triangular rod-like Ni nanoparticles, triangular plate-like Ni nanoparticles are more liable to form protective layer. Compared to Ni-based nanolubricants, as-synthesized Cu@Ni nanolubricants exhibit better friction-reducing, antiwear, and extreme pressure properties. It is because the highly active Ni nanocores and O- and N-containing organic modifying agents can readily form boundary lubricating film on sliding steel surfaces, while Cu nanocores can easily deposit on sliding steel surface to form a protective layer (self-healing film) thereon. Ni nanoparticles as nanoadditives in solid-liquid lubricating system significantly reduce the friction in all lubrication regimes: As a nanolubricant, Ni nanoparticles exhibit popular and effective friction-reducing, antiwear, and extreme pressure properties.

Keywords: nickel nanoparticles, Cu@Ni nanolubricant, DLC-Ni solid liquid composite lubrication system, preparation, tribological behavior

1. Introduction

Recently, nanoparticles used as lubricant additives have received increased attention because of their excellent antiwear and friction-reducing properties [1, 2]. Several tribological mechanisms of nanolubricants have been proposed such as deposited film, surface alloying, cold welding, and rolling effect [3–7]. It has been well recognized that tribological properties of nanoparticles can be affected by their chemical composition, particle size, and morphology, in which, however, the exact tribological mechanisms of nanolubricants still remain unknown. On the other hand, their application in the field of tribology is still hindered by their poor dispersibility because of their high surface energy and activity leading to agglomeration in lubricant base oils. Therefore, it is vital to improve the dispersion capacity of nanoparticles in base oil through *in situ* surface modification with a variety of organic molecules [1, 2]. Compared with the *in situ* surface-modification technology established in the early 1990s [3, 7–14], current surface-modification technique for nanolubricants is inconvenient and less cost-effective, which involves several steps including synthesis of nanoparticles and follow-up dispersion of nanoparticles in lubricant base oils. In order to overcome the low productivity of nanoparticles and the difficultly in storing and dispersing the nanoparticles, it becomes significant to develop one-step method for preparing *in situ* surface-capped nanolubricants so as to promote their industrialization.

Unfortunately, current synthetic approaches for Ni nanoparticles with controllable size and shape are usually complicated and need to use a large number of solvents, such as polyol synthesis [15], electrochemical reduction [16], microwave heating, reduction of organometallic precursors [17–19], and thermal decomposition of organic complexes [20–22]. More badly, even after surface modification with organic molecules, currently available Ni-based nanolubricants are still liable to aggregate and can hardly be monodispersed in lubricant base oil [23, 24]. On the other hand, it was more ideal to study the relationship between particle size, morphology of metal nanolubricants, and its tribological mechanism through one kind of metal nanoparticle.

In this chapter, a facile *in situ* one-step thermal decomposition route is used to prepare size and morphology-controllable monodispersed Ni nanoparticles with a narrow size distribution, which can be dispersed in PAO6 base oil and used as nickel-based nanolubricants directly. The tribological properties of Ni nanolubricants in base oil demonstrated strong dependence on the size, morphology, and composition. Meanwhile, the Ni nanoparticle significantly reduces the friction of diamond-like carbon (DLC)/di-iso-octyl sebacate (DIOS) liquid-solid composite lubricating system in all lubrication regimes: boundary, mixed, and elasto-hydrodynamic lubrication. This *in situ* one-step thermal decomposition route probabilizes popular, effective antiwear and friction-reducing properties by applying Ni-based nanolubricants.

2. The effect of size of Ni nanolubricant on its tribological properties

The chemical reaction for *in situ* preparation of Ni nanoparticles is expressed in **Figure 1**. OAM(oleylamine) is a lot in comparison to nickel, but actually, OAM is not only coordination ligand of nickel acetate by amidocyanogen in early solution, but also reducing agent and surface-capping agent for Ni nanoparticles. Upon completion of the reaction, decomposition byproducts of $Ni(HCOO)_2 \cdot 2H_2O$ are evaporated under a current of inert gas at high temperature and only surface-capped Ni nanoparticles by oleylamine and oleic acid remain in PAO6 base oil, and excess oleylamine and oleic acid in PAO6 do not matter because these are also lubricant additives, so nickel-based nanolubricants are obtained directly by this *in situ* one-step route. This should be critical to the tribological application of nickel-based nanolubricants. As shown in **Figure 2**, there are three characteristic peaks at 44.5°, 51.8°, and 76.4° appearing in the XRD patterns of Ni nanoparticles with different sizes. The three peaks can be assigned to the diffractions of (111), (200), and (220) crystal planes of pure face-centered cubic (FCC) Ni (JCDPS No.7440-02-0). The diffraction peaks are significantly broadened, which indicates that the crystalline size of Ni nanoparticles is very small.

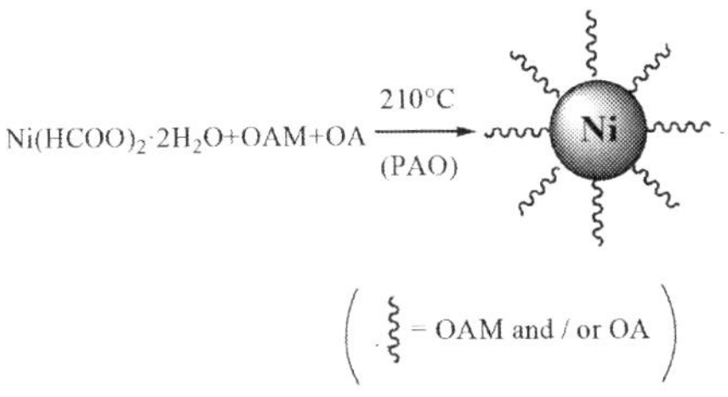

Figure 1. Chemical reaction for preparing Ni nanoparticles.

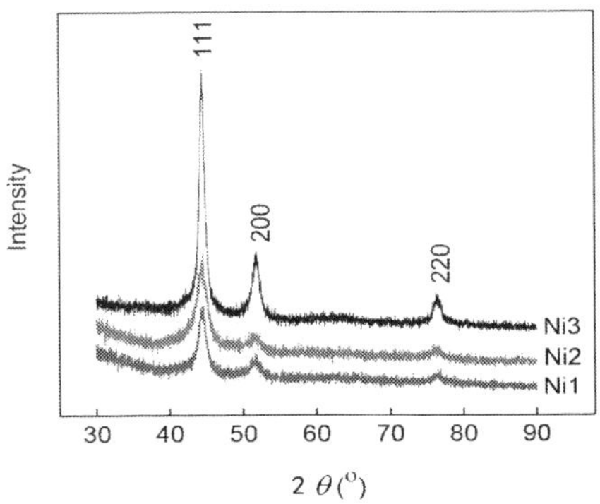

Figure 2. XRD patterns of Ni nanoparticles with different size.

The TEM images of Ni nanoparticles with different size as well as the size distribution are shown in **Figure 3**. It can be seen that Ni nanoparticles have sphere-like morphology, and through varying dosage of PAO6 from 80 to 40 and 20 mL their average diameter is tunable from about 7.5 to 13.5 and 28.5 nm. As-synthesized Ni nanoparticles show no sign of obvious aggregation. It can be deduced that the nucleation and growth process of Ni nanocrystalline can be adjusted by OAM and OA through controlling chemical reaction driving force and affording monodispersed Ni nanoparticles with different sizes. The simultaneous presence and relative amounts of binary modifiers OAM and OA are paramount. OAM is the reductant and at the same time controls the nucleation rate. OA as an efficient surface-modification agent are favorable for probing capping mechanism and affording size-tunable monodispersed Ni nanoparticles. As is the experimental result, the higher ligands concentrations led to the formation of the particles with larger sizes, because the higher the concentration of the ligands, the lower the monomer reactivity would be; therefore, there would be fewer nucleuses formed and resulted in larger nanoparticles [25, 26].

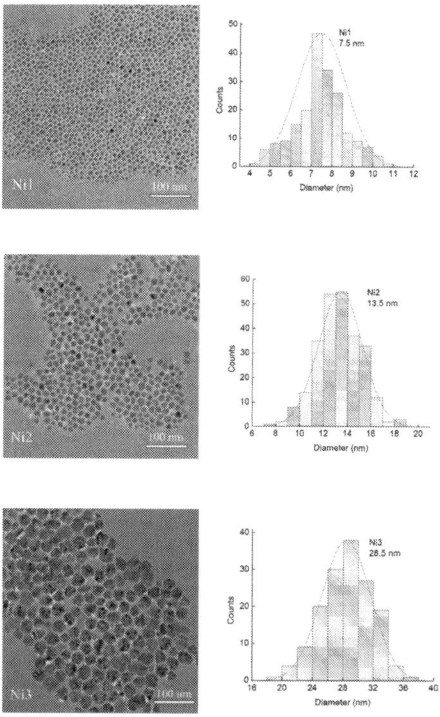

Figure 3. TEM images and size distribution of surface-capped Ni nanoparticles with different size.

Figure 4 shows the optical picture of the nickel nanoparticles in PAO6 after a month. The results indicate that the nickel nanoparticles are still equably distributed in PAO6 after a month and there is no obvious agglomeration behavior, which fits in quite with the TEM analysis.

FTIR(Fourier transform infrared spectrometer) analysis was conducted to study the interaction between organic surface modifier and Ni nanocore. As shown in **Figure 5**, the characteristic absorption bands of pure OA at 1711 and 939 cm^{-1} correspond to the stretching vibrations of C=O and the bending vibration of —OH of carboxyl group. As to pure OAM, the characteristic absorption bands of amine group at 3367 and 3295 cm^{-1} can be assigned to the symmetric and asymmetric stretching vibrations of primary amine N—H, the peak at 1583 cm^{-1} can be assigned to the scissoring mode of $-NH_2$, and the one at 967 cm^{-1} can be assigned to the bending vibration mode of $-NH_2$ [27]. The characteristic peaks of oleic group at 2922 and 2852 cm^{-1} appeared in both pure OAM and OA, the bending vibration mode of C—H at 1464 cm^{-1}, and the methylene rocking vibration of a minimum of four methyl groups ($(CH_2)_4$) in the 720 cm^{-1} region [28]. As to surface-modified Ni nanoparticles, the presence of OAM and OA as capping agents on the surface are confirmed by the characteristic peaks of oleic group at 2922, 2852, 1464, and 720 cm^{-1}. However, surface-capped Ni nanoparticles do not show the signal of primary amino group at 3367, 3295, 1583, and 967 cm^{-1} and the signal of free carboxyl group at 1711 and 939 cm^{-1}, at the same time some new absorption peaks of asymmetric and symmetric stretching vibrations of —COO— at 1543 and 1460 cm^{-1} and stretching vibration of C—N at 1071 cm^{-1} appeared, which indicates that two oxygen atoms in OA are symmetrically coordinated to the surface of Ni nanocore, and the OA chain is attached in a bidendate mode [29]. Therefore, it can be deduced that the organic surface modifiers OAM and OA chemically bond on the surface of Ni nanoparticles that prevents nanoparticles from oxidation and aggregation, which improved their compatibility with lubricating base stocks.

Figure 4. Picture of the nickel nanoparticles in PAO6 after a month.

TGA/DTA curves of surfaced-modified Ni nanoparticles are shown in **Figure 6**. As-synthesized Ni nanoparticles lose about 10% weight in the temperature range of 28~500°C. Within 28~250°C the slow weight loss corresponds to the volatilization of a small amount of OAM and OA physically adsorbed on the surface of Ni particles. Around 256°C the rapid weight loss is due to the decomposition of chemically adsorbed OAM and OA. Over 566°C the weight loss can be due to the elimination of decomposed residuals [30]. At the same time, in the DTA curve the exothermic peak around 271°C is due to the thermal decomposition of organic modification layer. It can be deduced that OAM and OA chemically bonded with Ni nanocores favor to improve the thermal stability of Ni nanoparticles.

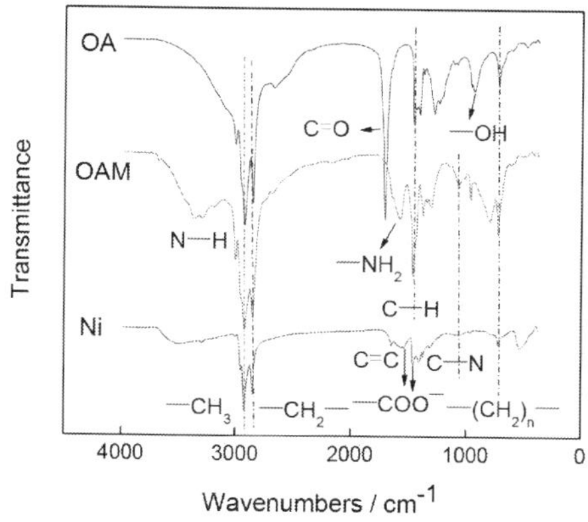

Figure 5. FTIR spectra of OA, OAM, and as-separated Ni nanoparticles.

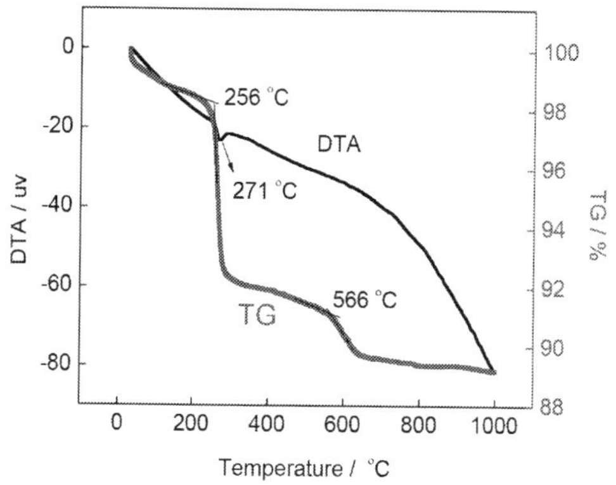

Figure 6. TG-DTA curves of as-separated Ni nanoparticles.

The variation of coefficient of friction (COF) (a) and wear scar diameter (WSD) (b) with the concentration of Ni nanoparticles in PAO6 are shown in **Figure 7** (four-ball friction and wear tester, load 300 N, rotary speed 1450 rpm, duration 30 min; for a comparison the data for PAO6 containing modifier are also given). It can be seen that the friction coefficient increases slightly with different Ni nanoparticles size and the surface-modifying agents, but the antiwear ability of base stock PAO6 is significantly improved. Moreover, their size and concentration is closely related with the antiwear ability of Ni nanoparticles in PAO6. At the additive concentration of 0.05% (mass fraction), sample Ni1 with smaller average diameter is more effective than samples Ni2 and Ni3 in reducing WSD of the steel-steel contact. It is consistent with the well-known Hall-Petch effect and Archard's law, which is mostly used in adhesion and abrasive wear conditions [31]. Besides, direct contact between the sliding steel surfaces can be avoided by the deposition and/or adsorption films of surface-modified Ni nanoparticle on rubbed steel surfaces thereby retarding adhesion wear and increasing the antiwear ability of PAO6 base oil.

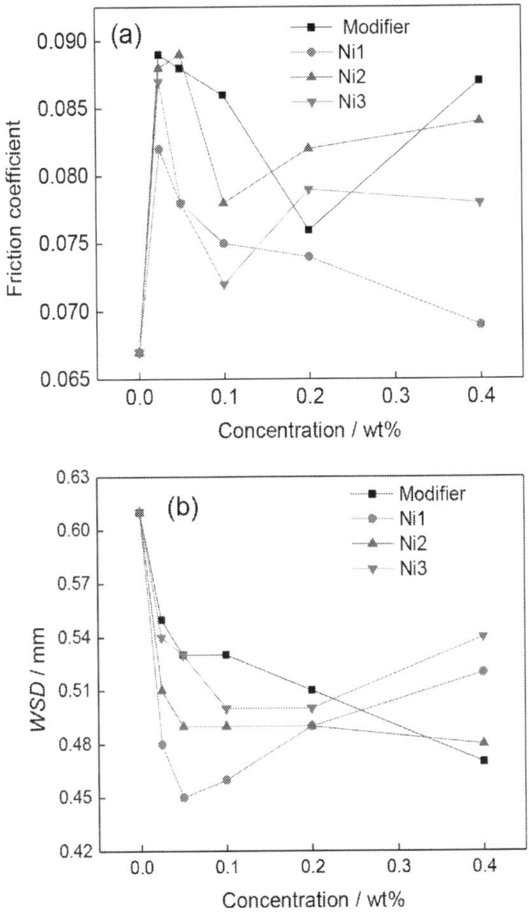

Figure 7. Variation of COF (a) and WSD (b) with the concentration of Ni nanoparticles.

The COF and WSD as functions of load under the lubrication of PAO6, PAO6 + modifier, and PAO + 0.05% Ni nanoparticles are shown in **Figure 8**. It can be seen that with rising load the COF decreases initially and increases later, and WSD increases monotonously therewith. Besides, under lower load, the applied force concentrates on a small number of large particles, which results in a higher friction coefficient owing to severe abrasion. Under higher load, large nanoparticles will undergo deformation, which leads to an even distribution of applied force on contact area thereby reducing COF. Besides, Ni nanoparticles with smaller size can improve the antiwear ability of PAO6 base oil, possibly because they are more easily chemically adsorbed on steel sliding surfaces to form boundary lubricating films.

Figure 9 shows the P_B and P_D values of various lubricant systems. PAO6 doped by Ni nanoparticles with different size has a higher P_B value than PAO6 and PAO6 + surface modifiers, and the P_B value of the lubricants is irrelevant to the size of the Ni nanoparticles. It can be deduced that Ni nanoparticles are able to improve the load-carrying capacity of PAO6.

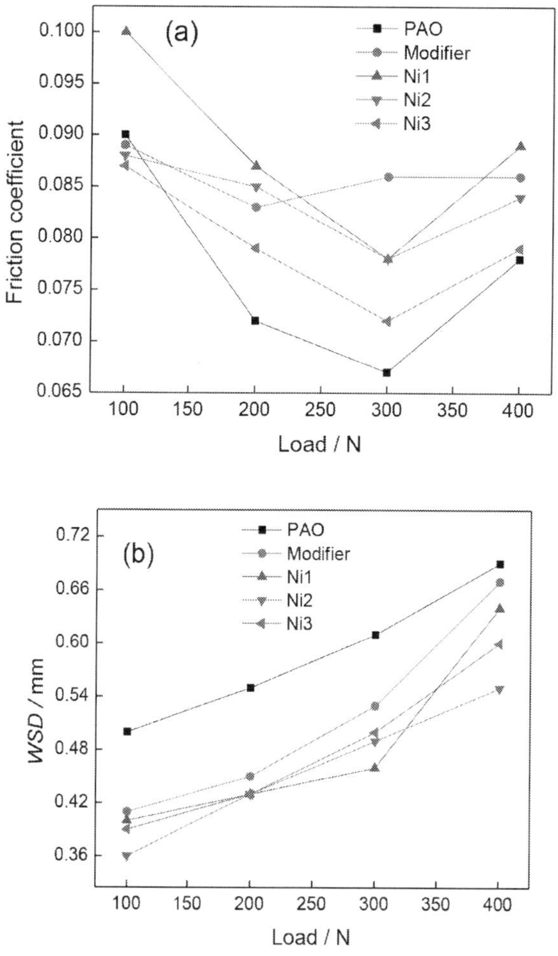

Figure 8. Friction coefficient (a) and wear scar diameter (b) versus applied load under the lubrication of PAO6, PAO6 + surface-modifiers, and nanolubricants containing 0.05% Ni nanoparticles with different size (four-ball friction and wear tester, 1450 rpm, 30 minutes).

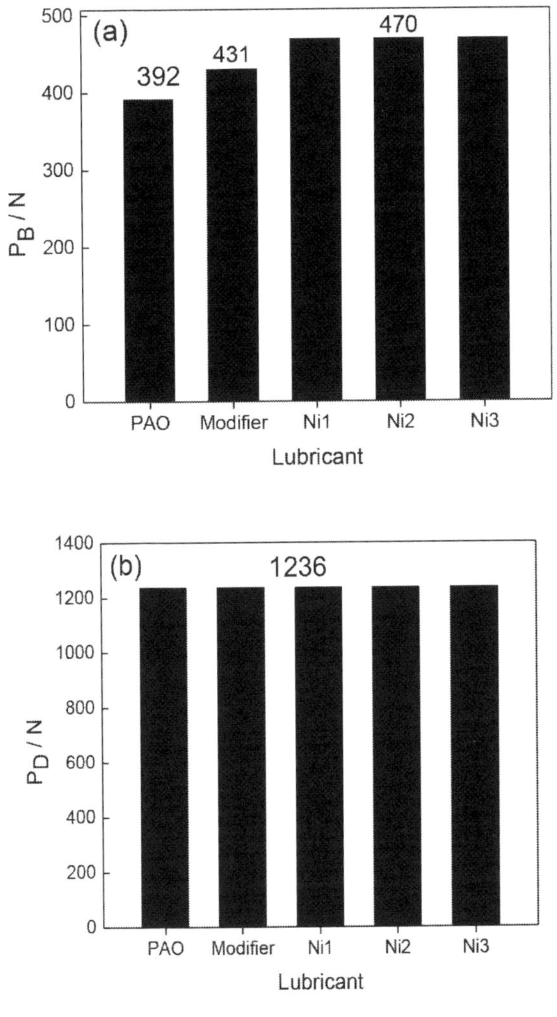

Figure 9. P_B(a) and P_D(b) values of various lubricant systems (four-ball friction and wear tester, 1450 rpm, 300 N).

SEM images of worn surfaces on steel balls lubricated by various lubricant systems are shown in **Figure 10**. It can be seen that the WSD of steel balls lubricated by PAO6 + Ni nanoparticles is smaller than that lubricated by pure PAO. Particularly, the smallest WSD is obtained under the lubrication of PAO6 + Ni1. Compared with PAO6 + Ni2 or PAO6 + Ni3, the worn surface on the steel ball lubricated by PAO6 + Ni1 is smooth and shows only shallow scratch grooves, which well conforms to the better antiwear ability of nanolubricants with smaller Ni nanoparticles (Ni1).

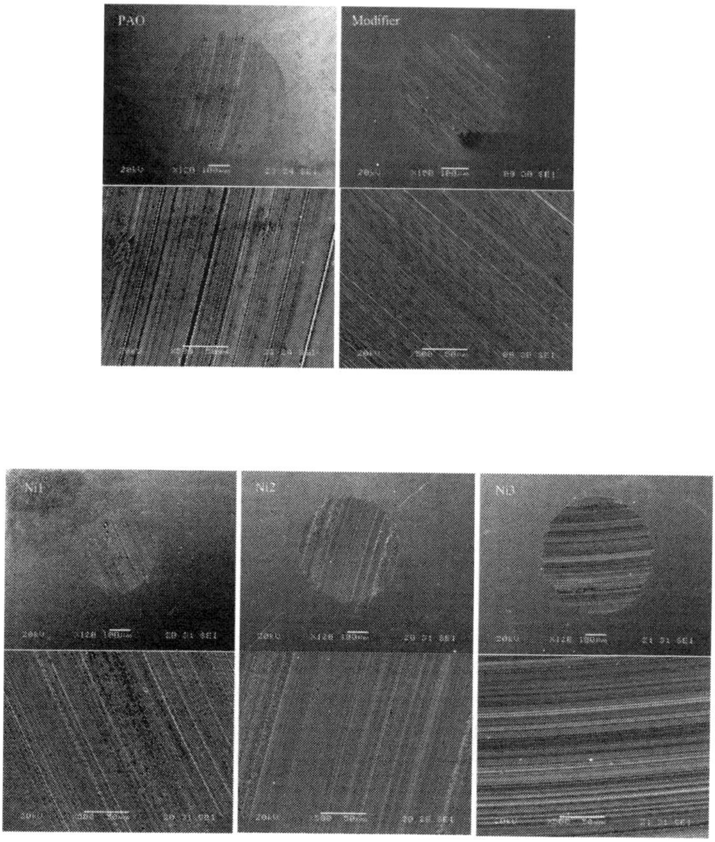

Figure 10. SEM micrographs of wear scars lubricated with PAO6 as well as PAO6 + surface-modifiers and nanolubricants containing 0.05% Ni nanoparticles (300 N, 30 minutes).

In order to reveal the friction-reducing and antiwear mechanism of Ni nanoparticles, chemical states of several typical elements on the worn surfaces of steel balls was analyzed by XPS. The XPS spectra of C 1s, O 1s, N 1s, Fe 2p, and Ni 2p on the worn surface of steel ball lubricated with PAO + 0.1% Ni1 nanoparticles are shown in **Figure 11**. Atomic concentrations of elements on worn surface are listed in **Table 2**. The C—N, COO—, C=C, $-CH_2-$, and —CH3 signals are attributed to chemically adsorbed and/or reacted surface modifier on sliding surfaces. The Fe 2p3/2 peaks at 710.90 and 713.16 eV are attributed to Fe_2O_3 [32]. It can be reasonably concluded that under selected experimental conditions ferrous oxides are formed on the worn surfaces of the steel balls. Moreover, Ni 2p3/2 peak at 852.58 eV and Ni 2p1/2 peak at 871.20 eV are attributed to Ni nanocores, which are released from surface-modified nanoparticles and transferred onto sliding surfaces during the rubbing process. Moreover, Ni 2p3/2 peak at 857.49 eV and Ni 2p1/2 peak at 876.20 eV (corresponding satellite peak is located at 862.40 eV and O

1s peak is located at 531.47 eV) indicated that a part of Ni nanocores can react with oxygen to form Ni_2O_3. In addition, as shown in **Table 2**, a higher atomic concentration of Ni, O, and N appeared on the worn surface lubricated by PAO6 + Ni1 than those lubricated by PAO6 + Ni2 and by PAO6 + Ni3, at the same time, the one lubricated by PAO6 + Ni3 has a higher atomic concentration of Fe. It can be deduced that nanolubricants with smaller size are able to form boundary lubricating film with good coverage on sliding surfaces thereby showing better antiwear ability.

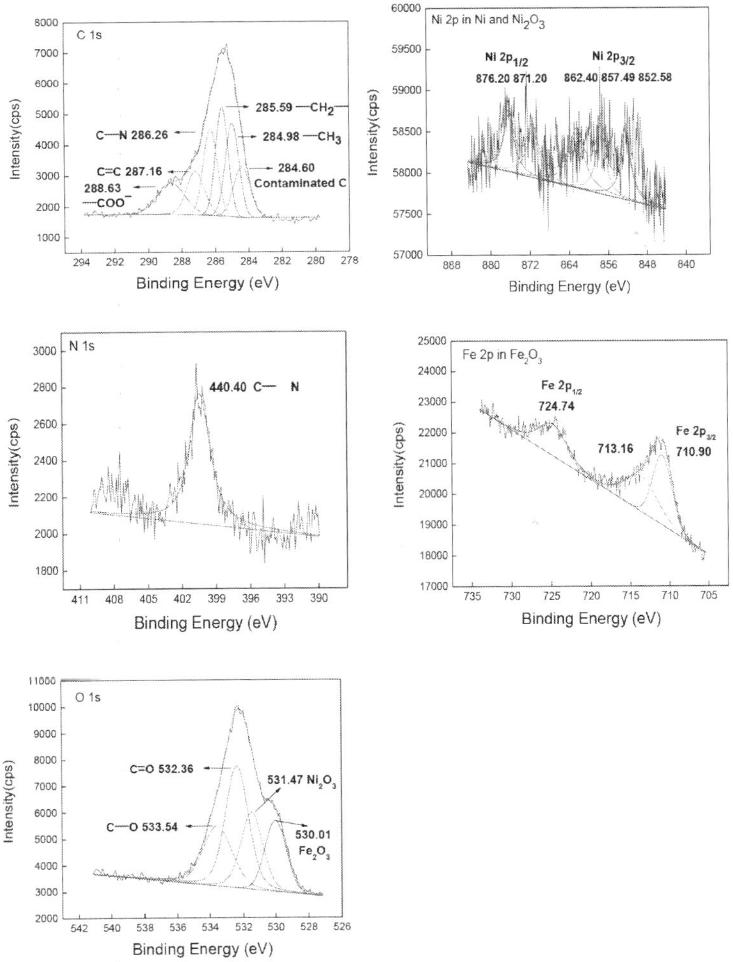

Figure 11. Curve-fitted XPS spectra of typical elements on rubbed surface of steel ball lubricated with nanolubricants containing 0.1% Ni1 (four-ball tribometer, 1450 rpm, 400 N, 30 minutes).

Based on the XPS results, it can be concluded that during sliding of the steel-on-steel contact nickel-based nanolubricants undergo tribochemical reaction and release organic surface modifier and highly chemically active Ni nanocores, which can be easily adsorbed on sliding surfaces and generate a boundary lubricating film thereby improving the antiwear ability [30]. Besides, smaller Ni nanoparticles are easier to fill up micropits and grooves on worn surfaces, and form a compact protective layer. As a result, the adsorbed and deposited Ni layer greatly contributes to improve the tribological properties of PAO6 base oil.

3. The effect of morphology Ni nanolubricant on its tribological properties

The properties of nanoparticle usually lie on the atom on surface, because the atomic bonds on the surface are unsaturated, and the chemical environment and electron cloud density of the surface atom is quite different from the atoms inside. The proportion and the type of atomic bonds of atom on the surface are decided not only by the size but also by the morphology of nanoparticles [33–36]. For example, Ag nanoparticle has a high catalytic activity, good electrical properties, and photocatalytic bactericidal activity, while flaky triangular Ag nanoparticles have great potential in biosensing and biomedical field. Ni nanoparticle with 15 nm diameter is superparamagnetic, while Ni nanoparticle with 85 nm diameter has high coercivity. The tribological mechanism of nanoparticle with different morphology should be better studied through one kind of nanoparticle. Surface modification can effectively improve the dispersion stability of nanoparticles in lubricating oil base [2–3], and surface modification can affect preferentially growing crystal plate. Nanoparticles with different shapes could be prepared through controlling conditions of surface modification. Ni nanoparticles with three different shapes were prepared through thermal decomposition of Ni $(HCOO)_2$ in decane modified with 12 alkyl sulfonate (DBS) and oleylamine (OAM).

Preparation conditions are listed in **Table 1**. Three morphologies of Ni nanoparticles were prepared by varying the concentration, molar ratio of reactants, and reaction time. XRD spectra of Ni nanoparticles with different shapes are shown in **Figure 12**. Three characteristic diffraction peaks: 44.5°, 51.8°, and 76.4° correspond to the crystal plate (111), (200), and (220) in face-centered cubic (FCC) structure of Ni (JCDPS No.7440-02). Diffraction peaks widening are significant for all samples. With the increase in modifiers content, widening phenomenon was much more obvious, which indicates that the Ni nanoparticle morphology, size, and crystallinity are affected in-depth by modifiers' contents.

Sample code	PAO volume (mL)	TEM diameter (nm)	Scherrer diameter (nm)
Ni1	80	7.5	6.0
Ni2	40	13.5	8.3
Ni3	20	28.5	10.0

Table 1. Size of as-prepared Ni nanoparticles determined by TEM observation and Scherrer equation.

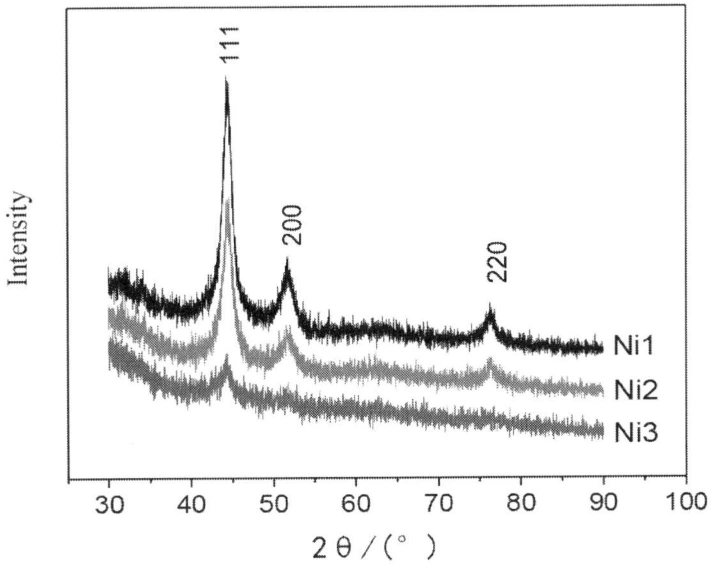

Figure 12. XRD patterns of Ni nanoparticles with different morphology.

The TEM images of Ni nanoparticles with different shapes are shown in **Figure 13** (low magnification on the left, high magnification on the right). Combined with the information of **Table 1**, it can be concluded that the shape of Ni nanoparticles is decided by the mol ratio of formic acid nickel to DBS. When the mol ratio is 5:1, the shape of Ni nanoparticles is spherical; when the mol ratio is 3:1, the shape is triangular plate; and when the ratio is 1:2, the shape of the Ni nanoparticles is triangular rod. Because of the inducement of surface modifier different crystal plate of Ni nanoparticles grows differently, which leads to different shapes. In addition, Ni nanoparticle shows narrow size distribution. No obvious aggregation appears because of the surface organic modification.

To further examine the interaction between the organic surface modifier and Ni nanocore, we conducted FTIR analysis of surface-modified Ni nanoparticles as well as pure OAM and OA. As shown in **Figure 14**, the characteristic absorption bands of pure OAM at 3367 and 3295 cm^{-1} correspond to the symmetric and asymmetric stretching vibration of N-H, and the absorption bands at 1583 and 967 cm^{-1} belong to the shear and flexural vibration mode of $-NH_2$ [27]. For a solid DBS, absorption peaks at 1183 and 1053 cm^{-1} correspond to asymmetric and symmetric stretching vibration of S=O. In addition, for pure OAM and DBS, characteristic peak of nonpolar chain appears, which is stretching vibration peak of CH_3 and CH_2 at 2922 and 2852 cm^{-1}, C–H bending vibration peak at 1464 cm^{-1}, and not less than four consecutive methylene rocking vibration peak at 720 cm^{-1} [28]. For surface modified Ni nanoparticles, the existence of nonpolar chain in surface modifiers OAM and DBS can be confirmed by characteristic peaks at 2922, 2852, 1464, and 720 cm^{-1}. For surface modified Ni nanoparticles, the primary amino characteristic peaks at 3367, 3295, 1583, and 967 cm^{-1} do not appear; however, the stretching vibrations peak of c=c at 1629 cm^{-1} and asymmetric and symmetric stretching vibrational peaks of S=O at 1183 and 1053 cm^{-1} appears. Therefore, it can be concluded that surface modifiers (OAM and DBS) and Ni nanoparticles are chemically combined, and this stable chemical bond effectively blocks reunification and oxide particles of Ni; meanwhile, nonpolar chain improves its compatibility with base oil.

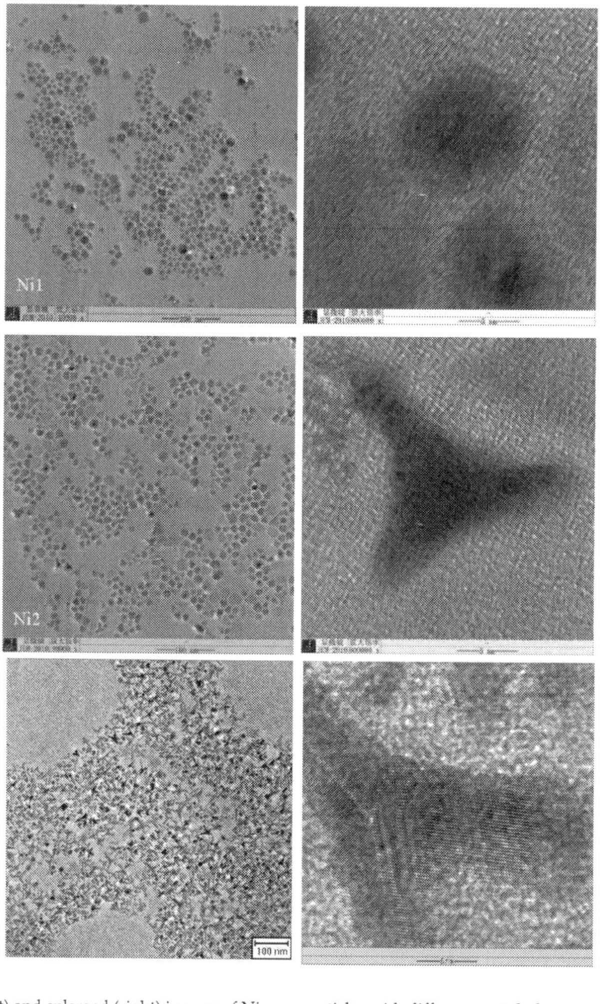

Figure 13. TEM (left) and enlarged (right) images of Ni nanoparticles with different morphology.

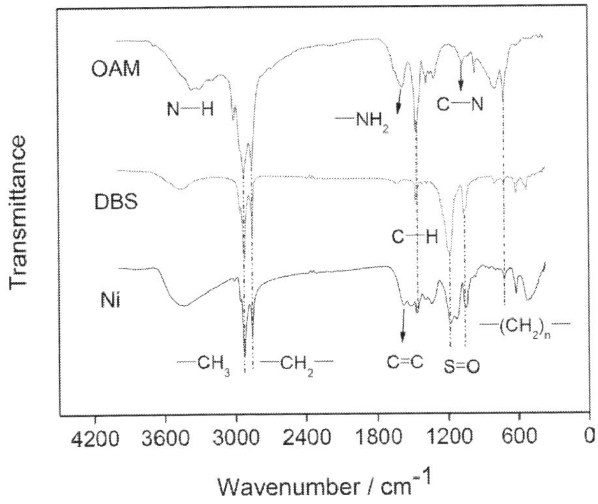

Figure 14. FTIR spectra of DBS, OAM, and as-separated Ni nanoparticles.

The variation of friction coefficient and wear scar diameter with the concentration of Ni is shown in **Figure 15** (four ball machine, 300 N, 1450 rpm, 30 minutes). It can be seen that the friction reduction and antiwear properties of Ni nanoparticles is closely related with its shape and concentration. Compared with the base oil PAO, within a suitable concentration range, Ni1 and Ni3 nanoparticles present a certain friction reduction property, while Ni2 nanoparticles significantly improve the wear resistance of PAO. In the experiment concentration range, Ni2 (triangular plate) is more effective than Ni1 (spherical shape) and Ni3 (triangular rode) in reducing WSD of steel partner. Therefore, it can be deduced that Ni nanoparticles with triangular plate shape easily forms a continuous film on the sliding surface, which prevents direct contact between friction interface so as to prevent the adhesion, thus showing a better abrasion resistance.

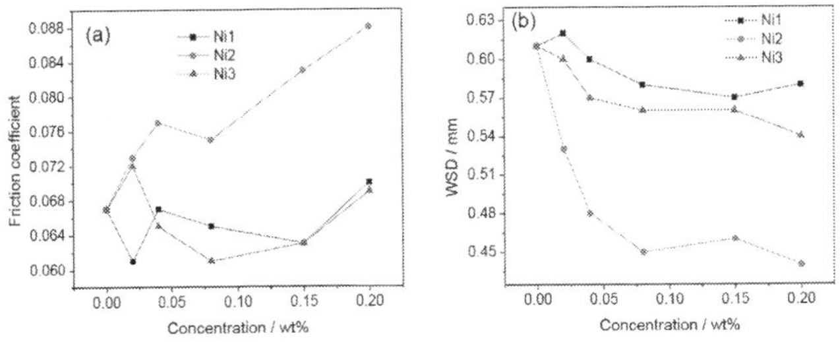

Figure 15. Variation of friction coefficient (a) and wear scar diameter (b) with the concentration of Ni nanoparticles (four-ball friction and wear tester, 1450 rpm, 300 N, 30 minutes).

Figure 16 showed variation of coefficient under (a) and the wear scar diameter (b) with the load (four-ball, 1450 rpm, 30 minutes). For pure PAO6 and PAO6 added with 0.2% Ni nanoparticles, the coefficient of friction decreases with the increasing of load. Under higher load, addition of Ni nanoparticles can reduce friction, while for the whole test load, three shapes of Ni nanoparticles can effectively improve antiwear properties of PAO6 base oil. Especially for triangular plate-like Ni2, the WSD can be reduced by 35%. It may be due to its easy adsorption on steel sliding surface and formation of lubricating film, which leads to better antiwear capacity.

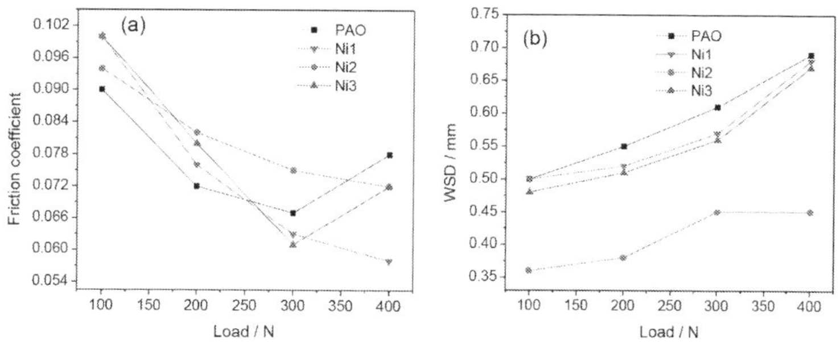

Figure 16. Friction coefficient (a) and wear scar diameter (b) versus applied load under the lubrication of PAO6 and 0.2% Ni nanoparticles with different morphology (four-ball friction and wear tester, 1450 rpm, 30 minutes).

Figure 17 shows the P_B and P_D of PAO6 and Ni nanoparticles of three morphologies. As shown in the figure, P_D values show no difference, but addition of Ni nanoparticles can improve extreme pressure properties of PAO6, especially for triangular plate Ni, its EP is the best. This suggests that surface modified Ni nanoparticles can effectively improve the bearing capacity of PAO6.

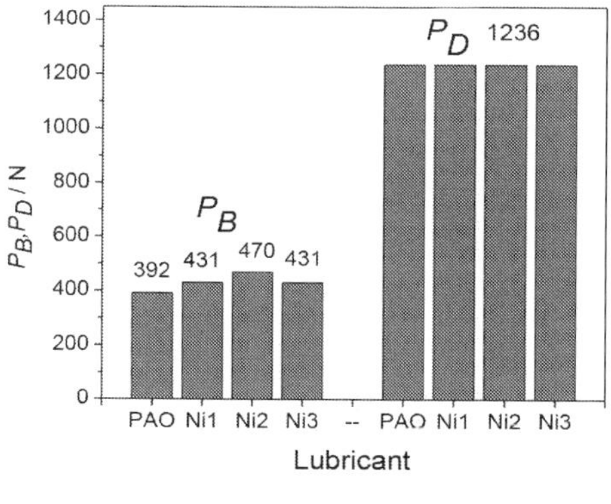

Figure 17. $P_B$and $P_D$values of various lubricant systems (four-ball friction and wear tester, 1450 rpm, 10 seconds).

SEM micrographs of wear scars lubricated with PAO6 as well as PAO6 containing 0.2% Ni nanoparticles are shown in **Figure 18**. Deep plow ditch and marks of tearing between friction pairs appears on the worn surface lubricated with PAO6. Lubricating effect of different Ni nanoparticles is obviously different. Compared with Ni1 and Ni3, WSD of steel ball lubricated with Ni2 is the least, and the worn surface is the smoothest. For the nanoparticle Ni1 and N3, there is some designated marks and stick. Once again, it is proved that triangular plate Ni nanoparticles have better abrasion resistance.

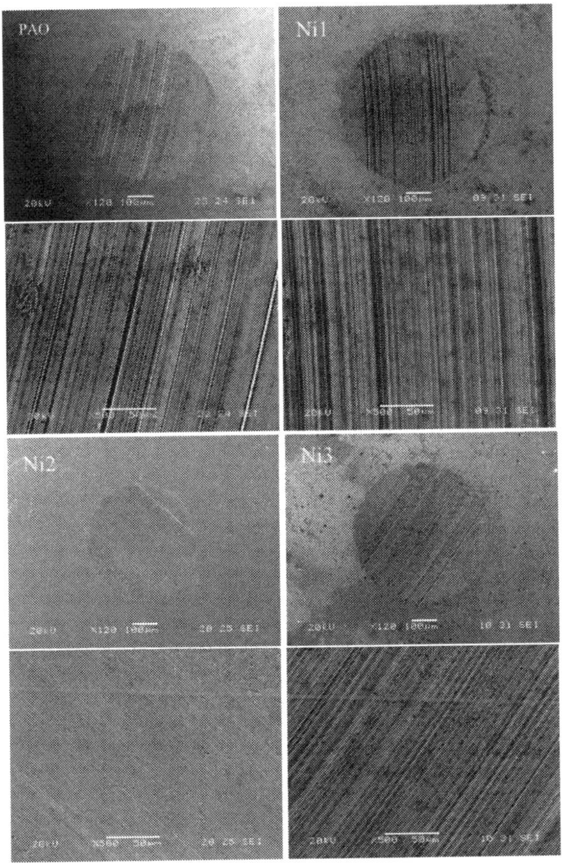

Figure 18. SEM micrographs of wear scars lubricated with PAO6 as well as PAO6 containing 0.2% Ni nanoparticles (four-ball friction and wear tester, 1450 rpm, 300 N, 30 minutes).

The worn surfaces of the steel balls lubricated with different lubricant are analyzed by mean of XPS so as to acquire more information about the tribochemical reactions involved during the friction process. **Figure 19** shows the curve-fitted XPS spectra of C1s, N1s, O1s, S2p, Fe2p, and Ni2p on the worn surface of steel ball lubricated with nanolubricants containing 0.2% Ni (four-ball, 1450 rpm, 400 N, 30 minutes). It can be noticed that S, O, N, and C are detected on the worn surface. This indicates that surface modifier DBS and OAM are adsorbed on the worn surface or takes part in the tribochemical reaction during the sliding process.

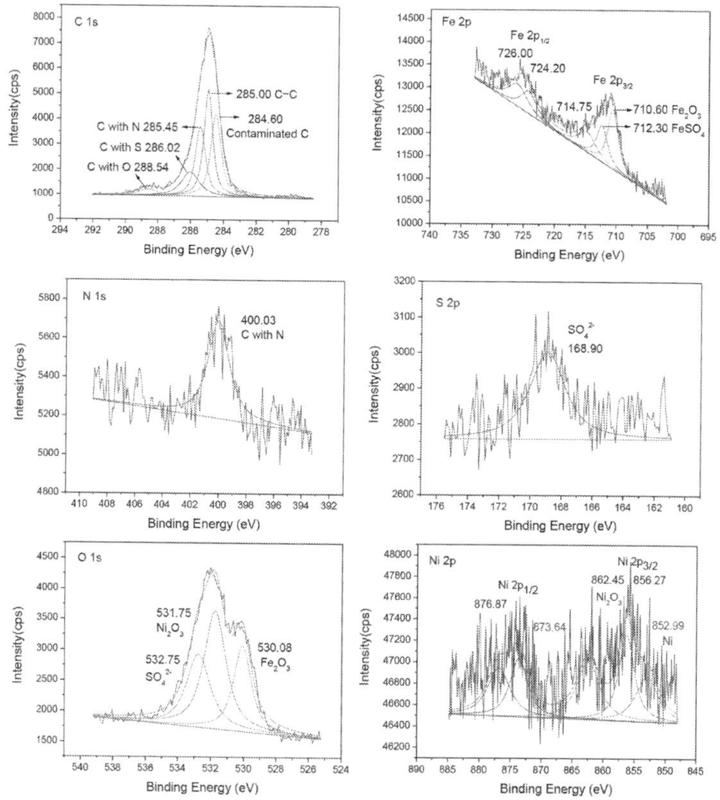

Figure 19. Curve-fitted XPS spectra of typical elements on rubbed surface of steel ball lubricated with nanolubricants containing 0.2% Ni (four-ball friction and wear tester, 1450 rpm, 400 N, 30 minutes).

It can be seen that the Fe2p shows XPS peaks at the binding energies 710.60, 712.30, 724.2, and 726.00 eV. The peak at a binding energy of 710.60 and 724.20 eV corresponds to the Fe_2O_3 [32]. The peak at 712.30 and 726.00 eV corresponds to the $FeSO_4$. This indicates that ferrous oxides have been formed on the worn surface, and the surface modifier DBS reacted with steel ball during the sliding process.

Ni 2p shows XPS peaks at the binding energies 852.99 and 856.27 eV. The peak at a binding energy of 852.99 eV corresponds to the Ni 2p3/2, and the peak at 856.27 eV corresponds to

Ni_2O_3. This indicates that Ni nanoparticles transferred to the worn surface of steel ball, and part of it was oxidized.

When the element O is mentioned, the binding energy of O1s is about 530.08, 531.75, and 532.75 eV. The binding energy at 530.08 eV corresponds to the Fe_2O_3 while the peak at 531.75 eV corresponds to the Ni_2O_3. The binding energy at 532.75 eV corresponds to the sulfate radical.

Based on the analysis of XPS, it can be inferred reasonably that, during the process of sliding between steel and steel, under low load, the organic surface modifiers outside the Ni nanoparticles could easily be adsorbed to the surfaces of friction pairs generating the layer of organic lubricating film. Under high load, the friction generated by mechanical, chemical, and electrochemical action of friction, and nano-Ni exchange of energy and material exchange between, to form a metal on the friction surface protective film, metal oxide protective film for metal sulfate, separated from the friction surface, play as antiwear and friction. For, the lubricating effect of Ni nanoparticles with different shapes, the Ni nanoparticle with shape of triangular sheet shows the best lubricating effect, which could be attributed to its larger contact area with sliding surface and, more easily absorbing on the worn surface and forming boundary lubricating film.

4. Tribological properties of Ni nanolubricant in DIOS/DLC solid-liquid composite lubricating system

Diamond-like carbon (DLC) coatings are becoming attractive protective films for many automotive parts as they offer high hardness, ultralow friction, and good wear resistance under dry or lubricated contacts [37–42]. However, the key problem with their lubrication is a low reactivity toward existing additives and oils [43, 44]. Although it was confirmed that DLC coatings react with these metal-tailored additives, their interactions are much weaker and the formation of tribochemical films was much lower and difficult to control. On the other hand, today's most effective additives will need to be reduced or phased out in the near future for environmental reasons and replaced with greener lubrication technologies.

Kalin' research group reported the lubrication behaviors of MoS_2 nanotubes as a green, physically based additive for DLC-based solid-liquid synergetic system [45]. The addition of MoS_2 nanotubes (with diameters of 100 nm and length of 20 μm) to PAO oil significantly improved the friction behavior in the boundary- and mixed-lubrication regimes. However, for the elasto-hydrodynamic lubrication (EHL) regime, the effectiveness disappeared completely. The study results of EHL suggested that the classical EHL modes are accurate down to films, as thin as 10–20 nm [46]. Hamed Ghaednia and Hasan Babaei reported the friction reduction of nanoparticle (with diameter of 7 nm) in EHL regime, and proposed that the obstructed flow induced by nanoparticles of friction reduction mechanism reduced the friction between layers of lubrication molecules. So, nanoparticle additives with size bigger than the oil film thickness would prevent the occurrence of EHL. The research about nanoparticles additives for DLC-based, solid-liquid synergetic system is considered rare. Moreover, the study about the lubrication of solid-liquid synergetic system constructed from DLC coatings and real nanofluid (nanoparticle with diameter of less than 10 nm) has not been reported.

We show in this study that the Ni nanoparticle (7 nm in diameter) significantly reduces the friction in the DLC/steel contacts lubricated by fully formulated oil di-iso-octyl sebacate (DIOS) in all lubrication regimes: boundary, mixed, and elasto-hydrodynamic lubrication. Analysis of the composite and topography of worn surface was operated systematically in order to reveal the friction reduction mechanism.

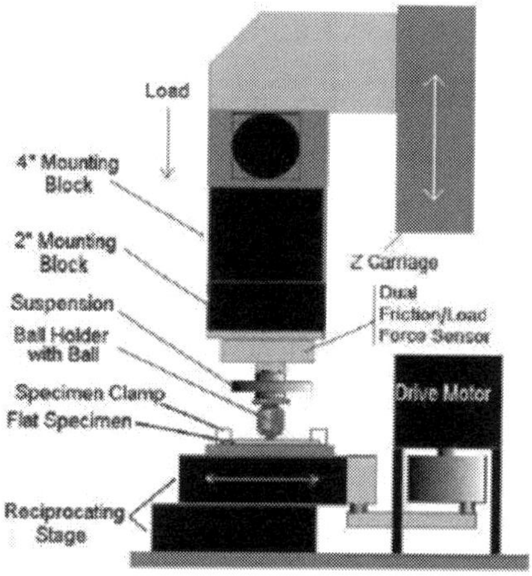

Figure 20. Schematic diagram of UMT tribometer.

The tribological properties of DLC and DLC/DIOS solid-liquid composite system were measured on a ball-on-disk tribometer shown in **Figure 20**. GCr 15 steel ball was used as friction pair with a diameter of 4 mm. Its elastic modular and Poisson ratio was 205 GPa and 0.28, respectively. In order to study the lubricating mechanism of Ni nanoadditive in DLC/DIOS solid-liquid system, three lubricating conditions were used including dry friction with DLC, solid-liquid composite lubricating by DLC/DIOS, and DLC/DIOS added with Ni nanoparticles. All of the friction tests were performed through a reciprocating sliding test mode at ambient temperature with a load range from 0.5 to 4 N. The relative humidity was 20–40%. Each test was repeated three times with a stroke of 5 mm. In order to obtain the Stribeck curves, a wide range of average contact velocities from 60 to 480 rpm were applied to achieve different lubrication regimes, i.e., boundary, mixed, and elastro-hydrodynamic. Because of the characteristics of reciprocating motion, within the stroke some variation in regimes may occur as a consequence of varied velocity. However, this has negligible effect in boundary and mixed lubrication. In elastro-hydrodynamic lubrication (EHL) regime, it is clear that EHL cannot be achieved at end-stroke. Accordingly, in this work for the region claimed as the EHL regime, these conditions are fully valid only within the central part of the stroke. Moreover, in the central part of the stroke where velocity is higher than the average, the conditions are even more beneficial toward EHL than "average" calculations for the lambda parameter.

The coefficient of friction under the three lubricating conditions is shown in **Figure 21** (4 N load and sliding speed 240 rpm). It can be seen that with the increasing of friction time, the coefficient of friction lubricated by DLC and DLC/DIOS fluctuated slightly, while that of DLC/DIOS added with Ni nanolubricant is very stable. The coefficient of friction decreased greatly under DLC/DIOS and DLC/DIOS added with Ni nanolubricant lubricated conditions, and DLC/DIOS with Ni nanolubricant composite lubrication system shows better antifriction effect.

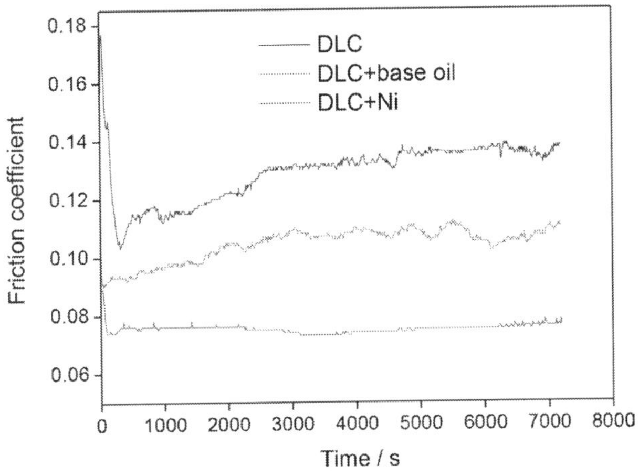

Figure 21. Coefficient of friction versus time under three lubricating conditions.

The effect of sliding speed on coefficient of friction under the three lubricating conditions is shown in **Figure 22**. As can be seen, coefficient of friction lubricated by DLC/DIOS and that added with Ni nanolubricant decreases rapidly with the sliding speed, while that of dry friction with DLC film is almost unchanged. With the increasing of sliding speed, lubricating regime changes from boundary lubrication to mixed lubrication. DLC/DIOS solid-liquid composite lubricating system and that added with Ni nanolubricant greatly reduces the coefficient of friction of the DLC film, and at low speed the antifriction effect of DLC/DIOS with Ni nanolubricant is more outstanding.

Based on the graphite transfer film theory of DLC, the metastable carbon DLC films overcome energy barrier during the friction process and transform into graphite structure [47–49]. In graphite structure, the weak interlayer binding force between layers leads to lower shear strength that reduces the coefficient of friction, which results in good antifriction effect of DLC films. The Ni nanolubricant added in DLC/DIOS lubricating system could absorb on the sliding surface and form a lower shear strength film, and then reduce the friction between friction surfaces.

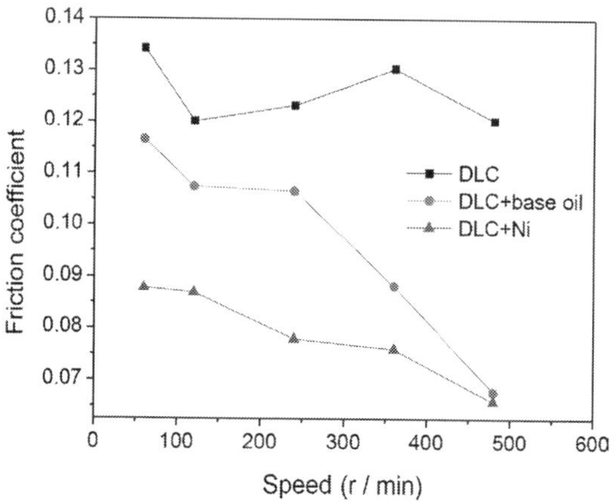

Figure 22. Coefficient of friction versus sliding speed under three lubricating conditions.

The effect of load on coefficient of friction under the three lubricating conditions is shown in **Figure 23**. With the increase of load, the coefficient of friction under dry friction and Ni nanolubricant lubricating conditions were slowly declining, and for those lubricated with DLC/DIOS, the COF fluctuated. The effect of load on the friction coefficient is less, but the existence of base oil lubrication and Ni nanolubricant greatly reduces the friction coefficient of DLC films, and the Ni nanolubricant makes the coefficient of friction more stable.

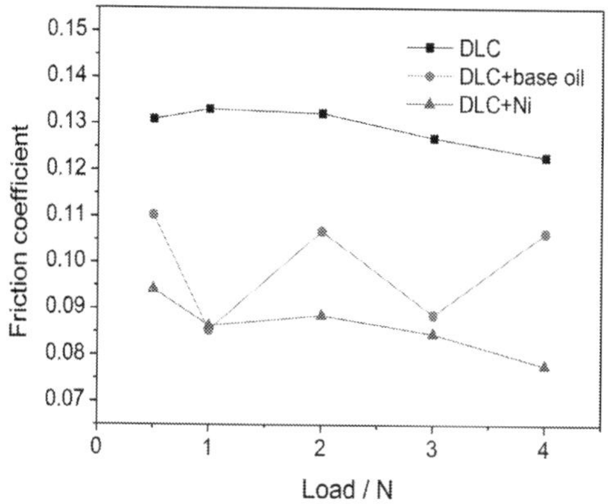

Figure 23. COF versus load under three lubricating conditions.

The wear rate of DLC films under the three lubricating conditions is shown in **Figure 24** (4 N load, sliding speed of 240 rpm time was 4 h). Under the experimental conditions, the wear rate of DLC decreases by 72% and 86%, respectively, because of lubricating of DIOS and Ni nanolubricant. Thus, antiwear effect of Ni nanolubricant composite lubricating is better.

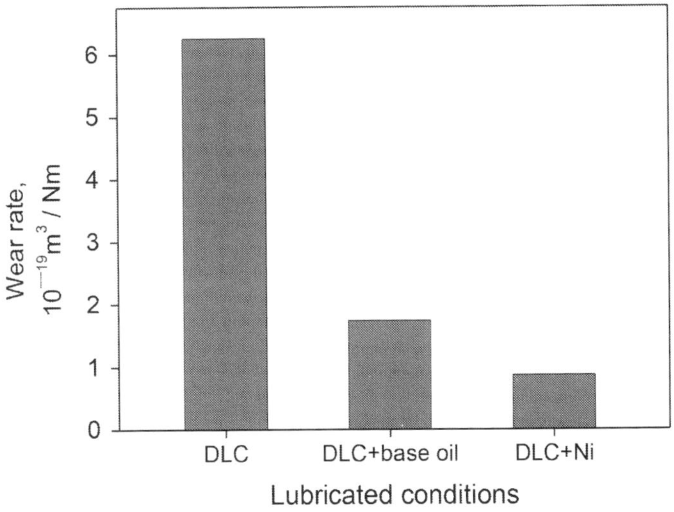

Figure 24. Wear rate of DLC films under three lubricating conditions.

The SEM images of wear scar on DLC films under the three lubrication conditions is shown in **Figure 25** (4 N load, sliding speed of 240 rpm, 4 h). As can be seen, under dry friction, the wear mark of DLC film is widest with grooved surface strip, and obviously has certain adhesive wear; while under the lubricating of DLC/DIOS, the wear mark is significantly smaller and shallow. For the DLC/DIOS with Ni nanolubricant, the wear mark is the most narrow (<100 m) and the surface is very smooth. Therefore, Ni nanolubricant greatly increases the antiwear effect of composite lubricating system, which is consistent with the above analysis of wear rate.

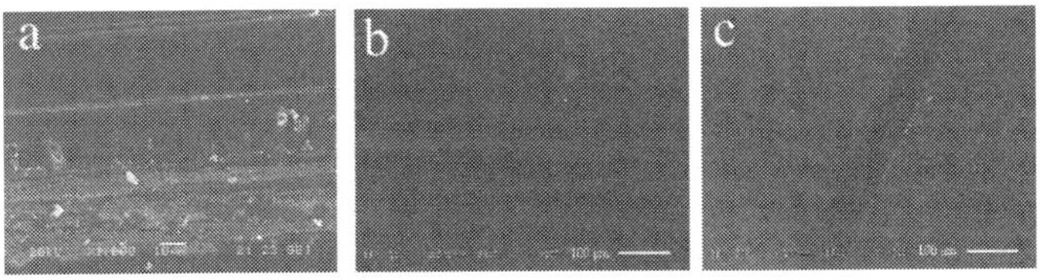

Figure 25. SEM images of wear scar on DLC films under three lubrication conditions.

For comparison, the SEM images of wear scar on steel balls under the three lubrication conditions are also shown in **Figure 26**. The wear scars under different conditions are quite different. Under dry friction the wear scar is the largest, with obvious edge debris in which the black substance may be graphitized transfer film. For DLC/DIOS and that added with Ni nanolubricant, wear scars are obviously smaller with less debris on their edges, the worn edge grinding on grinding ball mill is less. There are some shallow scratch grooves on the steel ball under DLC/DIOS lubrication condition; for that with Ni nanolubricant, the wear scar surface is smooth without any grooves and debris. Therefore, it can be concluded that DLC/DIOS with Ni nanolubricant composite lubricating system shows good synergistic effect of antiwear and antifriction, which is consistent with the above analysis.

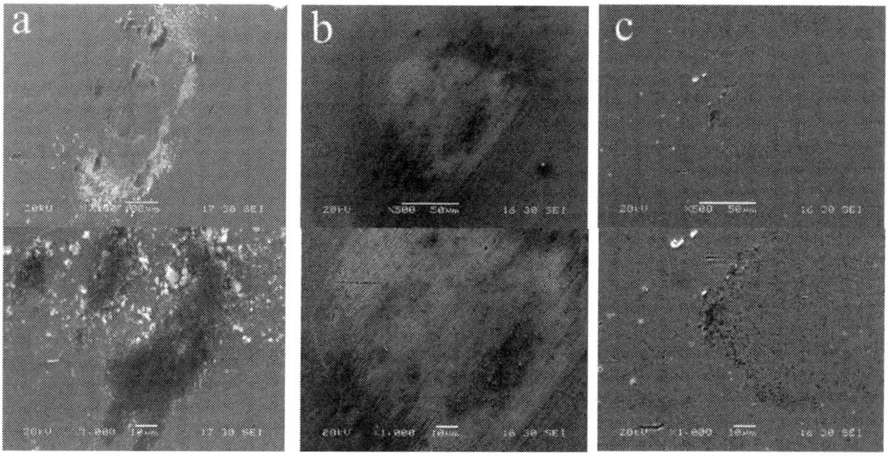

Figure 26. SEM images of wear scar on steel balls under three lubrication conditions.

The Stribeck curve describes the variation of friction coefficient with the contact conditions and shows lubricating regimes conversion between elasto-hydrodynamic lubrication, mixed lubrication, and boundary lubrication, which is a simple and effective method for judging lubricating regimes and predicting coefficient of friction [50–55]. Through comparing coefficient of friction under different lubrication regime, for example, the difficulty into EHL lubrication regime and the bottom width of it, the lubricating performance of lubricant can be valued comprehensively. Therefore, Stribeck curve was measured under lubrication of DLC/DIOS and that with Ni lubricant, which is shown in **Figure 27** (Sommerfield parameters was calculated according to [54]). In all lubrication regimes, the coefficient of friction of Ni nanolubricant is far better than that of base oil. Compared with lubrication condition of DLC/DIOS, the mixed lubrication regime is more stable and wider under lubrication condition of Ni nanolubricant, which can be attributed to the high chemical activity of Ni nanolubricant forming a solid adsorption film easily on the friction surface, so as to reduce friction resistance and exhibiting excellent tribological properties in the boundary and mixed lubrication regimes.

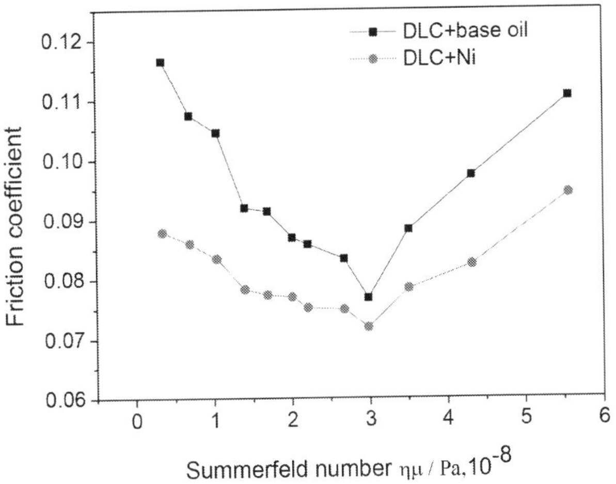

Figure 27. Stribeck curves of solid-liquid composite lubricating systems.

5. Tribological properties of Cu@Ni nanolubricants

In recent years, Cu nanoparticles as metal nanoadditives shows good antiwear and dynamic self-repairing performance [2–5, 7–10, 33–36]. For example, in this group, two alkyl two phosphorothioate (DDP) surface-modified Cu nanoparticles with average diameter of 5 nm was produced through *in situ* surface modification and liquid phase chemical reduction technology. As lubricating additives this Cu nanoparticles exhibit excellent energy saving, lubrication, antiwear, and self-repairing performance [34], which have been achieved in industrial production, and achieved considerable economic benefits. However, the modifier contains S, P elements, which easily leads to catalyst poisoning and environmental hazard, so is restricted for use in application process only.

If modifier does not contain S, P elements, Cu nanoparticles are easily oxidized [56–59]. How to produce stable Cu nanoparticles without S, P elements? Preliminary work shows that Ni nanoparticles are relatively stable and shows good antiwear performance. Then synthesis of Cu@Ni nanoparticles is going to be an effective method. Ni as shell can increase the stability of the center particles of Cu, which will further improve the tribological properties of nanoparticles [60, 61].

In situ preparation synthesis process of Cu@Ni nanoparticle is shown in **Figure 28**. Based on the difference of decomposition temperature of Cu $(HCOO)_2$ and Ni $(HCOO)_2$, the two compounds decomposes successively in basic oil of PAO6, Cu (HCOO) 2 decomposes first and forms Cu nucleus. Ni $(HCOO)_2$ decomposes subsequently. Because of the lower nucleation energy for heterogeneous nucleation, Ni forms nucleus on the surface of Cu nanoparticles and then forms

Ni coating on it. Therefore, a simple one-step synthesis method is set up for preparing Cu@Ni nanolubricant in basic oil, which is crucial for the application of nanolubricant.

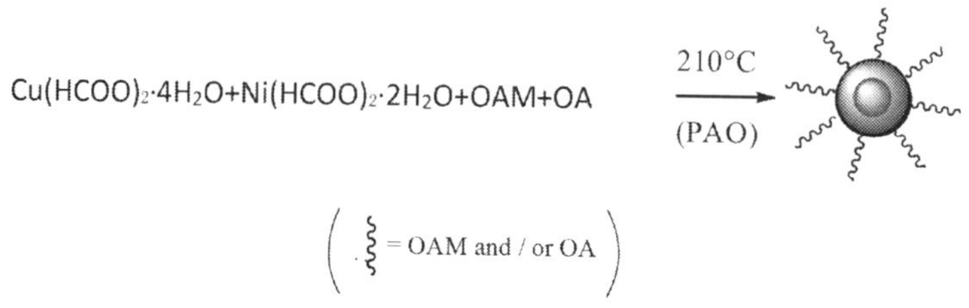

Figure 28. Synthetic procedure for preparing Cu@Ni nanoparticles.

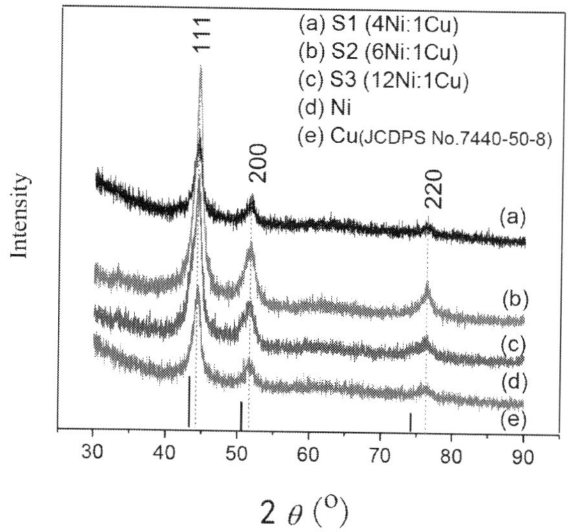

Figure 29. XRD patterns of Cu@Ni nanoparticles with different Cu content.

XRD maps of Cu@Ni nanoparticles with different Cu content are shown in **Figure 29**. For comparison, XRD map of pure Ni nanoparticles under the same conditions is also shown. It can be seen that three peaks at 44.5, 51.8, and 76.4 degrees corresponds to the (111), (200), and (220) crystal face of face-centered cubic (FCC) Ni (JCDPS No. 7440-02), respectively. There are no obvious characteristic peaks of Cu (JCDPS No. 7440-02). It may be that the grain size of Cu nucleus in Cu@Ni nanoparticles is too small (as shown in **Table 2**), and the XRD peak of Cu nucleus is remarkably broadened. In order to prove the existence of Cu nucleus, XPS analysis is used, which is shown in **Figure 30**. Cu@Ni nanoparticles Cu, Ni XPS, the binding energy at peaks of 932.85 and 952.67 eV corresponds to Cu 2p1/2 and Cu 2p3/2, signs of oxidized Cu do not appear. So the stability of nano-Cu can be improved.

Sample code	Ni:Cu	Solution A (mL)	Scherrer diameter (nm)
S1	4:1 (12:3)	1.5	8.4
S2	6:1 (12:2)	1.0	7.3
S3	12:1	0.5	6.2

Table 2. Size of as-prepared Cu@Ni nanoparticles determined by Scherrer equation.

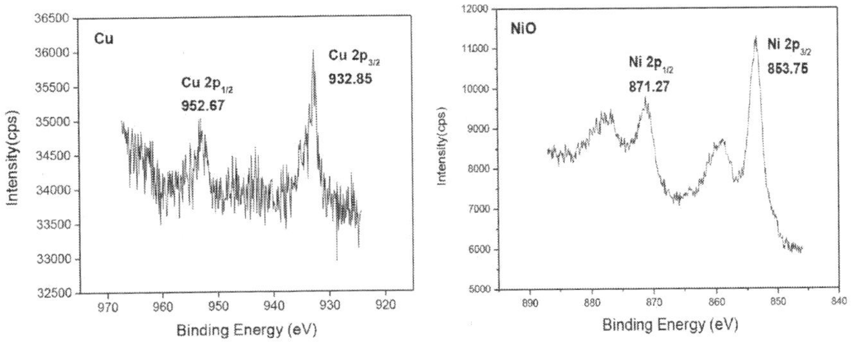

Figure 30. XPS spectra of Cu 2p, Ni 2p in Cu@Ni nanoparticles.

TEM images of Cu@Ni nanoparticles containing different amounts of Cu are shown in **Figure 31** (the panorama is on the left, the partial enlarged detail is on the right). It can be seen that Cu@Ni nanoparticles are uniform, spherical particles without obvious agglomeration, the average diameter is 12~15 nm. The coating outside of samples S1 and S2 is obvious, in which slightly shallow shell is Ni coating, the black core is Cu; for S3, Cu content may be too little to see. At the same time, binary modifier and controlling relative content of OAM and OA is crucial for synthesis, in which OAM is used as reducing agent controlling nanocrystal nucleation rate, high effect modifier of OA block the growth of nanocrystals, and the synergistic reaction of OAM and OA results in the monodispersed Cu@Ni nanoparticles.

The optical image of Cu@Ni nanoparticles stored for a month is shown in **Figure 32**. It can be seen that, after a month, the Cu@Ni nanoparticles still disperses in PAO6 uniformly, no obvious sedimentation behavior is observed, which is crucial for the application of nanoparticles.

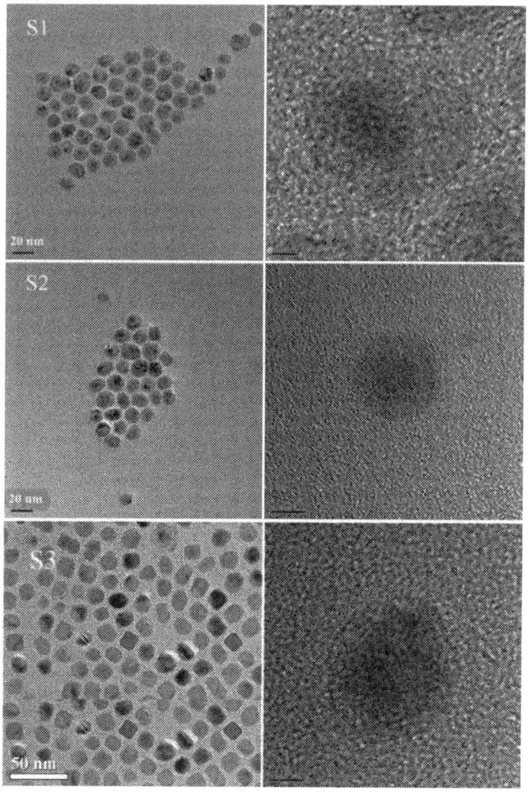

Figure 31. TEM images of surface-capped Cu@Ni nanoparticles with different content of Cu.

Figure 32. Photograph of the Cu@Ni nanoparticles stored in PAO6 after one month.

In order to further study the interaction between organic modifier and inorganic particles of Cu@Ni, FTIR is performed to analyze the modified Cu@Ni nanoparticles, pure OAM and OA. As shown in **Figure 33**, the characteristic peaks at 1711 and 939 cm^{-1} corresponds to the stretching vibration of C=O and bending vibration of —OH, respectively, in pure OA. As for OAM, the amine absorption peaks at 3367 and 3295 cm^{-1} corresponds to the symmetric and asymmetric stretching vibration of primary amine N-H, peaks at 1583 and 967 cm^{-1} corresponds to the shear mode and bending vibration modes of —NH2 [27]. For pure OAM and OA, characteristic peaks of nonpolar chain at 2922 and 2852 cm^{-1} are observed, bending vibration of C—H at 1464 cm^{-1} and rocking vibration of not less than four consecutive methylene at 720 cm^{-1} is also observed. For surface-modified Cu@Ni nanoparticles, the existence of nonpolar chain in surface modifiers can be certified through the presence of characteristic peak at 2922, 2852, 1464, and 720 cm^{-1}. At the same time, characteristic peaks of primary amine at 3367, 3295, 1583, and 967 cm^{-1} and the free carboxyl group at 1711 and 939 cm^{-1} is not observed, while the asymmetry and symmetry stretching vibration of —COO— peak at 1543 and 1460 cm^{-1} and stretching vibration of C—N at 1071 cm^{-1} appears [28]. It can be concluded reasonably that the interaction between organic surface modifiers and Cu@Ni nanoparticles is a kind of chemical combination, which is essentially different from surface wetting of general dispersant. Because of the more stable chemical coating, oxidation and agglomeration of Cu@Ni particles is effectively prevented, and the nonpolar chain can improve its compatibility with base oil.

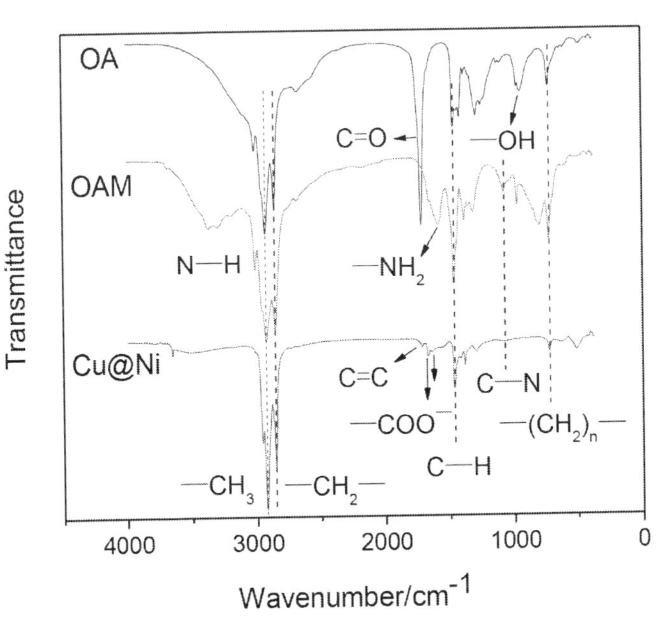

Figure 33. FTIR spectra of OA, OAM, and as-separated Cu@Ni nanoparticles.

The coefficient of friction (COF) (a) and wear scar diameter (WSD) (b) with different Cu@Ni nanoparticles (here represented by S2) concentration is shown in **Figure 34** (four ball machine, 300 N, 1450 rpm, 30 minutes; at the same time data of PAO6 and PAO6 only containing modifier are also given). As can be seen, compared with base oil PAO6, the coefficient of friction of surface modifier increased, but the wear scar diameter is greatly reduced. COF and WSD of Cu@Ni nanoparticles decrease significantly compared with surface modification agent, especially antiwear ability of it in the experimental concentration range. Compared with PAO6, the wear scar diameter of Cu@Ni nanoparticles decrease by 42.6% with mass fraction of 0.025%, by 47.5% with mass frication of 0.80%. In the last chapter, under the lubrication of Ni nanolubricant, the wear scar diameter reduced by 30%. As a result, compared with Ni nanoparticles, the antiwear ability of Cu@Ni composite nanoparticles increases by 17.5%. Therefore, it can be concluded reasonably that because of the synergistic effect of Ni and Cu lubricating and self-repairing film deposited on friction pair avoids direct contact between the interface friction, delays the adhesive wear and timely repairs worn parts, and greatly improves the antiwear ability of base oil PAO6.

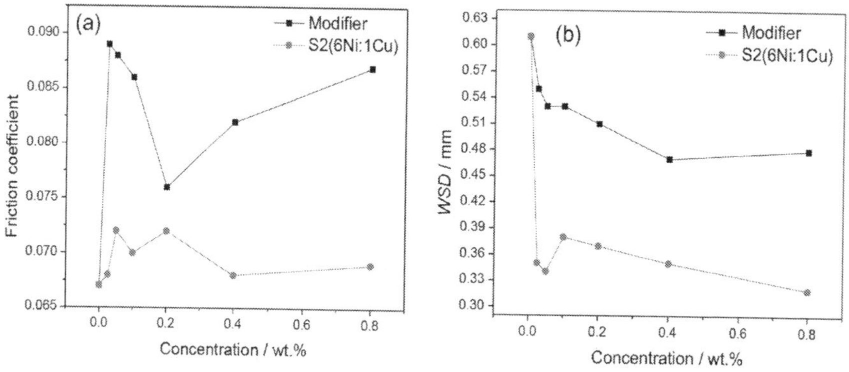

Figure 34. Variation of friction coefficient (a) and wear scar diameter (b) with the concentration of Cu@Ni nanoparticles.

The COF (a) and WSD (b) of PAO6, PAO6 containing only modifiers, and PAO6 containing 0.8% Cu@Ni nanoparticle versus load are shown in **Figure 35** (four ball machine, 1450 rpm, 30 minutes; at the same time, that of PAO6 lubricant containing Ni nanoparticles surface modified under the same condition is also given for comparison). It can be seen that with the increasing of load, COF of Cu@Ni nanoparticles decreases rapidly, and WSD increases slowly. COF of other lubrication conditions sharply decreases at first and then increases WSD. In addition, overall the COF and WSD of Cu@Ni nanoparticles decreases greatly compared to the rest of the lubrication conditions, especially under the load of 400 N, which reveals once again that the Cu@Ni composite nanoparticles has better antifriction, antiwear, and bearing performance than Ni nanoparticles, which may be due to the formation of boundary lubricating film on the sliding surface of the steel ball because of the synergistic action of Ni and Cu.

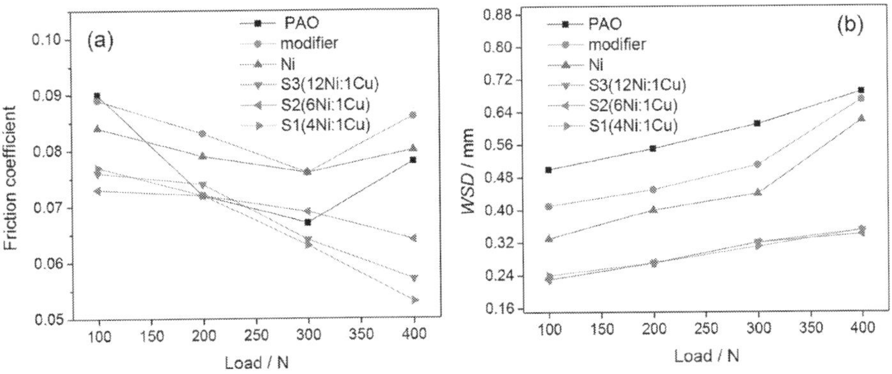

Figure 35. Friction coefficient (a) and wear scar diameter (b) versus applied load under the lubrication of PAO6, PAO6 + surface-modifiers, and nanolubricants containing 0.8% Cu@Ni nanoparticles with different content of Cu.

P_B(a) and P_D (b) value of PAO6, PAO6 containing only modifiers, and PAO6 containing Cu@Ni nanoparticle are given in **Figure 36** (that of PAO6 containing Ni nanoparticles is also given for comparison). As shown in **Figure 36**, $P_B$of Cu@Ni nanoparticles is obviously higher than that of PAO6 and PAO6 containing only modifiers, however, is similar with that of Ni nanoparticle. For the P_D of all lubricants, Cu@Ni nanoparticles shows the highest values compared with the rest of the lubricants (see **Figure 36 b**). Once again, it is proved that Cu@Ni nanoparticles can effectively ameliorate the extreme pressure ability of PAO6.

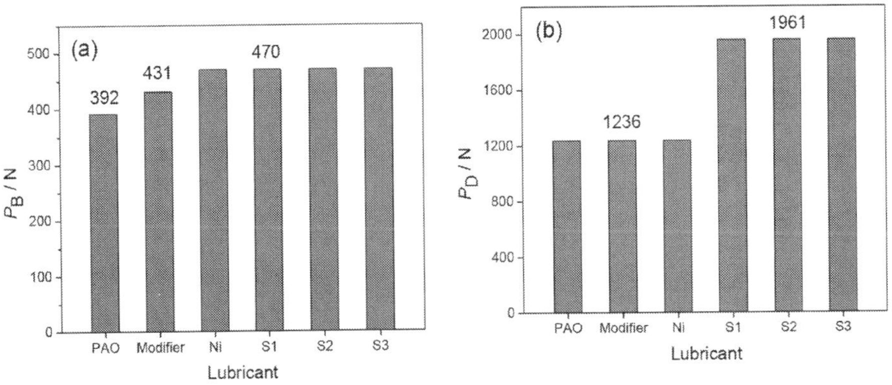

Figure 36. P_B (a) and P_D (b) values of various lubricant systems.

The SEM images of worn surface on steel ball under the three lubrication conditions are shown in **Figure 37** (partial enlarged details are given). It can be seen that the WSD of steel ball lubricated by Cu@Ni of nanoparticles is obviously less than that of pure PAO6 and PAO6 containing modifier, and the worn surface is smooth with minor scratches. For PAO6, obvious scratches and adhesion wear is observed, which once again proves that the Cu@Ni nanoparticles shows better antiwear performance.

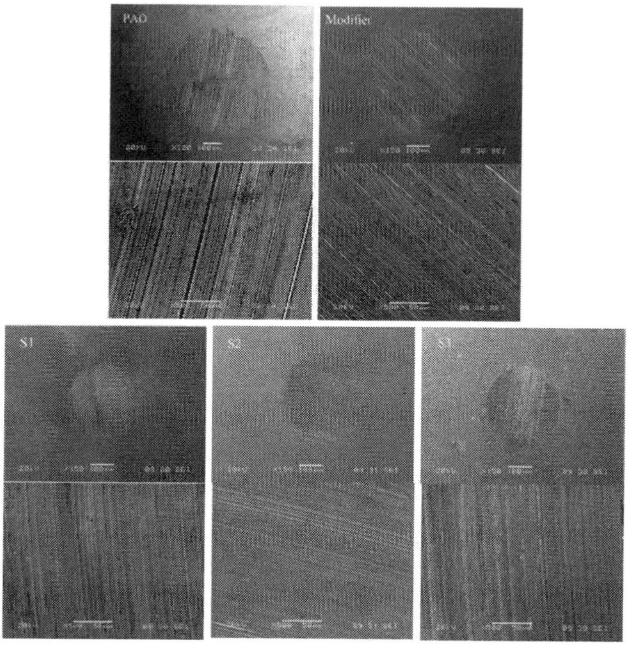

Figure 37. SEM micrographs of wear scars lubricated with PAO6 as well as PAO6 + surface-modifiers and nanolubricants containing 0.8% Cu@Ni nanoparticles.

In order to reveal the antifriction and antiwear mechanism of Cu@Ni nanoparticles, XPS is used to analyze the chemical state of several typical elements on worn surface of steel ball. **Figure 38** shows the curve-fitted XPS spectra of C 1s, O 1s, N 1s, Fe 2p, Ni 2p, and Cu 2p on wear scar surface under the lubrication conditions of PAO6 with 0.8% Cu@Ni nanoparticles (S2) (four ball mode, 1450 rpm and 400 N, 30 minutes). It can be noticed that O, N elements are detected which come from surface modifiers containing C—N, COO—, C=C, and C—H absorbed or reacted with worn surface of steel ball. The Fe 2p3/2 peak at 710.81 eV (corresponding satellite peak is located at 713.00 eV) is attributed to Fe_2O_3 (corresponding O 1s peak is located at 529.99 eV) [31]. We can therefore reasonably conclude that ferrous oxides are formed on the worn surfaces of steel balls under selected experimental conditions. Moreover, Ni 2p3/2 peak at 855.94 eV and Ni 2p1/2 peak at 873.64 eV indicate that Ni nanocores can be released from surface-modified Ni nanoparticles and transfers onto sliding steel surface during the rubbing process. Furthermore, a part of Ni nanocores can react with oxygen to form Ni_2O_3 (corresponding O 1s peak is located at 531.66 eV). Moreover, Cu 2p3/2 peak at 932.90 eV and Cu 2p1/2 peak at 953.05 eV indicate that metal Cu released from Cu@Ni nanoparticles transfers to the worn surface of steel ball, from which it can be concluded that the synergistic effect of Cu nanoparticles with self-repairing performance and Ni nanoparticles with high activity results in forming boundary lubricating film covering sliding surface during friction process, thus showing better tribological properties.

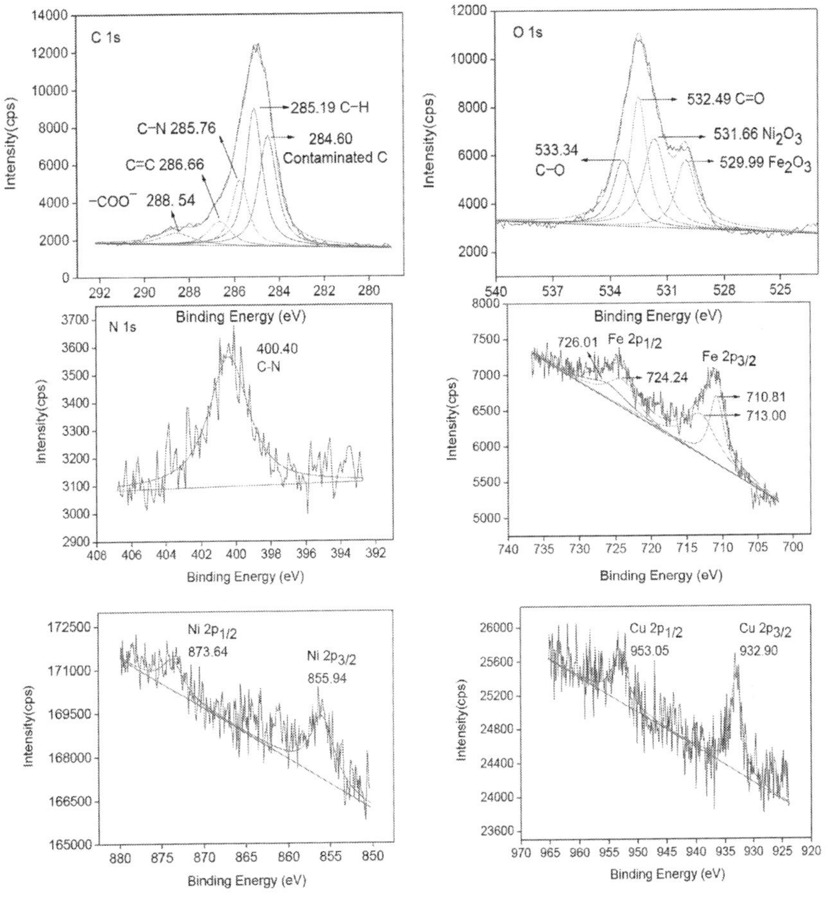

Figure 38. Curve-fitted XPS spectra of typical elements on rubbed surface of steel ball lubricated with nanolubricants containing 0.8% CuNi(S2).

According to the XPS analysis it can be inferred that during the direct contact sliding process, tribochemical reaction occurs between Cu@Ni nanoparticles and steel ball, in which organic modifiers and Ni, Cu nanocores with high activity are released and transferred to worn surface easily to form continuous and uniform boundary lubricating film. Especially under the synergistic effect of Ni and Cu nanoparticles, an antifriction and self-repairing film deposited on friction pair can reduce direct contact between sliding surface, delay adhesion wear, and timely repair wear parts, thus greatly improving the antifriction, antiwear, and extreme pressure performance of PAO.

6. Conclusion

In this chapter, nano-Ni and the Cu@Ni nanoparticles with different size and shape are prepared through *in situ* thermal decomposition. Their properties as additives in base oil and DLC solid-liquid composite lubrication system are systematically studied. It was found that Ni and Cu@Ni nanoparticles showed outstanding antifriction and antiwear properties in base oil and DLC solid-liquid composite lubrication system. And the size and shape of nanoparticles affects their tribological properties greatly. The results are as follows:

1. Ni nanoparticles with three type of particle size can significantly improve the wear resistance as lubricating oil additives, and its wear resistance is closely related to particle size. The smaller the particle size is, the better the wear resistance is. When Wt 0.05% Ni particle was added in base oil, the WSD of steel ball can be reduced by 30%.
2. Nano-Ni particles with three different morphologies including triangular rod, triangular plate, and spherical shape were added in lubricating oil. Triangle sheet-like Ni nanoparticles' performance is better than the other ones due to its larger contacting area with steel ball. The WSD of steel ball can be reduced by 35%.
3. The Ni nanoparticles (7 nm diameter) significantly reduce the friction in the DLC/steel contacts lubricated by fully formulated oil di-iso-octyl sebacate (DIOS) in all the lubrication regimes: boundary, mixed, and elasto-hydrodynamic lubrication.
4. Compared with pure Ni nanolubricant, Cu@Ni nanolubricant shows better antimill, friction reduction, and bearing capacity. The antiwear capacity is enhanced by 17.5%. The mechanism can be deduced that the collaborative work of Ni nuclear deposition film and Cu transfer film can effectively reduce direct contact between friction interface and timely repair wear parts.

Author details

Yujuan Zhang, Shengmao Zhang*, Pingyu Zhang, Guangbin Yang and Zhijun Zhang

*Address all correspondence to: zsm@henu.edu.cn

Engineering Research Center for Nanomaterials, Henan University, Kaifeng, China

Collaborative Innovation Center of Nano Functional Materials and Applications of Henan Province, Henan University, Kaifeng, China

References

[1] Bakunin, V.N., Suslov, A.Y., Kuzmina, G.N., Parenago, O.P., Topchiev, A.V. Synthesis and application of inorganic nano-particles as lubricant components-a review. J. Nanopart. Res. 2004;6:273–284.

[2] Wei Dai, Bassem Kheireddin, Hong Gao, Hong Liang, Roles of nanoparticles in oil lubrication, Tribology internationaly, 2016;102:88–98.

[3] Zhou, J., Wu, Z., Zhang, Z., Liu, W., Xue, Q. Tribological behavior and lubricating mechanism of Cu nanoparticles in oil. Tribol. Lett. 2000;8:213–218.

[4] Tarasov, S., Belyaev, S. Alloying contact zones by metallic nanopowders in sliding wear. Wear. 2004;257:523–530.

[5] Lee, K., Hwang, Y., Cheong, S., Choi, Y., Kwon, L., Lee, J. Understanding the role of nanoparticles in nano-oil lubrication. Tribol. Lett. 2009;35:127–131.

[6] Herna´ndez Battez, A., Viesca, J.L., Gonza´lez, R., Blanco, D., Asedegbega, E., Osorio, A. Friction reduction properties of a CuO nanolubricant used as lubricant for a NiCrBSi coating. Wear. 2010;268:325–328.

[7] Zhang, Z., Zhang, J., Xue, Q. Synthesis and characterization of a molybdenum disulfide nanocluster. J. Phys. Chem. 1994;98:12973–12977.

[8] Zhang, Z., Xue, Q., Zhang, J. Synthesis, structure and lubricating properties of dialkyldithiophosphate-modified Mo-S compound nanoclusters. Wear. 1997;209:8–12.

[9] Xue, Q., Liu, W., Zhang, Z. Friction and wear properties of a surface-modified TiO2 nanoparticle as an additive in liquid paraffin. Wear. 1997;213:29–32.

[10] Zhou, J., Yang, J., Zhang, Z., Liu, W., Xue, Q. Study on the structure and tribological properties of surface-modified Cu nanoparticles. Mater. Res. Bull. 1999;34:1361–1367.

[11] Zhou, J., Wu, Z., Zhang, Z., Liu, W., Dang, H. Study on an antiwear and extreme pressure additive of surface coated LaF3 nanoparticles in liquid paraffin. Wear. 2001;249:333–337.

[12] Dang, H., Sun, L., Zhou, J., Zhang, Z. Synthesis, structure and tribological properties of stearic acid coated (NH4)3PMo12O40 nanoparticles. Tribol. Lett. 2004;17:311–316.

[13] Li, X., Cao, Z., Zhang, Z., Dang, H. Surface-modification in situ of nano-SiO2 and its structure and tribological properties. Appl. Surf. Sci. 2006;252:7856–7861.

[14] Xiong, X., Kang, Y., Yang, G., Zhang, S., Yu, L., Zhang, P. Preparation and evaluation of tribological properties of Cu nanoparticles surface modified by tetradecyl hydroxamic acid. Tribol. Lett. 2012;46:211–220.

[15] Tzitzios, V., Basina, G., Gjoka, M., Alexandrakis, V., Georgakilas, V., Niarchos, D. Chemical synthesis and characterization of hcp Ni nanoparticles. Nanotechnology. 2006;17:3750–3755.

[16] Sun, Y., Rollins, H., Guduru, R. Preparations of nickel, cobalt, and iron nanoparticles through the rapid expansion of supercritical fluid solutions (RESS) and chemical reduction. Chem. Mater. 1999;11:7–9.

[17] Han, M., Liu, Q., He, J., Song, Y., Xu, Z., Zhu, J.M. Controllable synthesis and magnetic properties of cubic and hexagonal phase nickel nanocrystals. Adv. Mater. 2007;19:1096–1100.

[18] Carenco, S., Labouille, S., Bouchonnet, S., Boissie`re, C., Le Goff, X.-F.L. Revisiting the molecular roots of a ubiquitously successful synthesis: nickel(0) nanoparticles by reduction of [Ni(acetylacetonate)2]. Chem. Eur. J. 2012;18:14165–14173.

[19] Carenco, S., Boissie`re, C., Nicole, L., Sanchez, C., Le Floch, P., Me´zailles, N. Controlled design of size-tunable monodisperse nickel nanoparticles. Chem. Mater. 2010;22:1340–1349.

[20] Lu, A.-H., Salabas, E.L., Schüth, F. Magnetic nanoparticles: synthesis, protection, functionalization, and application. Angew. Chem. Int. Edit. 2007;46:1222–1244.

[21] Winnischofer, H., Rocha, T.C.R., Nunes, W.C., Socolovsky, L.M., Knobel, M., Zanchet, D. Chemical synthesis and structural characterization of highly disordered Ni colloidal nanoparticles. ACS Nano. 2008;2:1313–1319.

[22] Chen, Y., Peng, D., Lin, D., Luo, X. Preparation and magnetic properties of nickel nanoparticles via the thermal decomposition of nickel organometallic precursor in alkylamines. Nanotechnology. 2007;18:505703.

[23] Qiu, S., Zhou, Z., Dong, J., Chen, G. Preparation of Ni nanoparticles and evaluation of their tribological performance as potential additives in oils. J. Tribol. 2001;123:441–443.

[24] Chou, R., Battez, A.H., Cabello, J.J., Viesca, J.L., Osorio, A., Sagastume, A. Tribological behavior of polyalphaolefin with the addition of nickel nanoparticles. Tribol. Int. 2010;43:2327–2332.

[25] Liang, X., Wang, X., Zhuang, J., Chen, Y., Wang, D., Li, Y. Synthesis of nearly monodisperse iron oxide and oxyhydroxide nanocrystals. Adv. Funct. Mater. 2006;16:1805–1813.

[26] Sun, S., Zeng, H., Robinson, D.B., Raoux, S., Rice, P.M., Wang, S., Li, G. Monodisperse MFe2O4 (M = Fe, Co., Mn) nanoparticles. J. Am. Chem. Soc. 2004;126:273–279.

[27] Li, N., Zhang, X., Chen, S., Hou, X., Liu, Y., Zhai, X. Synthesis and optical properties of CdS nanorods and CdSe nanocrystals using oleylamine as both solvent and stabilizer. Mater. Sci. Eng. B. 2011;176:688–691.

[28] Shukla, N., Liu, C., Jones, P.M., Weller, D. FTIR study of surfactant bonding to FePt nanoparticles. J. Magn. Magn. Mater. 2003;266:178–184.

[29] Söderlind, F., Pedersen, H., Petoral, R.M., Käll, P.-O., Uvdal, K. Synthesis and characterisation of Gd2O3 nanocrystals functionalised by organic acids. J. Colloid Interface Sci. 2005;288:140–148.

[30] Shah, F.U., Glavatskih, S., Antzutkin, O.N. Synthesis, physicochemical, and tribological characterization of S-Di-n-octoxyboron-O, O′-di-n-octyldithiophosphate. ACS Appl. Mater. Interfaces. 2009;1:2835–2842.

[31] Wang, L., Gao, Y., Xu, T., Xue, Q. A comparative study on the tribological behavior of nanocrystalline nickel and cobalt coatings correlated with grain size and phase structure. Mater. Chem. Phys. 2006;99:96–103.

[32] Cai, M., Liang, Y., Zhou, F., Liu, W. Tribological properties of novel imidazolium ionic liquids bearing benzotriazole group as the antiwear/anticorrosion additive in poly(ethylene glycol) and polyurea grease for steel/steel contacts. ACS Appl. Mater. Interfaces. 2011;3:4580–4592.

[33] Bradley, J.S., Tesche, B., Busser, W. Surface spectroscopic study of the stabilization mechanism for shape-selectively synthesized nanostructured transition metal colloids. J. Am. Chem. Soc. 2000;122:4631–4636.

[34] Yamamoto, T., Yin, H., Wada, Y. Morphology-control in microwave-assisted synthesis of silver particles in aqueous solutions. Bull. Chem. Soc. 2004;77:757–761.

[35] Wang, F., Zhang, Z.C., Chang, Z.Q. Effects of magnetic field on the morphology of nickel nanocrystals prepared by gamma-irradiation in aqueous solutions. Mater. Lett. 2002;55(1–2):27–29.

[36] Zhang, C., Zhang, S., Yu, L. Size-controlled synthesis of monodisperse Ag_2S nanoparticles by a solventless thermolytic method. Mater. Lett. 2012;85:77–80.

[37] Ronkainen, H., Varjus, S., Holmberg, K. Friction and wear properties in dry, water-and oil-lubricated DLC against alumina and DLC against steel contacts. Wear. 1998;222(2): 120–128.

[38] Miklozic, K.T., Lockwood, F., Spikes, H. Behaviour of boundary lubricating additives on DLC coatings. Wear. 2008;265(11–12):1893–1901.

[39] Kalin, M., Viintin, J. A comparison of the tribological behaviour of steel/steel, steel/DLC and DLC/DLC contacts when lubricated with mineral and biodegradable oils. Wear. 2006;261(1):22–31.

[40] Jia, Z., Xia, Y., Li, J. Friction and wear behavior of diamond-like carbon coating on plasma nitrided mild steel under boundary lubrication. Tribol. Intl. 2010;43(1–2):474–478.

[41] Sanche-Lopez, J.C., Erdemir, A., Donnet, C. Friction-induced structural transformations of diamond like carbon coatings under various atmospheres. Surf. Coat. Tech. 2003;163–164:444–450.

[42] Kalin, M., Viintin, J. Real contact temperatures as the criteria for the reactivity of diamond-like carbon coatings with oil additives. Thin Solid Films. 2010;518(8):2029–2036.

[43] Neville, A., Morina, A., Haque, T., Voong, M. Compatibility between tribological surfaces and lubricant additives—how friction and wear reduction can be controlled by surface/lube synergies. Tribol. Intl. 2007;40(10–12):1680–1695.

[44] Haque, T., Morina, A., Neville, A., Kapadia, R., Arrowsmith, S. Study of the ZDDP antiwear tribofilm formed on the carbon coating using AFM and XPS. J. ASTM Intl. 2007;4(11):http://dx.doi.org/10.1520/JAI100937.

[45] Kalin, M., Kogovšek, J., Remškar, M. Nanoparticles as novel lubricating additivesin a green, physically based lubrication technology for carbon coatings. Wear. 2013;303:480–485.

[46] Ghaednia, H., Babaei, H., Jackson, R.L., Bozack, M.J., Khodadadi, J.M. The effect of nanoparticles on thin film elasto-hydrodynamic lubrication. Appl. Phys. Lett. 2013;103:263111.

[47] Liu, Y., Erdemir, A., Meletis, E.I. A study of the wear mechanism of diamond-like carbon films. Surface Coatings Technol. 1996;82(1–2):48–56.

[48] Liu, Y., Erdemir, A., Meletis, E.I. An investigation of relationship between graphitization and frictional behavior of DLC coatings. Surface Coatings Technol. 1996;86–87(1–3):564–568.

[49] Erdemir, A., Bindal, C., Fenske, G.R. Characterization of transfer layers forming on surface sliding against diamond-like carbon. Surface Coatings Technol. 1996;86–87:692–697.

[50] Kogovšek, J., Remškar, M., Kalin, M. Lubrication of DLC-coated surfaces with MoS2 nanotubes in all lubrication regimes: Surface roughness and running-in effects. Wear. 2013;303:361–370.

[51] Gelinck, E.R.M., Schipper, D.J. Calculation of Stribeck curves for line contacts. Tribol. Intl. 2000;33:175–181.

[52] Huang, P. Physical-mathematical model and numerical analysis of nano thin film lubrication. Tribology. 2003;23(1):60–64.

[53] Kalin, M., Velkavrh, I. Non-conventional inverse-Stribeck-curve behaviour and other characteristics of DLC coatings in all lubrication regimes. Wear. 2013;297:911–918.

[54] Wang, W.Z., Huang, P. Study on the lubrication state of frictional pairs with different surface roughness based on Stribeck curves. Tribology. 2004;24(3):254–257.

[55] Nilsson, D., Prakash, B. Influence of different surface modification technologies on friction of conformal tribopair in mixed and boundary lubrication regimes. Wear. 2011;273:75–81.

[56] Chen, S., Sommers, J.M. Alkanethiolate-protected copper nanoparticles: spectroscopy, electroche-mistry, and solid-state morphological evolution. J. Phys. Chem. B. 2001;105:8816–8820.

[57] Engels, V., Benaskar, F., Jefferson, D.A. Nanoparticulate copper-routes towards oxidative stability. Dalton T. 2010;39:6496–9502.

[58] Tsarfati, T., Zoethout, E., Bijkerk, F. Growth and sacrificial oxidation of transition metal nanolayers. Surf. Sci. 2009;603:1041–1045.

[59] Honkanen, M., Vippola, M., Lepisto, T. Oxidation of copper alloys studied by analytical transmission electron microscopy cross-sectional specimens. J. Mater. Res. 2008;23(5): 1350–1357.

[60] Guo, H., Chen, Y., Ping, H. One-pot synthesis of hexagonal and triangular nickel-copper alloy nanoplates and their magnetic and catalytic properties. J. Mater. Chem. 2012;22:8336–8344.

[61] Zhang, S., Zeng, H. Solution-based epitaxial growth of magnetically responsive Cu@Ni nanowires. Chem. Mater. 2010;22:1282–1284.

Chapter 11

Lubrication in Chemical and Mechanical Planarization

Changhong Wu and Xiaoyan Liao

Additional information is available at the end of the chapter

http://dx.doi.org/10.5772/64484

Abstract

Chemical mechanical planarization (CMP) has been widely used in integrated circuit (IC) processing to achieve both local and global surface planarity through combined chemical and mechanical actions. The lubrication plays a significant role in CMP and can be determined by the Stribeck curve since it provides direct evidence of the extent of contact among wafer, pad asperities, and slurry particles. The advancements in the construction of the Stribeck curve are highlighted in this chapter. Traditionally, the procedure for constructing the Stribeck curve is as follows: (1) polish wafers at various pressures and sliding velocities to obtain the coefficient of friction (COF) values; (2) plot the experimental data as COF vs. Sommerfeld number; (3) construct the Stribeck curve via curve fitting. Recently, an alternative method was presented to construct the Stribeck curve via only performing one wafer polishing experiment. Pressure and sliding velocity are varied separately or together for a desired length of time, so that multiple measurements can be taken within one run. In this study, a back-propagation (BP) neural network is proposed to construct the Stribeck curve. Results show that the BP neural network could construct a more accurate Stribeck curve and thus could better provide insight into the lubrication mechanism of CMP processes.

Keywords: chemical mechanical planarization, lubrication, Stribeck curve, neural network

1. Introduction

Integrated circuit (IC) technology plays a critical role in today's advanced industries. Since transistors were invented in the 1940s [1], IC technology has been growing nonstop. The circuit density and complexity keep on increasing driven by Moore's Law [2]. To date, the transistor count in a central processing unit (CPU) can reach up to several billions [3]; the commercial technology node scales down to 14 nm and the maximum wiring layer is 13 [4]. High volume manufacturing of ICs relies heavily on the advancements in each unit process such as chemical mechanical planarization (CMP), lithography, etching, deposition, and cleaning. Among these processes, CMP has been an enabling technology in the realization of IC manufacturing. Presently, CMP is the only planarization technique that can offer excellent local and global planarity. For the most advanced ultra-large-scale integration (ULSI), CMP is the only choice for global surface planarization [5]. In this chapter, the lubrication of CMP is reviewed. Particularly, the advancements in the construction of Stribeck curve are highlighted.

2. What is CMP?

CMP typically refers to chemical mechanical planarization. As its name indicates, CMP simultaneously employs both chemical and mechanical actions to selectively remove the exposed material from elevated features for improved planarization [6–14]. Driven with the planarization challenge in IC manufacturing for smaller minimum feature size and more wiring levels, CMP was invented by International Business Machines (IBM) in the mid-1980s [15].

Figure 1 schematically shows a generalized CMP process on a rotary polisher. The wafer to be polished is secured by a retaining ring, which is attached to the carrier head. The rotating wafer is pressed against the rotating polishing pad under a certain pressure from the carrier head [5, 7]. Slurry, containing chemicals and nanoparticles, is injected onto the pad surface and transported to the pad-wafer interface through platen-wafer rotation, pad surface structures as well as the retaining ring. Slurry chemicals react with the wafer surface to form a softer porous layer, which is then removed by mechanical forces generated among wafer surface, pad asperities, and slurry nanoparticles. In addition, a pad conditioner is typically used to refresh pad asperities and keep material removal rate stable [5, 7, 13, 14].

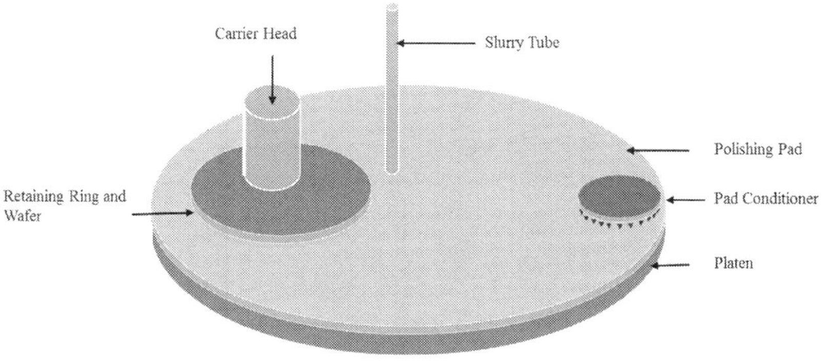

Figure 1. Schematic of a generalized CMP process on a rotary polisher [7].

http://dx.doi.org/10.5772/64484

3. Stribeck curve

Tribology is generally understood as the science and technology of interacting surfaces in relative motion [16]. As schematically shown in **Figure 2**, three-body (i.e., wafer, pad, and slurry particles) contact occurs in CMP. Tribology, including friction, lubrication, and wear, plays a crucial role in CMP [16]. In this chapter, we will focus on the lubrication in CMP.

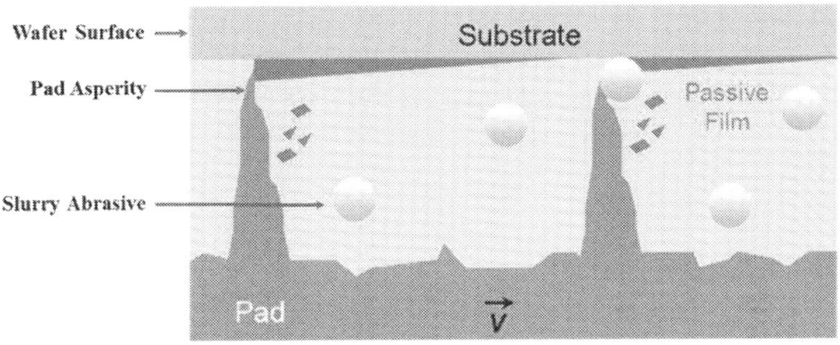

Figure 2. Schematic of three-body contact in CMP [17].

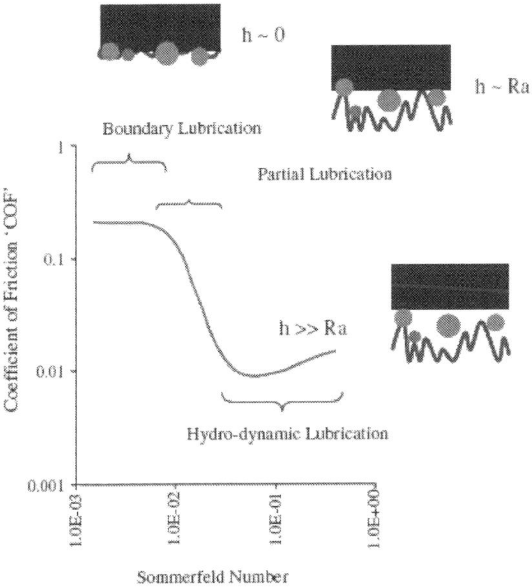

Figure 3. Generic Stribeck curve based on Sommerfeld number [18].

The lubrication mechanism in CMP processes can be determined by the Stribeck curve [18]. **Figure 3** shows a generic Stribeck curve based on Sommerfeld number [18]. In this figure, Y axis is coefficient of friction (COF), and X axis represents the Sommerfeld number. COF is defined as the ratio of the shear force to the down force. During CMP, shear force is generated among the wafer, pad asperities, and slurry nanoparticles. The shear force and down force can be measured using appropriate instruments such as CETR CP4 and Araca APD-800 [5, 19]. For example, using Araca APD-800, **Figure 4** shows the transient shear force and down force as a function of polishing time, and **Figure 5** shows the transient COF. Since COF is defined as the ratio of the shear force to the down force, the transient COF can be expressed as:

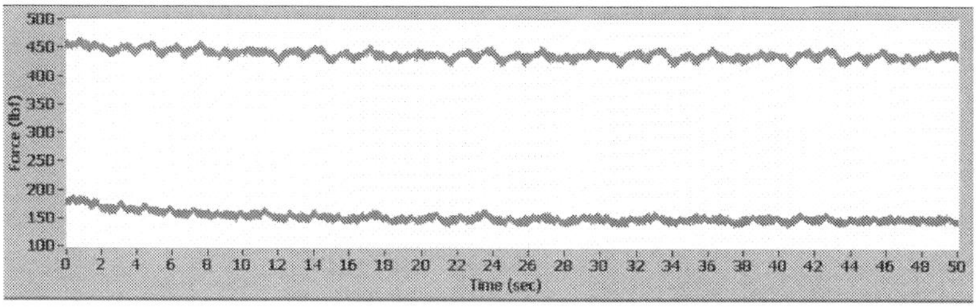

Figure 4. The transient shear force and down force as a function of polishing time [5].

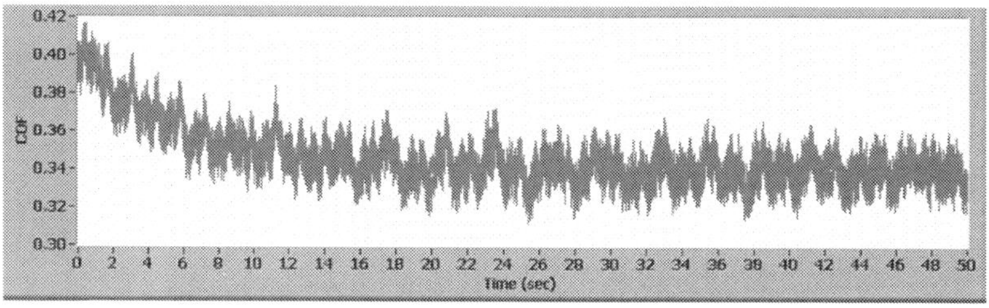

Figure 5. Transient COF based on **Figure 4** [5].

$$COF_i = \frac{\text{Shear force}_i}{\text{Down force}_i} \tag{1}$$

In **Figure 3**, X axis is the Sommerfeld number (*So*), which is defined as follows [18]:

$$So = \frac{\mu V}{p\delta_{eff}} \tag{2}$$

where μ is the viscosity of slurry, V is the pad-wafer sliding velocity, p is the wafer pressure, and δ_{eff} is the effective slurry film thickness at the wafer and pad interface.

Slurry viscosity μ can be measured experimentally, and the pad-wafer sliding velocity V can be calculated based on the tool geometry and the relative angular velocity of the wafer and platen [18]. The wafer pressure p is the applied down force divided by the contact area between the wafer and pad. Finally, the effective slurry film thickness δ_{eff} can be assumed as the arithmetic average roughness of the pad based on slurry film thickness measurement results by dual emission laser induced fluorescence (DELIF) [20–22]. In some cases, the pseudo Sommerfeld number (V/p) is used to replace Sommerfeld number for simplicity [23].

As mentioned previously, the lubrication mechanism in CMP processes can be determined by the Stribeck curve since it provides direct evidence of the extent of contact among wafer, pad asperities, and slurry particles [18]. As evident from **Figure 3**, there are three contact modes: boundary lubrication, partial lubrication, and hydro-dynamic lubrication.

Boundary lubrication occurs at smaller values of the Sommerfeld number. In this regime, the slurry film is quite thin, and thus, the wafer, pad, and slurry particles are intimate contact with one another [18]. COF is high and independent of the Sommerfeld number. From a process control point of view, the boundary lubrication is preferable because of its high and stable COF [7].

The partial lubrication occurs at intermediate values of the Sommerfeld number. In this regime, the slurry film thickness is similar to that of the roughness of the pad [18]. The slurry film layer partially separates the wafer and pad, and COF sharply drops as increase in the Sommerfeld number. A small change in process parameter such as polish pressure and velocity may cause a significant change in COF.

Hydro-dynamic lubrication occurs at larger values of the Sommerfeld number. In this regime, the slurry film thickness is larger than that of the roughness of the pad [18]. As such, very little contact exists between the wafer and pad. COF is small and increases slightly as increase in the Sommerfeld number [5].

Since the Stribeck curve can determine the lubrication mechanism in CMP processes, it has many practical applications. For example, it can help screen certain consumable sets (pad, slurry, wafer, retaining ring, etc.) by determining whether and how they contact one another during CMP. It can also help determine the optimal polish parameters (wafer and retaining ring pressure, pad/wafer velocity, slurry flow rate, etc.) [24].

4. Advancements in Stribeck curve construction

Traditionally, the procedure of constructing the Stribeck curve for a given consumables set is as follows [25]: (1) polishing wafers at various pressures and sliding velocities. One or more wafers typically need to be polished to obtain the COF at each given polish pressure and sliding velocity; (2) plotting experimental data as COF vs. Sommerfeld number or pseudo Sommerfeld number (V/p); and (3) performing curve fitting or simply connecting data points to obtain the Stribeck curve.

Figure 6 shows Stribeck curves where oxide wafers were polished on different types of pads using Fujimi PL-4217 slurry with 25% abrasives. As evident from **Figure 6**, the relationship between COF and Sommerfeld number is complicated. First, for a given consumables set, this relationship needs to be described using different mathematical functions depending on the lubrication regions. Second, as shown in **Figure 6**, this relationship varies with consumables set (pad in this case). A process could fall in different lubrication regions for different consumable sets at the same Sommerfeld number. For example, when Sommerfeld number ranges from 0.0002 to 0.0008, IC 1400 k-groove pad and IC 1000 perforated pad are in the boundary and partial lubrication region, respectively. As such, it is difficult to describe the relationship between COF and Sommerfeld number using a general explicit function. This renders challenges to curve fitting for the Stribeck curve construction, which needs to predetermine the form of function. On the other hand, if we just simply connect data points to obtain the Stribeck curve, it could be not accurate if the data points are not enough.

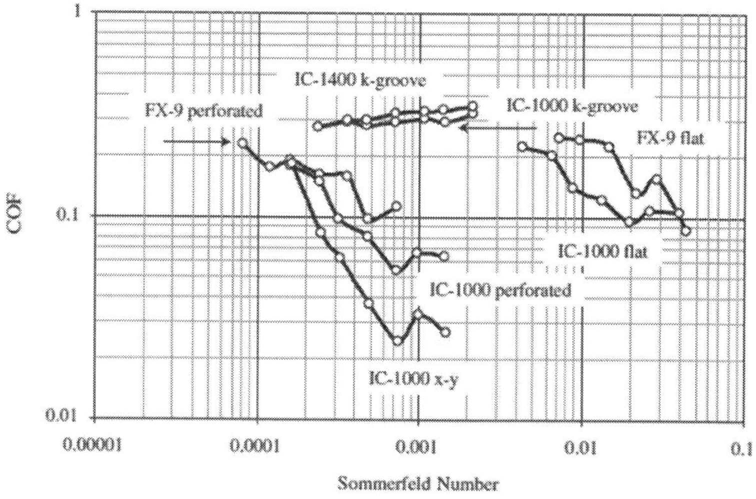

Figure 6. Stribeck curves at 25% abrasives with Fujimi PL-4217 slurry [18].

Zhuang and Jiao et al. proposed an alternative method to construct the Stribeck curve by only performing one wafer polishing experiment [23, 24]. Pressure and sliding velocity are varied separately or together for a desired length of time, so that multiple measurements (COF, V, p) can be taken within one run. This is attributed to their polisher (i.e., Araca APD-800), which is capable of measuring shear force, down force, and rendering a value for COF in real time.

Figure 7(a) shows the 'traditional' Stribeck curve. Three hundred-millimeter blanket copper wafers were polished on a Cabot Microelectronics Corporation (CMC) D100 pad with CMC

EP-C600Y-75 slurry at different polish pressures (1.0–2.5 psi) and sliding velocities (0.6–1.5 m/s). One wafer was polished for each particular V and p combination, and totally, nine wafers were polished. In comparison, **Figure 7(b)** shows the Stribeck curve obtained by the new method where only one wafer was polished in the range of V/p investigated. As observed in **Figure 7**, the CMP process is in the boundary lubrication. Also, it should be noted that the new method can collect enough data by only polishing one wafer and thus does not need to perform curve fitting to obtain the Stribeck curve.

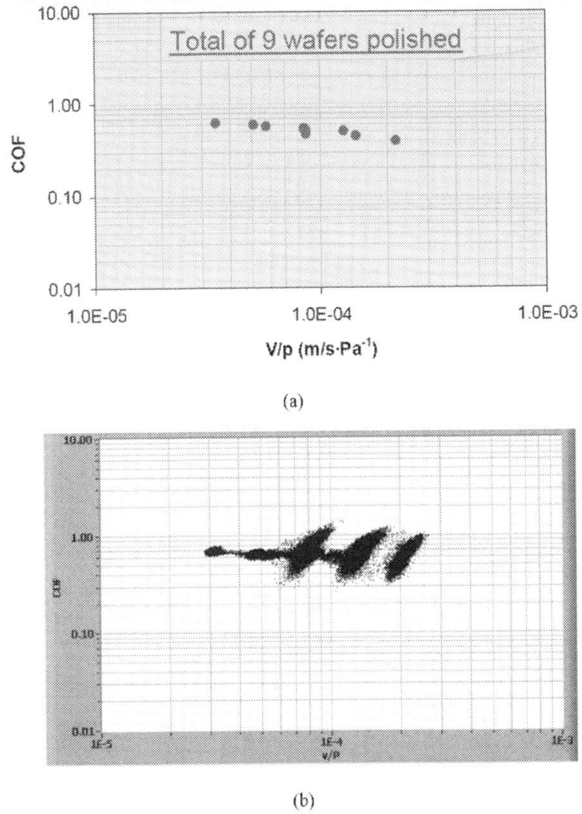

Figure 7. Stribeck curves obtained by the (a) traditional and (b) new method on D100 pad [24].

Figure 8(a) and **(b)** shows another example of Stribeck curves, which were obtained by the traditional and new method, respectively. Polishing conditions and consumables in **Figure 8** were same as **Figure 7** except that Dow IC 1000 k-groove pad was employed in **Figure 8**. As observed in **Figure 8**, the CMP process is in the partial lubrication. As we can see, the change in the type of the pad, the lubrication mechanisms of the CMP processes may change.

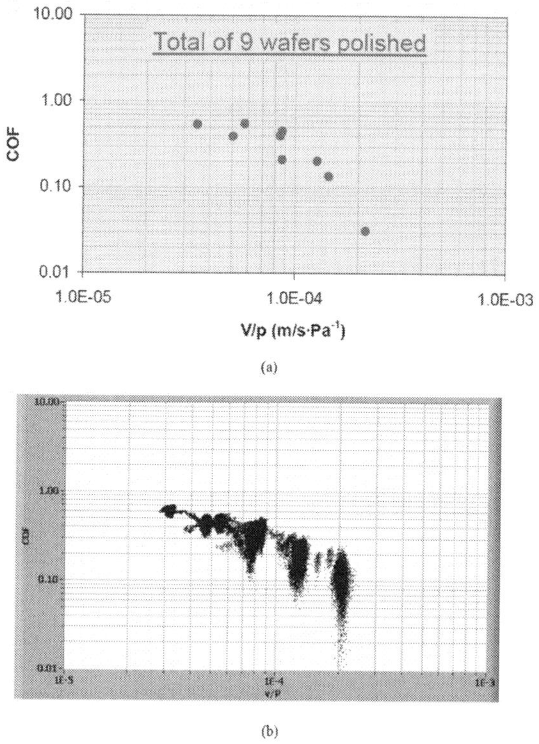

Figure 8. Stribeck curves obtained by the (a) traditional and (b) new method on IC 1000 pad [24].

5. Artificial neural network for Stribeck curve

The above new method proposed by Zhuang and Jiao et al. can provide an accurate Stribeck curve. As mentioned previously, this is attributed to their polisher which is capable of measuring shear force, down force, and rendering a value for COF in real time. Due to the limitation of the experimental apparatus, the traditional method (i.e., polishing one wafer at each given pressure and velocity) is still commonly used. To construct an accurate Stribeck curve based on the limited data points (COF vs. V/p), which are obtained by the traditional method, a new neural network-based method is presented here.

Artificial neural networks have been widely used for nonlinear function approximation, pattern identification, system modeling, and control [25–33]. For example, Wu *et al.* employed the neural network to predict molten temperature of blended coal ash and estimate the degree of slagging of the coal-fired boiler in power plant [29, 30]. Furthermore, Wu *et al.* optimized power coal blending based on neural network and genetic algorithm [28, 32]. These successful work provided a method for the construction of Stribeck curve in this study.

There are various types of neural networks such as back-propagation (BP), linear, and RBF neural networks. Among these networks, BP neural network is the most widely used one. In this chapter, a BP neural network is employed to construct the Stribeck curve.

Figure 9 schematically shows the architecture of a 3-layer BP neural network. The first layer (i.e., input layer) and the last layer (i.e., output layer) consist of the independent and dependent variables, respectively. The dependent variables in the output layer are determined by the independent variables in the input layer [26]. In Stribeck curve, COF is shown as a function of Sommerfeld number or pseudo Sommerfeld number (V/p). Therefore, the input layer of the neural network is V/p and the output layer is COF.

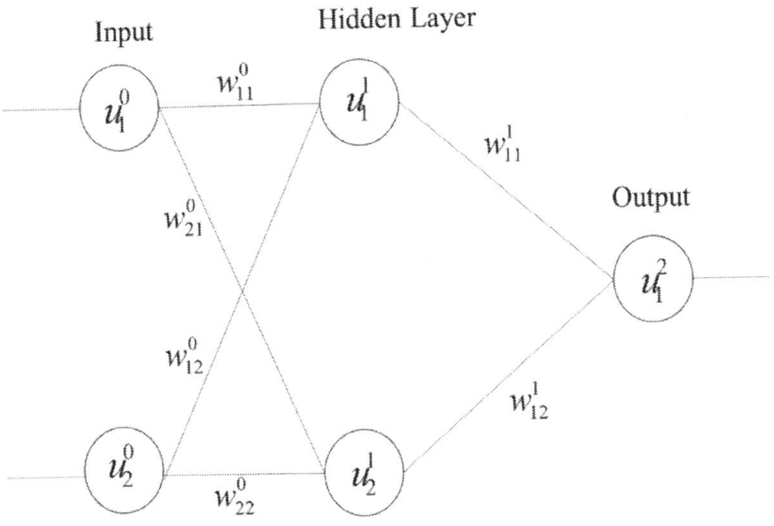

Figure 9. Architecture of a three-layer BP neural network.

For BP neural network, transfer functions ('tansig,' 'purelin,' 'logsig,' etc.) are used to transfer data from one layer to the next layer. The BP algorithm for neural network training was described in detail in Ref. [34]. The main procedure is [28–34]: at first, the weights (w_{ik}) and biases of the network are initialized. A certain amount of samples is then used for training. The values of the input and output of the training samples are already known. Outputs of the training samples are calculated based on the inputs, weights, and biases. Then, the network outputs and the real outputs of the training samples are compared to calculate the error. If the error meets the target, the training is completed. If not, the weights and biases of the network need to be adjusted and the outputs and errors are recalculated. Repeat the above procedure until the error goal is met. After training, it is expected that the neural network can predict the value of the output when a new value of the input is given.

A case study is presented here. Three-layer BP neural network is employed to construct the Stribeck curve. Both input layer and output layer have only one neuron, and the number of neuron in the hidden layer is selected as 4. The nine data points in **Figure 7(a)** are used as the training samples. For the input, V/p = [0.000035 0.000051 0.000058 0.000084 0.000086 0.000087 0.00012 0.00014 0.00021]; for the output, COF = [0.64 0.6 0.57 0.54 0.51 0.47 0.5 0.44 0.4]. The error goal is set as 10^{-5}, the transform function from the input layer to hidden layer is 'tansig,' and the one from the hidden layer to output layer is 'purelin.'

After training, the neural network is used to predict COF for a particular combination of pressure and velocity. In this case, the COF values are predicted with a V/p interval of 10^{-5} from 0.000035 to 0.00021. The prediction COF values obtained by the neural network are plotted as a function of V/p, as shown in **Figure 10** (see the blue curve). As we can see, the prediction COF values agree well with the actual COF values (see the circles) of the training samples. If we compare the Stribeck curve obtained by the neural network in **Figure 10** with the Stribeck curve obtained by the experimental approach in **Figure 7(b)**, they are consistent . Therefore, after training, the neural network can successfully capture the nonlinear relationship between the input and output. COF for a particular combination of pressure and velocity can be accurately predicted by the neural network. We believe that the successful construction of the Stribeck curve by the BP neural network could better provide insight into the lubrication mechanism of CMP processes.

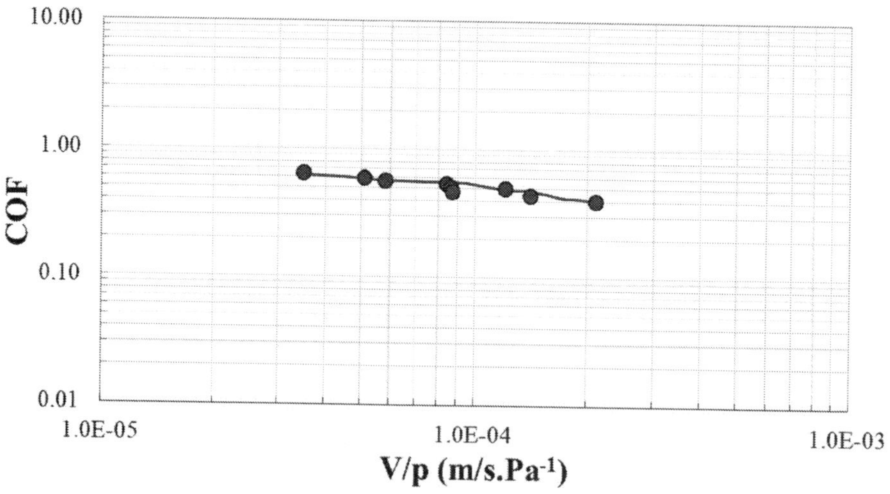

Figure 10. Stribeck curves obtained by the neural network.

http://dx.doi.org/10.5772/64484

Author details

Changhong Wu[1] and Xiaoyan Liao[2*]

*Address all correspondence to: xiaoyan@email.arizona.edu

1 Advanced Technology Development (ATD), GLOBALFOUNDRIES, Malta, NY, USA

2 School of Chemical Engineering and Energy, Zhengzhou University, Zhengzhou, Henan, China

References

[1] W. Shockley, *"The path to the conception of the junction transistor"*, IEEE Trans. Elec. Dev., ED-31, 1523 (1984).

[2] G. Moore, "Cramming more components onto integrated circuits", Electronics, 38 p. 114-117 (1965) .

[3] V. Krishnaswamy, L. Shin, and S. Turullols, 9th IEEE Asian Solid-State Circuits Conference Proceedings of Technical Papers, Sentosa Island, Singapore; p. 17 (2013).

[4] http://www.itrs.net/

[5] C. Wu, "Control of slurry flow, temperature and aggressive diamonds in chemical mechanical planarization", Ph.D. Thesis, University of Arizona, Tucson, AZ (2015).

[6] J. Steigerwald, S. Murarka, and R. Gutmann, Chemical Mechanical Planarization of Microelectronic Materials, Wiley, NY (1997).

[7] X. Liao, "Process optimization and fundamental consumables characterization of advanced dielectric and metal chemical mechanical planarization", Ph.D. Thesis, University of Arizona, Tucson, AZ (2014).

[8] X. Liao, Y. Zhuang, L. Borucki, J. Cheng, S. Theng, T. Ashizawa, and A. Philipossian, "Effect of pad surface micro-texture on removal rate during interlayer dielectric chemical mechanical planarization process", Jpn. J. Appl. Phys., 52, 018001 (2013).

[9] X. Liao, Y. Zhuang, L. Borucki, J. Cheng, S. Theng, T. Ashizawa, and A. Philipossian, "Effect of pad surface micro-texture on wafer topography during shallow trench isolation chemical mechanical planarization process", Jpn. J. Appl. Phys., 53, 086501 (2014).

[10] X. Liao, Y. Zhuang, L. Borucki, S. Theng, X. Wei, T. Ashizawa, and A. Philipossian, "Effect of pad surface micro-texture on coefficient of friction and removal rate during copper CMP process", *Electrochem. Solid-State Lett.*, 14, H201 (2011).

[11] X. Liao, Y. Sampurno, Y. Zhuang, and A. Philipossian, "Effect of slurry application/ injection schemes on slurry availability during chemical mechanical planarization (CMP)", *Electrochem. Solid-State Lett.*, 15, H118 (2012).

[12] C. Wu, Y. Sampurno, X. Liao, Y. Jiao, S. Theng, Y. Zhuang, L. Borucki, and A. Philipossian, "Pad surface thermal management during copper chemical mechanical planarization", *ECS J. Solid State Sci. Technol.*, 4(7), P206-P212 (2015).

[13] C. Wu, Y. Sampurno, X. Liao, Y. Zhuang, L. Borucki, S. Theng, and A. Philipossian, "Effect of pad groove design on slurry injection scheme during interlayer dielectric chemical mechanical planarization", *ECS J. Solid State Sci. Technol.*, 4(7), P272-P276 (2015).

[14] C. Wu, Y. Zhuang, X. Liao, Y. Jiao, Y. Sampurno, S. Theng, F. Sun, A. Naman, and A. Philipossian, "Aggressive diamond characterization and wear analysis during chemical mechanical planarization", *ECS J. Solid State Sci. Technol.*, 2(1), P36-P41 (2013).

[15] D. Moy, M. Schadt, C. Hu, F. Kaufman, A. Ray, N. Mazzeo, E. Baran, and D. Pearson, Proceedings 1989 VMIC Conference, Santa Clara, CA, US; p. 26 (1989).

[16] H. Liang and D. Craven, Tribology in Chemical Mechanical Planarization, Taylor & Francis, Boca Raton, FL (2005).

[17] D. DeNardis ., "Evaluation and modeling of alternative copper and inter-layer dielectric chemical mechanical planarization technologies", Ph.D. Thesis, University of Arizona, Tucson, AZ (2006).

[18] A. Philipossian and S. Olsen, "Fundamental tribological and removal rate studies of inter-layer dielectric chemical mechanical planarization", *Jpn. J. Appl. Phys.*, 42, 6371-6379 (2003).

[19] http://www.cetr.com/

[20] A. Lawing, Proceedings of 7th International Chemical-Mechanical Planarization for ULSI Multilevel Interconnection Conference (CMP-MIC), Fremont, CA, US; p. 1 (2002).

[21] J. Lu, J. Coppeta, C. Rogers, V. Manno, L. Racz, and A. Philipossian, The Effect of Wafer Shape on Slurry Film Thickness and Friction Coefficients in CMP *Mater. Res. Soc. Symp. Proc.*, 613, E1.2.1 (2000).

[22] C. Rogers, J. Coppeta, and L. Racz, Analysis of Flow Between a Wafer and Pad During CMP Processes *J. Electron. Mater.*, 27, 1082 (1998).

[23] Y. Jiao, X. Liao, C. Wu, S. Theng, Y. Zhuang, Y. Sampurno, M. Goldstein, and A. Philipossian, "Tribological, thermal and kinetic attributes of 300 vs. 450 mm chemical mechanical planarization processes", *J. Electrochem. Soc.*, 159, H255 (2012).

[24] Y. Zhuang, ERC TeleSeminars, May 29, 2010.

[25] J. Yi, Y. Sheng, and C. Xu, "Neural network based uniformity pro-file control of linear chemical-mechanical planarization", *IEEE Trans. Semicond. Manuf.*, 16, 609-620 (2003).

[26] X. Liao, C. Wu, and J. Zheng, "New neural network-based method for Stribeck curve construction during chemical mechanical planarization", The International Semiconductor Technology Conference/China Semiconductor Technology International Conference (ISTC/CSTIC 2016), Shanghai, China, March 13–14 (2016).

[27] G. Wang and C. Yu, "Developing a neural network-based run-to-run process controller for chemical-mechanical planarization", *Int. J. Adv. Manuf. Technol.*, 28, 899-908 (2006).

[28] Y. Liao, C. Wu, and X. Ma, "New hybrid optimization model for power coal blending", Proceedings of 2005 International Conference on Machine Learning and Cybernetics, Guangzhou, August 18, vol. 7, pp. 4023-4027 (2005).

[29] C. Wu, X. Ma, and Y. Liao, "Forecasting slagging of coal-fired boiler in power plant based on fuzzy network", *J. Combust. Sci. Technol.*, 2, 17 (2006).

[30] C. Wu and X. Ma, "Experimental research and forecast of the softening temperature of blended coal ash", *J. Eng. Therm. Power.*, 21, 179 (2006).

[31] C. Wu, X. Ma, Z. Dong, Z. Zhao, H. Li and Y. Chen, "Study on NO emission from co-combustion of plastic powder and rice hull in CFB", *Trans. Chin. Soc. Agr. Machin.*, 38, 92-97 (2007).

[32] Y. Liao, C. Wu, and X. Ma, "Research on optimization model for power coal blending based on genetic algorithm", ASME 2005 Power Conference, Guangzhou, China pp. P203-P207 (2005).

[33] X. Liao, H. Zou, and C. Wu, "Research on NOx emissions from co-combustion of wood powder and plastic powder in CFB", *Boiler Technol.*, 37, P72-P76 (2006).

[34] X. Wen, L. Zhou, and X. Li. MATLAB Neural Network Simulation and Application, Chinese Science Press, Beijing, China (2003).